Current Topics in Microbiology

120 and Immunology

Parasite Antigens in Protection, Diagnosis and Escape

Edited by R.M.E. Parkhouse

With 17 Figures

Springer-Verlag Berlin Heidelberg GmbH

R.M.E. PARKHOUSE, MD
National Institute for Medical Research
The Ridgeway, Mill Hill
London NW7 1AA, United Kingdom

ISBN 978-3-662-09199-9   ISBN 978-3-662-09197-5 (eBook)
DOI 10.1007/978-3-662-09197-5

© by Springer-Verlag Berlin Heidelberg 1985
Originally published by Springer-Verlag Berlin Heidelberg New York in 1985.
Softcover reprint of the hardcover 1st edition 1985
Library of Congress Catalog Card Number 15-12910

2123/3130-543210

Preface

An estimated 2–3 billion people in the less developed countries suffer from infections, often multiple, caused by a variety of parasitic organisms. These infections are frequently debilitating rather than fatal, and the toll in human misery is fearsome. To this may be added the prevalence of similar diseases in domestic animals, which diminish supplies of animal protein. As the world population increases, the already enormous problem also continues to grow. The resources of the less developed nations are inadequate for solving the problem, and in the developed countries a lack of interest in tropical diseases has meant low priority for research. Two recent methodological advances now raise the real possibility of a systematic and effective attack upon these diseases – hybridoma and recombinant nucleic acid technologies. The combination of these with the still necessary clinical, parasitological and immunological information permits a logical, planned and realistic approach to diagnosis and treatment. The central aim of these modern techniques is to define antigens with regard to diagnosis, protection and pathology. In the case of some diseases, work has already commenced along these lines; in the case of others, knowledge lags a long way behind.

This volume represents a summary of current knowledge about a wide, representative spectrum of tropical diseases. There is considerable common ground between the different infections as regards objectives and the methods for achieving them. Some of the more sophisticated investigations associated with, for example, malaria and schistosomes, are thus relevant to the less explored coccidia or nematodes. The book is not intended to be a "current state-of-the-art" volume. Rather, it presents the approach of modern biologists to the understanding, diagnosis and elimination of tropical diseases. The range of experience with the widely different models should therefore provide a unifying intellectual basis. Although not all the parasites reviewed are equally well understood and investigated, the intellectual and technical approaches, focussing as they must on chemically defined antigens, contribute a linking theme and constitute the basis for the elimination of these terrible diseases.

Autumn 1985   R.M.E. PARKHOUSE

Table of Contents

Indexed in Current Contents

List of Contributors

ALEXANDER, J., Department of Bioscience and Biotechnology, University of Strathclyde, The Todd Centre, Glasgow G4 0NR, Scotland, United Kingdom

ALMOND, N.M., Division of Immunology, National Institute for Medical Research, The Ridgeway, Mill Hill, London NW7 1AA, United Kingdom

AVRON, B., Department of Biophysics and Unit for Molecular Biology of Parasitic Diseases, The Weizmann Institute of Science, 76 100 Rehovot, Israel

CHAYEN, A., Department of Biophysics and Unit for Molecular Biology of Parasitic Diseases, The Weizmann Institute of Science, 76 100 Rehovot, Israel

HARRISON, L.J.S., University of Edinburgh, Centre for Tropical Veterinary Medicine, Easter Bush, Roslin, Midlothian EH25 9RG, Scotland, United Kingdom

HUGHES, H.P.A., Department of Immunology, St. George's Hospital Medical School, Cranmer Terrace, London SW17 0RE, United Kingdom

HUGHES, D.L., AFRC Institute for Research on Animal Diseases, Compton, Newbury, Berkshire RG16 0NN, United Kingdom

NEWBOLD, C.I., Nuffield Department of Clinical Medicine, University of Oxford, John Radcliffe Hospital, Headington, Oxford OX3 9DU, United Kingdom

PARKHOUSE, R.M.E., National Institute for Medical Research, The Ridgeway, Mill Hill, London NW7 1AA, United Kingdom

ROSE, M.E., Houghton Poultry Research Station, Houghton, Huntingdon, Cambs. PE17 2DA, United Kingdom

RUSSELL, D.G., Max-Planck-Institut für Biologie, Correns-strasse 38, D-4700 Tübingen

SIMPSON, A.J.G., Division of Parasitology, National Institute for Medical Research, Mill Hill, London NW7 1AA, United Kingdom

SMITHERS, S.R., Division of Parasitology, National Institute for Medical Research, Mill Hill, London NW7 1AA, United Kingdom

TERRY, R.J., Department of Applied Biology, Brunel University, Uxbridge, Middlesex UB8 3PH, United Kingdom

TURNER, M.J., Merck Sharp & Dohme, Research Laboratories, 126 E. Lincoln Avenue, P.O.Box 2000, Rahway, NJ 07065, USA

Introduction

R.J. TERRY

When Michael Parkhouse first approached me concerning this book, I asked him whether I could delay writing the Introduction until I had seen what there was to introduce. I am glad that he agreed, because it not only makes my task easier, but it also allows me to congratulate my fellow contributors on their most scholarly and extremely useful reviews. Although I have tried to keep up with developments in as many areas as possible, I was startled by the amount of information we now have on the antigens of some parasites, and could not help contrasting our present knowledge with that available to me when I first became interested in the immunology of parasites, exactly 30 years ago.

There were at that time two books dealing with the subject, that by TALIAFERRO (1929) and the more recent review by CULBERTSON (1941). They were both extremely well-written books, full of interest and speculation, but without much hard information. KABAT and MAYER (1948), in their extremely important treatise on immunochemistry, never once mentioned parasite antigens. BOYD (1956) accurately summarized the state of play in his sentence – "Parasites of course contain proteins but the chemistry of those proteins is very imperfectly known." (!)

I should perhaps add that our general understanding of immunology was very imperfect at this time. The studies of Porter and Edelman, which were to lead to the elucidation of antibody structure, were just getting started, and it was to be some years before Gowans was to show conclusively that the small lymphocyte was the cellular basis of the immune response. And yet, the techniques which were going to lead directly to our present level of understanding were in many instances already developed.

TISELIUS and KABAT (1939) had applied the technique of electrophoresis to immune sera and purified antibodies, and had shown that most antibodies had γ-mobility and could be separated from other serum proteins. KUNKEL and TISELIUS (1951) and SMITHIES (1955) had demonstrated that electrophoresis could be carried out on paper and starch gels, respectively, which provided permanent records of the separations. SVEDBERG and PEDERSEN (1940) had described ultracentrifugation and how it might be used to separate macromolecules, whilst OUDIN (1946) and OUCHTERLONY (1949) had described their techniques for immunoprecipitation in agar gels. The techniques of electrophoresis and diffusion in agar gel had recently been combined by GRABAR and WILLIAMS (1955) to provide a tool with great analytical properties, whilst COONS et al. (1942) and EISEN and KESTON (1949) had begun to label proteins with fluorescent markers and radiolabels, respectively.

Department of Applied Biology, Brunel University, Uxbridge, Middlesex UB8 3PH, United Kingdom

Science advances as much by new techniques as by new concepts, and the new methodology was enthusiastically applied to a number of immunological problems, including the immunology of infectious diseases and even parasite immunology. For example, SMITHERS and TERRY (1959) used ^{131}I-labelled proteins to show that the increase in "γ-globulins" seen in experimental trypanosomiasis was the result of greatly increased synthesis rather than of reduced catabolism. But such studies were rare, and in general the application of the new technology to the study of parasites was very slow. At a symposium organized by the international organizations of medical sciences under the joint auspices of UNESCO and WHO and reported in a volume of over 600 pages (ACKROYD 1964), parasites were poorly represented. Ten pages were devoted to *Trichomonas foetus* of cattle, an interesting parasite but not one in the first rank when related to human welfare. There was a brief mention of antigen-coated bentonite particles in the immunodiagnosis of *Trichinella* and *Echinococcus* and a couple of lines on a passive haemagglutination test for *Toxoplasma* diagnosis.

There was nothing on the major tropical parasitic diseases of man–malaria, schistosomiasis and filariasis; the parasites mentioned in the symposium can all be found in temperate countries. Herein lies one of the reasons for slow progress in understanding the immunology of the major parasitic diseases: lack of money and consequently lack of effort. But there were other factors which bore upon this problem besides a cynical "I'm all right Jack" attitude. Parasitic diseases were largely chronic, persisting in some cases for many years. Clearly then, there was no effective immunity! Perhaps parasites were antigenically inert or, alternatively, they might be just too big to be affected by antibodies or even by antibodies in combination with phagocytic cells. Immune responses perhaps bounced off them, like rifle bullets off a tank. Furthermore, parasites had complex life cycles, often requiring vectors, and were difficult to maintain in the laboratory and to obtain in any quantity. Most, and all the important, parasites could not be cultured in vitro. And, to be fair, chemotherapy seemed to be winning the battle against parasites. Judicious combinations of antimalarial chemotherapy and the destruction of mosquitos with insecticides were certainly reducing the extent of the malarious areas of the world, and there were high hopes of equally effective drugs in African trypanosomiasis, schistosomiasis and filariasis.

A number of events changed our viewpoint drastically. Firstly, antiparasite chemotherapy and "vectoricides" did not prove as effective as had been hoped, resistance of both parasites and vectors developing rapidly in some parts of the world. Around the same time a means of vaccinating against a parasite, *Dictyocaulus viviparus*, lungworm of cattle, was devised, using irradiated infective larvae (JARRETT et al. 1958). This concept was developed into a commercial vaccine, which proved to be very effective and has been widely used (POYNTER et al. 1960). Unfortunately, irradiated larvae of other important nematode parasites, such as *Haemonchus contortus*, proved to be ineffective in young animals, although they did protect adults (MANTON et al. 1962), thus limiting the general use of irradiated antiparasite vaccines.

But there were other promising signs of immunity to parasites: the demonstration that antibodies could at least partially protect against malaria of man (COHEN et al. 1961) and experimental animals (TERRY 1956); and experimental demonstrations of possible anti-malarial vaccines (TARGETT and FULTON 1965); TARGETT and VOLLER 1965).

The reverse of these partial successes was the clear demonstration that parasites had evolved ways of evading or otherwise subverting host immunity. It had long been known that infections with African trypanosomes followed a cyclical pattern, but it was not until the work of GRAY (1965a, b) in West Africa that we began to comprehend the true nature of antigenic variation. BROWN and BROWN (1965) discovered a similar phenomenon in the monkey malaria parasite *Plasmodium knowlesi*. Somewhat later, VICKERMAN and LUCKINS (1969) localized the variant antigens to the trypanosome surface coat and LE PAGE (1967) and CROSS (1975) began the biochemical studies so well described by Mervyn Turner in this volume.

An equally interesting evasion mechanism was also demonstrated in a helminth parasite *Schistosoma mansoni*. SMITHERS and TERRY (1965) showed that following a small initial infection, rhesus monkeys developed a powerful resistance to reinfection with large numbers of *S. mansoni* cercariae. It was further demonstrated that the presence of adult worms provided a major stimulus to this anti-cercarial immunity, but paradoxically the adults were unaffected by this immunity (SMITHERS and TERRY 1967). This phenomenon was later described as "concomitant immunity" (SMITHERS and TERRY 1969) and clearly pointed to the adult worms being able to evade the immune responses that they evoke. SMITHERS et al. (1969) made and tested the hypothesis that the adult worms escaped destruction by acquiring molecules of host origin, these "host antigens" adhering firmly to the parasite surface. Young invading stages that had not yet acquired host antigens were susceptible to destruction by host immune responses, providing that these could act with sufficient rapidity.

The decade of the 1970s thus began with a somewhat pessimistic view of parasite control through chemotherapy and vector destruction, and a rather more optimistic view of control through vaccination. Perhaps somewhat paradoxically, the evasion mechanisms that had been uncovered proved to be attractive to young workers looking for a suitable challenge. There was a feeling that perhaps parasites were "too clever by half" but once their evasion mechanisms had been delineated they might be disabled.

And then the missing ingredient became increasingly available – money! Government and quasi-government bodies, such as the Medical Research Council of the United Kingdom, had always given some support for research into "Third World Diseases", and independent bodies such as the Wellcome Trust had also allocated funds for tropical medical research. Countries most active in these areas tended to be the ex-colonial powers, but as the decade progressed other governments, and particularly that of the United States, began to provide more funding. Additionally, private foundations, such as the Rockefeller and the Edna McConnell Clark Foundations, began to fund work on parasites and, of major importance, the Special Programme on Research and Training in Tropical Diseases (TDR) was launched by the World Health Organization. Five of the six diseases identified by WHO for special attention are caused by parasites. More recently still, special European Economic Community funds have become available for research on "Health in the Tropics".

This combination of increased funding, increased knowledge of basic immunology, and increased interest in parasite immunology led to many advances during the first half of the 1970s. We learned a great deal about the complex nature of the immune responses requires to destroy parasites, and also a great deal about the sophisticated devices evolved by parasites to combat these responses. But our detailed understanding of the biochemical nature, distribution, and immunogenicity of individual parasite antigens was far from adequate.

A number of technical advances have combined to change this situation, allowing the investigations which have led to extremely detailed knowledge of some parasite antigens, as reported in the various chapters which follow. Some of these advances have related to the in vitro culture of parasites which has enabled us not only to obtain relatively large quantities of parasite antigens but also to study their biochemistry and, in some instances, their functions, divorced from the host. Particularly useful in this respect have been the partial culture of schistosomes (CLEGG 1965); the cultivation of bloodstream forms of African trypanosomes (HIRUMI et al. 1977); and the culture of the bloodstream stages of the most important human malaria parasite, *Plasmodium falciparum* (TRAGER and JENSEN 1976).

Of the greatest importance to the study of parasite antigens, particularly surface antigens, has been the advent of hybridoma technology and the production of monoclonal antibodies (KOHLER and MILSTEIN 1975). The following pages are plentifully sprinkled with references to this technology, and there is no doubt that these highly specific probes will continue to be of value in the identification, purification and characterization of key antigens.

Finally, a beginning has been made in applying DNA recombinant technology to a study of parasite antigens, in some instances with the explicit aim of producing appropriate antigens in large quantities as a basis for vaccination. In this respect we are furthest advanced in our studies on malaria, and whilst being sensibly cautious, we can be hopeful of some success in other parasites.

Turning now to the chapters in this volume, there is evident variation, not only in length, but in the detailed understanding of the antigens of the several parasites. We know most about the antigens of malaria (Newbold), schistosomiasis (Simpson and Smithers), and trypanosomiasis (Turner), although our knowledge of African trypanosome antigens is largely confined to the variant specific glycoproteins. We know least about the antigens of *Eimeria* (Rose), *Babesia* and *Theileria* (Allison) and *Entamoeba* (Chayen and Avron); our understanding of the antigens of other parasites – *Toxoplasma* (Huw Hughes), taeniid cestodes (Harrison and Parkhouse), *Fasciola* and other trematodes (Denys Hughes), and nematodes (Almond and Parkhouse) – is intermediate.

I do not believe that this variation in understanding reflects varying difficulties with the different systems, much less varying abilities of the investigators concerned. There is however a strong positive correlation between the depth of our understanding and the number of workers concerned with a particular parasite, and, at one step removed, with the amount of money being spent. The message is reasonably clear: the tools for improving our understanding have been developed and are appropriate for all parasites – but people must be paid to use them.

One aspect which is common to all chapters is the very clear way in which they have been written. Not only that, but all have obeyed the instructions of Michael Parkhouse, as Editor, to include enough parasitology, immunology and pathology to permit the reader to appreciate the special aspect of parasite antigens against a meaningful general background. The result is a volume which reviews the major areas of interest in the whole field of parasite immunology and is not merely a catalogue of more or less well-defined parasite antigens.

Finally,what use can we make of all the information already gathered and about to be gathered? There is general optimism that at least some of the defined parasite antigens may provide candidate vaccines, and this optimism is particularly strong in the field of malaria. It is foreign to my nature to be a wet blanket, but as one who took part in the development of the *Dictyocaulus* vaccine, I know that even given a candidate vaccine, there are still many problems to solve in production, standardization, and the delivery of the vaccine before it makes much impact against the disease. It is often easier and quicker to make use of novel antigens in improved diagnostic techniques, although the lessons of a recent conference held in Stockholm on this subject (WHO/International Association for Biological Standards Meeting, February 1985) indicated that the problems lay in making the techniques simpler and cheaper, rather than more sensitive and specific. But in any event, the more we know about those parasite macromolecules that we designate "antigens", the more chance we shall have of controlling these parasitic diseases by one means or another.

And I can well imagine some middle-aged researcher, in the year 2015, beginning an introductory chapter to a volume on parasite antigens, with the words: "When I first became interested in parasite immunology, now almost 30 years ago, I found that the volume edited by Parkhouse . . .".

References

Ackroyd JF (ed) (1964) Immunological methods. Blackwell, Oxford

Boyd WC (1956) Fundamentals of immunology, 3rd edn. Interscience, New York

Brown KN, Brown IN (1965) Immunity to malaria: antigenic variation in chronic infections of *Plasmodium knowlesi*. Nature 208:1286–1288

Clegg JA (1965) In vitro cultivation of *Schistosoma mansoni*. Exp Parasitol 16:133–147

Cohen S, McGregor IA, Carrington S (1961) Gamma globulin and acquired immunity to human malaria. Nature 192:733–737

Coons AH, Creech HJ, Jones N, Berliner E (1942) The demonstration of pneumococcal antigen in tissues by the use of fluorescent antibody. J Immunol 45:159–170

Cross GM (1975) Identification, purification and properties of clone-specific glycoprotein antigen constituting the surface coat of *Trypanosoma brucei*. Parasitology 71:393–417

Culbertson JT (1941) Immunity against animal parasites. Columbia University Press, New York

Eisen HN, Keston AS (1949) The immunologic reactivity of bovine serum albumin labelled with trace amounts of radioactive iodine (^{131}I). J Immunol 63:71–80

Grabar P, Williams CA (1955) Méthode immunoélectrophorétique d'analyse de mélanges de substances antigénique. Biochim Biophys Acta 17:67–74

Gray AR (1965a) Antigenic variation in clones of *Trypanosoma brucei*: 1. Immunological relationship of the clones. Ann Trop Med Parasitol 59:27–36

Gray AR (1965b) Antigenic variation in a strain of *Trypanosoma brucei* transmitted by *Glossina morsitans* and *G. palpalis*. J Gen Microbiol 41:195–214

Hirumi H, Doyle JJ, Hirumi K (1977) African trypanosomes: cultivation of animal-infective *Trypanosoma brucei* in vitro. Science 196:992–994

Jarrett WFH, Jennings FW, McIntyre WIM, Mulligan W, Urquhart GM (1958) Irradiated helminth larvae in vaccinations. Proc R Soc Med 51:743–744

Kabat EA, Mayer MM (1948) Experimental immunochemistry. Thomas, Springfield

Kohler G, Milstein C (1975) Continous culture of fused calls secreting antibody of defined specificity. Nature 256:495–497

Kunkel HG, Tiselius A (1951) Electrophoresis of proteins on filter paper. J Gen Physiol 35:89–118

Le Page R (1967) Studies on the variable antigens of *Trypanosoma brucei*. Trans R Soc Trop Med Hyg 61: 139–140

Manton VJA, Peacock R, Poynter D, Silverman PH, Terry RJ (1962) The influence of age on naturally acquired resistance to *Haemonchus contortus* in lambs. Res Vet Sci 3:308–314

Ouchterlony O (1949) An in vitro test of the toxin-producing capacity of *Corynebacterium diphtheriae*. Lancet I:346–348

Oudin J (1946) Méthode d'analyse immunochimique par précipitation spécifique en milieu gelifié. CR Acad Sci 222:115–125

Poynter D, Jones BV, Nelson AMR, Peacock R, Robinson J, Silverman PH, Terry RJ (1960) Symposium on husk: IV. Recent experiences with vaccinations. Vet Rec 72:1078–1090

Smithers SR, Terry RJ (1959) Changes in the serum proteins and leucocyte counts of rhesus monkeys in the early stages of infection with *Trypanosoma gambiense*. Trans R Soc Trop Med Hyg 53:336–345

Smithers SR, Terry RJ (1965) Naturally acquired resistance to experimental infections of *Schistosoma mansoni* in the rhesus monkey *(Macaca mulatta)*. Parasitol 55:701–710

Smithers SR, Terry RJ (1967) Resistance to experimental infection with *Schistosoma mansoni* induced by the transfer of adult worms. Trans R Soc Trop Med Hyg 61:517–533

Smithers SR, Terry RJ (1969) Immunity in schistosomiasis. Ann NY Acad Sci 160:826–840

Smithers SR, Terry RJ, Hockley DJ (1969) Host antigens in schistosomiasis. Proc R Soc Lond [Biol] 171:483–494

Smithies O (1955) Zone electrophoresis in starch gels: group variations in the serum of normal human adults. Biochem J 61:629–641

Svedberg T, Pedersen KO (1940) The ultracentrifuge. Clarendon, Oxford

Taliaferro WH (1929) The immunity of parasitic infections. Century, New York

Targett GAT, Fulton JD (1965) Immunization of rhesus monkeys against *Plasmodium knowlesi* malaria. Exp Parasitol 17:180–193

Targett GAT, Voller A (1965) Studies on antibody levels during vaccination of rhesus monkeys against *Plasmodium knowlesi*. Br Med J II:1104–1106

Terry RJ (1956) Transmission of antimalarial immunity *(Plasmodium berghei)* from mother rats to their young during lactation. Trans R Soc Trop Med Hyg 50:41–46

Tiselius A, Kabat AE (1939) An electrophoretic study of immune sera and purified antibody preparations. J Exp Med 69:119–131

Trager W, Jensen JB (1976) Human malaria parasites in continuous culture. Science 193:673–675

Vickerman K, Luckins A (1969) Localization of variable antigens in the surface coat of *Trypanosoma brucei* using ferritin-labelled antibody. Nature 224:1125–1126

The *Eimeria*

M.E. ROSE

1 Introduction

Coccidiosis, i.e. infection with *Eimeria* and *Isospora* spp., is of great economic importance in agriculture. This is especially true of poultry production but, with the adoption of more intensive systems of husbandry for other domestic animals, the disease is beginning to assume greater proportions in these species also (GREGORY et al. 1980; JOYNER et al. 1981; FITZGERALD 1980). In the case of poultry it is unlikely that the broiler industry would have developed to its present extent in the absence of a means of controlling the parasite, i.e. effective medication (RYLEY 1982). Because of the ever-recurring problem of drug resistance this has necessitated a constant search for new compounds. The cost of this research, coupled with that of development and testing, has risen to such an extent that few pharmaceutical companies are now engaged in such work (RYLEY 1982). The need for an alternative means of control is, therefore, becoming more urgent and is arousing considerable interest in the possibility of vaccination against the disease. Hitherto, effort has been concentrated on the development of live attenuated strains of the various *Eimeria* spp. for use as prophylactic agents since the limited attempts to induce protective immunity by the injection of various crude preparations of non-viable parasite material have been unsuccessful (see ROSE and LONG 1980). However, the recent rapid advances made in the biochemical and molecular definition of the important antigens of some

Houghton Poultry Research Station, Houghton, Huntingdon, Cambs. PE17 2DA, United Kingdom

Current Topics in Microbiology and Immunology, Vol. 120
© Springer-Verlag Berlin · Heidelberg 1985

of the other parasitic protozoa, together with progress in genetic engineering, suggest that a similar approach should now be made in the case of the *Eimeria*.

Work on the immune responses of the host has been proceeding for some time but the antigens responsible for inducing protective immunity have not yet been characterised and their investigation is in its infancy. Nevertheless, there is information available which indicates likely areas for research and the aim of this review is to discuss the relevant aspects of the host-parasite relationship and to summarise the progress to date on the characterisation of antigens which may be involved in protection.

2 Life Cycle

The eimerian coccidia parasitise a very wide range of hosts, invertebrate and vertebrate, the latter including the common domesticated and laboratory animals. They are markedly host-specific being, with some exceptions, restricted to closely related species or subspecies of host (see JOYNER 1982). Any given host is usually parasitised by several species of eimeria, e.g. 7 in the chicken, at least 11 in the sheep and 14 in the mouse. In addition to host specificity there is a marked degree of tissue specificity and most eimeria develop in epithelial cells of the intestine. Within the intestine also there is a remarkable degree of site restriction: in the chicken, for example, *E. acervulina* grows in the duodenum whereas *E. tenella* is a parasite of the paired caeca.

The life cycle (Fig. 1) is a complicated one, of the sporozoan type, and consists of a fixed, small (often four) number of serial cycles of asexual reproduction (schizogony), followed by the formation of gametes (gametogony). These fuse to form a zygote which develops a thick wall and is passed out in the faeces as a resistant cyst (oocyst). A period of development (sporogony) outside the host ensues during which sporozoites are formed. The cyst becomes infective only when differentiation of sporozoites is completed. Infection destroys the intestinal epithelium, with effects varying from interference with the absorption of nutrients to haemorrhage and death, according to the species of parasite involved and the numbers of oocysts ingested. Infections are brief and self limiting due to the fixed and finite nature of schizogony.

3 Antigens Likely To Be Involved in Protection

In the case of most species of *Eimeria*, infection of the normal (immunocompetent) host induces an immune response which renders it resistant, in at least some degree, to subsequent attempts at infection. Little work has yet been done on the characterisation of any of the many antigens which could be involved. There is, however, information on the immunising abilities of different species and developmental stages, and their specificity in this respect, which should provide some indication of the most fruitful areas for investigation.

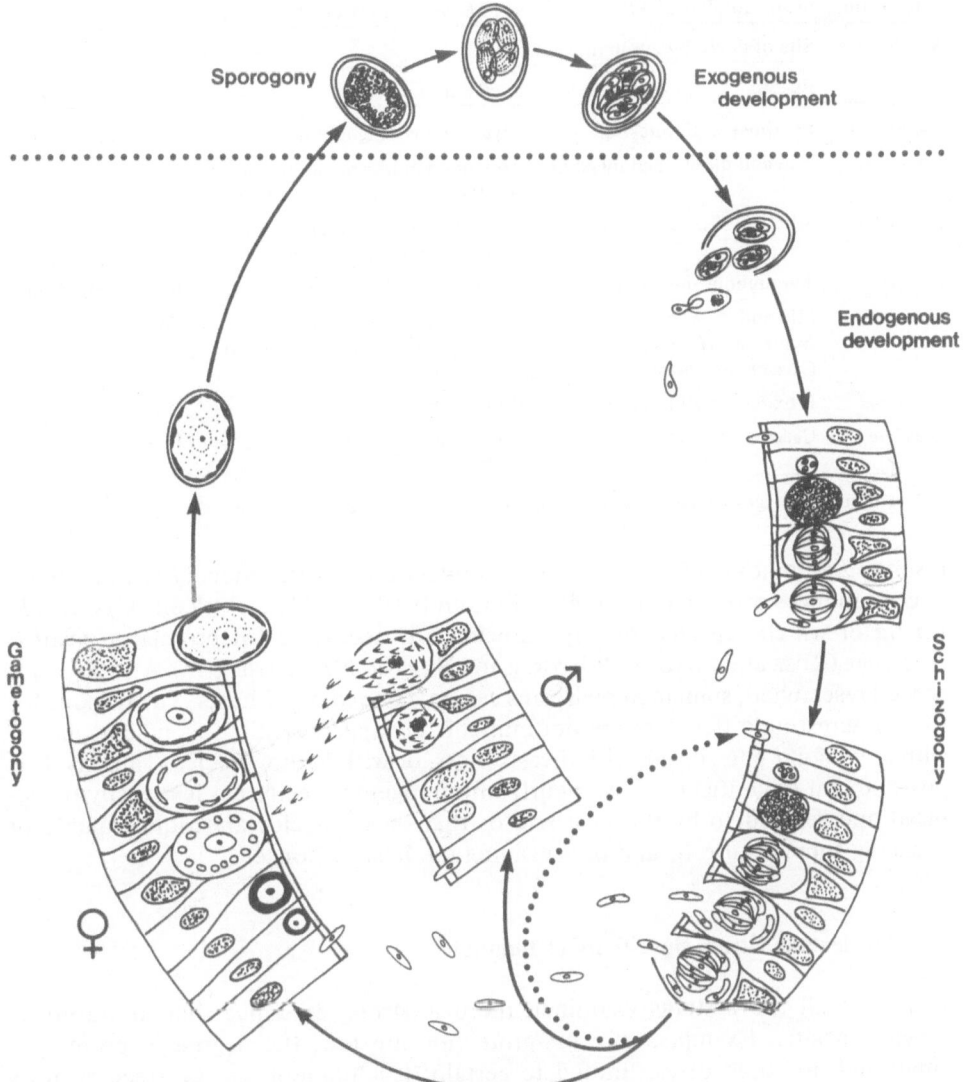

Fig. 1. Diagrammatic representation of the life cycle of *Eimeria* species

3.1 Species Differences in Immunogenicity

The degree of immunity induced by infection depends upon a number of factors (reviewed by ROSE 1978; ROSE and LONG 1980) which include the numbers of oocysts given, their mode of administration and, importantly, the species of parasite. There are remarkable differences in the immunising abilities of different species of *Eimeria* within the same host. For example, in the chicken infection with a single oocyst (JOYNER and NORTON 1976) or even sporocyst (LEE and FERNANDO 1978) of *E. maxima*

Table 1. Immunising abilities of different specias of *Eimeria* of the chicken

Species	Site of development in intestine		Immunising ability
	Region	Location in mucosa[a]	
E.acervulina	Proximal small intestine	Crypts and villi, epithelium	Moderately good
E.brunetti	Mid and distal small intestine	Villi, epithelium and lamina propria (second schizont, gametocytes)	Good
E.maxima	Mid and distal small intestine	Crypts and villi, epithelium and lamina propria (gametocytes)	Very good
E.mitis	Throughout small intestine	Crypts and villi, epithelium	Moderately good
E.necatrix	Mid and distal small intestine (schizogony) Caeca (gametogony)	Crypts and villi, epithelium and lamina propria (second schizont)	Poor
E.praecox	Proximal small intestine	Villi, epithelium	Good
E.tenella	Caeca	Crypts and surface epithelium, lamina propria (second schizont)	Poor

[a] Developmental stages which occur in the lamina propria are given in parentheses

results in a demonstrable resistance to reinfection and the inoculation of 50–100 oocysts induees prompt and complete immunity (ROSE 1974). This contrasts with *E. tenella* of which repeated or large inocula are necessary for protection against challenge (ROSE and LONG 1962; JOYNER and NORTON 1973; ROSE 1974). Although not so well researched, similar comparisons can be made in other hosts. The reasons for these interesting differences are not immediately apparent: there is no correlation with area of gut parasitised, with invasiveness, or with pathogenicity (Table 1). It is possible that the antigens of the poorly immunogenic species are less easily recognised and reacted to by the host and/or that these species are more capable of inducing suppressor cells and thus evading the host response.

3.2 Species and Strain Specificity of Immunity

In almost all the systems examined there is strong evidence that immunity is species specific. Examples of cross-protection are few, the degree of protection small and, in some cases, limited to certain developmental stages (see reviews ROSE 1973, 1978). In the more intensively studied coccidia of the fowl, immunological differences, as indicated in incomplete cross-protection, have also been demonstrated between different isolates of some, but not all, species. It is known to occur in *E. acervulina* (JOYNER 1969) and is particularly marked in the very immunogenic *E. maxima* (LONG 1974; NORTON and HEIN 1976; JEFFERS 1978; LONG and MILLARD 1979), but was not found amongst the isolates of *E. tenella* examined by JEFFERS (1978). More recently there have been indications of immunological differences in strains of *E. mitis* (SHIRLEY et al. 1983; MCDONALD, personal communication) and of *E. necatrix* (SHIRLEY, personal communication). The two species, *E. acervulina* and *E. maxima*, in which this trait is most marked, are the most prevalent in commercial broiler houses (JEFFERS 1974; LONG et al. 1975), despite their abilities to immunise very effectively. There is nevertheless no evidence (SHIRLEY and HOYLE 1981) for

Table 2. Antigenic cross-reactivity in ELISA between different species of *Eimeria*. (ROSE and MOKETT 1983)

Antigen[a]	Reactions of antisera[b] (% of homologous log 2 titre)			
	E.acervulina (chicken)	*E.maxima* (chicken)	*E.tenella* (chicken)	*E.nieschulzi* (rat)
E.acervulina	100	100	65	NT
E.maxima	<	100	67	<
E.tenella	52	95	100	<
E.nieschulzi	NT	82	50	100

[a] Soluble, prepared from disrupted oocysts
[b] From infected animals
<, negative at lowest dilution ($\log_2$ 5.64) tested; NT, not tested

antigenic modulation of the type which occurs in trypanosomes or plasmodia (see later). The specificity of immunity exhibited by the *Eimeria* suggests that the antigens important for protection should be sought amongst those that are unique to any given species (or even strain). It is possible, however, that the immunogenic differences between species may result from differences in the presentation of similar proteins and might therefore be overcome by the use of synthetic peptides (see LERNER 1982).

There is a great deal of antigenic cross-reactivity in the eimeria which is evident between different species which parasitise the same host and between species which parasitise different hosts. This has been demonstrated in vitro by Ouchterlony double diffusion in agar gel using soluble antigen prepared from oocysts (PIERCE et al. 1962; ROSE and LONG 1962), the indirect fluorescent antibody test (IFAT) using various stages of the parasite, both free and in tissues, as antigen (ČERNÁ 1970, 1974; SOKOLIĆ et al. 1974; ABU ALI et al. 1976; TANIELIAN et al. 1976; DANFORTH and AUGUSTINE 1982; TILAHUN and STOCKDALE 1982) and by the enzyme-linked -immunosorbent assay (ELISA) with soluble oocyst antigen (ROSE and MOCKETT 1983). Although cross-reactivity is common it is variable between different species in the same hosts and between species of different hosts – and there are some interesting anomalies. For example, antiserum from chickens infected with *E. maxima* reacted in the ELISA with antigens of *E. acervulina* and *E. nieschulzi* (of the rat), but the converse did not apply (Table 2). Differences in the presentation of antigen by different parasites and/or in their recognition by different hosts may account for this.

Cross-reactivity between *Eimeria* antigens is also evident in vivo from the results of tests for delayed-type hypersensitivity carried out in the wattles of chickens immunised by infection. Particulate antigens prepared from the oocysts of three species of *Eimeria* which infect the chicken (*E. maxima, E. necatrix* and *E. tenella*) were cross-reactive and positive results were also obtained with an antigen prepared from *E. bovis*, a parasite of cattle (GIAMBRONE and KLESIUS 1980).

3.3 Immunogenicity of Different Developmental Stages

There is good evidence to suggest that the sexual stages of eimeria are poorly immunogenic and that the immunising effect of infection is due, in great part, to the development of the asexual stages (see ROSE 1982). Animals which experience infection with gametogony stages only (by the inoculation intraintestinally of merozoites derived from the cycle of schizogony immediately preceding gamete formation) do not resist challenge with the homologous, sexual, or asexual stages (see HORTON-SMITH et al. 1963a; ROSE 1967; ROSE and HESKETH 1976). Interestingly, the final hosts of the related heteroxenous coccidium *Sarcocystis* spp. acquire little or no immunity to reinfection, and in these hosts the infection is confined to the gametogony stages of the parasite (see TADROS and LAARMAN 1976; FAYER 1977). Thus it is possible that the poor immunogenicity of sexual developmental stages is a feature of the coccidia. This is borne out by the results obtained in an IFAT of sera from mice immunised by infection with *E. falciformis* var. *pragensis*: fluorescence with sexual stages was weak, in contrast to the strong reactions obtained with sporozoites and mature asexual developmental stages (MESFIN and BELLAMY 1980). It is not known with certainty whether infection with asexual stages only can protect against challenge with gametocytes but this would appear to be likely (ROSE and HESKETH 1976; MESFIN and BELLAMY 1979).

Amongst the asexual developmental stages some appear to be more immunogenic than others, for example, the second-generation schizont of *E. tenella* compared with the first-generation schizont, or the sporozoite. Certainly a strain of *E. tenella* selected for the absence of the second-generation schizont (McDOUGALD and JEFFERS 1976) is very poorly immunogenic (JOHNSON et al. 1979) and relapses occur after the withdrawal of drugs which prevent the development of, but do not kill, sporozoites (LONG and MILLARD 1968). Although this relapse suggests that even prolonged contact with sporozoites does little to stimulate immunity, it must be recognised that these sporozoites are metabolically inhibited and possibly not expressing their normal range of antigens. It seems likely that the sporozoite does contain important protective antigens as its development is inhibited in the solidly immune animal (HORTON-SMITH et al. 1963b; LEATHEM and BURNS 1967; NIILO 1967; MESFIN and BELLAMY 1979; ROSE et al. 1984). In view of the relation between the magnitude of infection and its immunising effect (e.g. MUKKUR and BRADLEY 1969; LIBURD 1973) true comparisons between the immunising abilities of different developmental stages are difficult to make.

It is not known whether there is any specificity between the different asexual developmental stages in the immunity induced by them and no detailed antigenic analyses of any of the developmental stages have been carried out. There appear to be many common antigens, as indicated by Ouchterlony gel diffusion (ROSE 1959; ROSE and LONG 1962) but there is evidence for some stage-specific ones. Serum from mice infected with *E. pragensis*, after absorption with merozoites, reacted in IFAT with sexual stages only (ČERNÁ and ŽALMANOVÁ 1972) and antiserum raised in rabbits against sporozoites of *E. tenella* also reacted in IFAT with second-stage but, surprinsingly, not first-stage generation schizonts (KOUWENHOVEN and KUIL 1976). More recently some hybridoma-produced monoclonal antibodies have been shown by IFAT and ELISA to be both species and developmental stage specific whereas others cross-reacted with either different species of coccidia or different developmental stages (DANFORTH 1983a; DANFORTH and AUGUSTINE 1983). The extent of cross-reactivity of three monoclonal antibodies, induced against sporozoites of *E. tenella*, with asexual stages was variable but none reacted in IFAT with sexual stages grown in cultured chick kidney cells (DANFORTH 1983a).

   The information available suggests that the asexual developmental stages are those most likely to express protective antigens but convenience must dictate which of these stages should be examined initially by the newer immunochemical techniques. Hitherto, none of the various attempts to induce protective immunity by the administration of crude preparations of non-viable parasite material or "metabolic products" has been successful (see ROSE and LONG 1980) despite the elicitation, in many cases, of immune responses indistinguishable from those observed in infected (and subsequently immune) animals.

3.4 Immunochemical and Molecular Investigations

Investigation of the immunochemistry and molecular biology of eimeria antigens has barely started; consequently very little published information is available. Hybridoma-produced monoclonal antibodies with varying degrees of specificity have been raised to sporozoites and merozoites of a number of different species of avian eimeria (DANFORTH 1982, 1983a; DANFORTH and AUGUSTINE 1982, 1983; MH WISHER, unpublished work). Varying degrees of species and developmental stage-specificity, as demonstrated by IFAT and ELISA, have been reported and also different patterns of staining in IFAT. Some monoclonal antibodies had antiparasite effects, demonstrable by their ability to inhibit penetration and development of sporozoites in cultured chick kidney cells. Pretreatment of sporozoites and the continued presence of the antibodies in the culture fluid was necessary to maintain the inhibition of development (DANFORTH 1983a; AUGUSTINE and DANFORTH 1983).

   Research has now begun on the molecular aspects of *Eimeria* antigens. Preliminary results have shown that the surface membranes of several species of *Eimeria* have similar profiles of polypeptides, with molecular weights ranging from 20 000 to >200 000. Indirect precipitation of *E. tenella* material with serum from chickens immune to this species enabled at least seven of the ^{125}I-labelled surface polypeptides (MW 24 000–125 000) and five ^{35}S-labelled polypeptides (MW 35 000 to >200 000) to be identified as antigens (WISHER 1983). This work is continuing and is being extended to other developmental stages considered likely to possess important antigens, and to isolated infected host cells.

   The characterisation of antigens, in combination with the development of monoclonal antibodies possessed of significant function in immunity, should provide the information necessary for the eventual production of a vaccine by gene manipulation.

4 Diagnosis

It is unlikely that currently available immunologically based techniques could be useful in the diagnosis of coccidiosis. In its acute stage the disease is readily recognisable, on postmortem examination, from the often characteristic lesions and the developmental stages associated with them. In the live animal oocysts are detectable in the faeces after the completion of gametogony. In most cases the cessation of oocyst production is usually followed by immunity to the clinical effects of disease and diagnosis at this stage is not likely to be useful, unless it can be species specific. The cross-reactivity of antigen/antibody reaction in *Eimeria* systems, already discussed, make this appear unlikely and cross-reactivity may extend across other genera of the coccidia.

It is, however, possible that in the future there could be an application in diagnosis for fully characterised antigens, and monoclonal antibodies with species specificity. The latter could be useful in the laboratory to differentiate species, particularly when the amount of parasite material is too small for analysis of enzymes by electrophoresis, a method in current use (SHIRLEY and ROLLINSON 1979). Similarly, the identification and isolation of species-specific antigens might enable a diagnostic test, such as delayed-type hypersensitivity, to be used in the field to determine whether animals had experienced infection with any particular species. This would indicate their likely susceptibility to infection and the prevalence of any given species.

5 Escape

There is little to suggest that *Eimeria* spp. make use of the mechanisms for avoiding the response of the host which are such a feature of many other successful parasites, but immunosuppression and strain variation in immunogenicity are seen.

5.1 Immunosuppression

Infection with *Eimeria* spp. has been shown to exacerbate concomitant infections with unrelated parasites. Thus, infections with *E. acervulina* or *E. mitis* in chickens produced increased parasitaemia or recrudescence of latent *Plasmodium juxta-nucleare* or *P. gallinaceum* (AL-DABAGH 1961) and the rejection of *Trichinella spiralis* (DUSZYNSKI et al. 1978) and *Nippostrongylus brasiliensis* (BRISTOL et al. 1983) was delayed in rats infected with *E. nieschulzi*. More recently, responses of peripheral blood leucocytes and spleen cells to mitogens such as phytohaemagglutinin or bacterial lipopolysaccharide have been shown to be decreased during infection of the chicken with *E. tenella*. Concurrently, however, there were responses to antigens prepared from the parasite (ROSE and HESKETH 1984).

5.2 Antigenic Variation

The partial strain-specificity of immunity exhibited by some species of *Eimeria* of the chicken has already been mentioned (Sect. 3.2). It is interesting that those species (*E. acervulina* and *E. maxima*) in which strain variation is a feature are both comparatively highly immunogenic and prevalent in the field. This apparent contradiction might be explicable if the parasites were capable of antigenic variation, as seen in some other parasitic protozoa, notably the trypanosomes. There is, however, no indication that variation of this type occurs since isolates of *E. maxima* obtained from the litter after successive crops of broilers were found not to differ when tested by cross-protection (SHIRLEY and HOYLE 1981). Also, the successive passage of *E. acervulina* in the laboratory through partially immunised chickens did not result in a strain which was immunologically distinct (again judged by cross-protection in chickens) from that used for the original inoculum (ROSE, unpublished work). The finite nature of the eimerian life cycle would appear not to be conducive to antigenic modulation in its accepted sense and the variation seen is probably due to the natural heterogeneity of parasites in the field.

6 Concluding Remarks

From the foregoing it should be evident that very little is known about the antigens of the eimeria but that the identification and characterisation of the antigens which elicit protective immunity are necessary and, probably, attainable goals. Existing methods, i.e. non-immunological and cross-immunity tests, for diagnosis are reasonably satisfactory and, in view of the extensive cross-reactivity between species, it is doubtful whether other immunological tests could prove useful. Species specific monoclonal antibodies and antigens could, however, play a part, respectively in the identification of species and in determining their prevalence.

References

Abu Ali N, Movsesijan M, Sokolić A, Tanielian Z (1976) Circulating antibody response to *Eimeria tenella* oral and subcutaneous infections in chickens. Vet Parasit 1:309–316
Al-Dabagh MH (1961) Synergism between coccidia parasites (*Eimeria mitis* and *E. acervulina*) and malarial parasites (*Plasmodium gallinaceum* and *P. juxtanucleare*) in the chick. Parasitology 51:251–261
Augustine PC, Danforth HD (1983) Effect of serum and hybridoma antibodies on invasion of cultured cells by *Eimeria* sporozoites (Abstr). Poult Sci 62:1376
Bristol JR, Pinon AJ, Mayberry LF (1983) Interspecific interactions between *Nippostrongylus brasiliensis* and *Eimeria nieschulzi* in the rat. J Parasitol 69:372–374
Černá Ž (1970) The specificity of serous antibodies in coccidioses. Folia Parasitol (Praha) 17:135–140
Černá Ž (1974) Antigenic relationships between *Eimeria acervulina, E. tenella* and *E. maxima*. Proceedings of the Third International Congress of Parasitology (Munich 26–31 August 1974) p 1135
Černá Ž, Žalmanova Z (1972) An attempt to analyse antigens from sexual and asexual stages of coccidia by the indirect fluorescence antibody reaction. Folia Parasitol (Praha) 19:179–181
Danforth HD (1982) Development of hybridoma-produced antibodies directed against *Eimeria tenella* and *E. mitis*. J Parasitol 68:392–397
Danforth HD (1983a) Use of monoclonal antibodies directed against *Eimeria tenella* sporozoites to determine stage specificity and *in vitro* effect on parasite penetration and development. Am J Vet Res 44:1722–1727
Danforth HD (1983b) Ultrastructural study of monoclonal antibody interaction with surface antigen of avian sporozoites (Abstr). Poult Sci 62:1408
Danforth HD, Augustine PC (1982) Cross-reactivity of monoclonal antibodies directed against various species of avian coccidia (Abstr). Poult Sci 6:1446
Danforth HD, Augustine PC (1983) Development of hybridoma antibodies directed against avian coccidia (Abstr). Poult Sci 62:1408

Duszynski D, Russell D, Roy SA, Castro G (1978) Suppressed rejection of *Trichinella spiralis* in immunized rats concurrently infected with *Eimeria nieschulzi*. J Parasitol 64:83–88

Fayer R (1977) *Sarcocystis leporum* in cotton-tail rabbits and its transmission to carnivores. J Wildl Dis 13:170–173

Fitzgerald PR (1980) The economic impact of coccidiosis in domestic animals. Adv Vet Sci Comp Med 24:121–143

Giambrone JJ, Klesius PH (1980) Chicken coccidiosis: correlation between resistance and delayed hypersensitivity. Poult Sci 59:1715–1721

Gregory MW, Joyner LP, Catchpole J, Norton CC (1980) Ovine coccidiosis in England and Wales 1978–1979. Vet Rec 106:461–462

Horton-Smith C, Long PL, Pierce AE, Rose ME (1963a) Immunity to coccidia in domestic animals. In: Garnham PCC, Pierce AE, Roitt I (eds) Immunity to Protozoa. Blackwell Scientific, Oxford, pp 273–295

Horton-Smith C, Long PL, Pierce AE (1963b) Behavior of invasive stages of *Eimeria tenella* in the immune fowl (*Gallus domesticus*). Exp Parasitol 13:66–74

Jeffers TK (1974) *Eimeria acervulina* and *E. maxima*: incidence, distribution and anticoccidial drug resistance of isolants in major broiler-producing areas. Avian Dis 18:331–342

Jeffers TK (1978) Genetics of coccidia and the host response. In: Long PL, Boorman KN, Freeman BM (eds) Avian coccidiosis. Proceedings of the Thirteenth Poultry Science Symposium 1977 (British Poultry Science Ltd). Longman, Edinburgh, pp 57–125

Johnson J, Reid WM, Jeffers TK (1979) Practical immunization of chickens against coccidiosis using an attenuated strain of *Eimeria tenella*. Poult Sci 58:37–41

Joyner LP (1969) Immunological variation between two strains of *Eimeria acervulina*. Parasitology 59:725–732

Joyner LP (1982) Host and site specificity. In: Long PL (ed) The biology of the coccidia. University Park Press, Baltimore, pp 36–62

Joyner LP, Gregory MW, Norton CC, Done JT, Wells GWH (1981) Coccidiosis and coprophagy in pigs. Vet Rec 108:264–265

Joyner LP, Norton CC (1973) The immunity arising from continuous low-level infections with *Eimeria tenella*. Parasitology 67:333–340

Joyner LP, Norton CC (1976) The immunity arising from continuous low-level infections with *Eimeria maxima* and *Eimeria acervulina*. Parasitology 72:115–125

Kouwenhoven B, Kuil H (1976) Demonstration of circulating antibodies to *Eimeria tenella* by the indirect immunofluorescent antibody test using sporozoites and second-stage schizonts as antigen. Vet Parasitol 2:283–292

Leathem WD, Burns WC (1967) Effects of the immune chicken on the endogenous stages of *Eimeria tenella*. J Parasitol 53:180–185

Lee E-H, Fernando MA (1978) Immunogenicity of a single sporocyst of *Eimeria maxima*. J Parasitol 64:483–485

Lerner RA (1982) Tapping the immunological repertoire to produce antibodies of predetermined specificity. Nature 299:592–596

Liburd EM (1973) *Eimeria nieschulzi* infections in inbred and outbred rats: infective dose, route of infection and host resistance. Can J Zool 57:273–279

Long PL (1974) Experimental infection of chickens with two species of *Eimeria* isolated from the Malaysian jungle fowl. Parasitology 69:337–347

Long PL, Millard BJ (1968) *Eimeria*: effect of meticlorpindol and methyl benzoquate on endogenous stages in the chicken. Exp Parasitol 23:331–338

Long PL, Millard BJ (1979) Immunological differences in *Eimeria maxima*: effect of a mixed immunising inoculum on heterologous challenge. Parasitology 79:451–457

Long PL, Tompkins RV, Millard BJ (1975) Coccidiosis in broilers: evaluation of infection by the examination of broiler house litter for oocysts. Avian Pathol 4:287–294

McDougald LR, Jeffers TK (1976) *Eimeria tenella* (Sporozoa, Coccidia): gametogony following a single asexual generation. Science 192:258–259

Mesfin GM, Bellamy JEC (1979) Effects of acquired resistance on infection with *Eimeria falciformis* var. *pragensis* in mice. Infect Immunol 23:108–114

Mesfin GM, Bellamy JEC (1980) Immunogenicity of the different stages of *Eimeria falciformis* var. *pragensis*. Vet Parasitol 7:87–93

Mukkur TKS, Bradley RE (1969) *Eimeria tenella*: packed blood cell volume, hemoglobin and serum proteins correlated with the immune state. Exp Parasitol 26:1-16

Niilo L (1967) Acquired resistance to reinfection of rabbits with *Eimeria magna*. Can Vet J 8:201-208

Norton CC, Hein M (1976) *Eimeria maxima*: a comparison of two laboratory strains with a fresh isolate. Parasitology 72:345-354

Pierce AE, Long PL, Horton-Smith C (1962) Immunity to *Eimeria tenella* in young fowls (*Gallus domesticus*). Immunology 5:129-152

Rose ME (1959) Serological reactions in *Eimeria stiedae* infection of the rabbit. Immunology 2:112-122

Rose ME (1967) Immunity to *Eimeria tenella* and *Eimeria necatrix* infections in the fowl. 1. Influence of the site of infection and the stage of the parasite. II. Cross-protection. Parasitology 57:567-583

Rose ME (1973) Immunity. In: Hammond DM, Long PL (eds) The Coccidia, *Eimeria, Isospora, Toxoplasma* and related genera. University Park Press, Baltimore, pp 295-341

Rose ME (1974) The early development of immunity to *Eimeria maxima* in comparison with that to *E. tenella*. Parasitology 68:35-45

Rose ME (1978) Immune responses of chickens to coccidia and coccidiosis. In: Long PL, Boorman KN, Freeman BM (eds) Avian coccidiosis. Proceedings of the thirteenth poultry science symposium, 1977. (British Poultry Science Ltd) Longman, Edinburgh

Rose ME (1982) Host immune responses. In: Long PL (ed) The biology of the coccidia. University Park Press, Baltimore, pp 329-371

Rose ME, Hesketh P (1976) Immunity to coccidiosis: stages of the life-cycle of *Eimeria maxima* which induce, and are affected by, the response of the host. Parasitology 73:25-37

Rose ME, Hesketh P (1984) Infection with *Eimeria tenella*: modulation of lymphocyte blastogenesis by specific antigen and evidence for immunodepression. J Protozool 31:549-553

Rose ME, Long PL (1962) Immunity to four species of *Eimeria* in the fowl. Immunology 5:79-92

Rose ME, Long PL (1980) Vaccination against coccidiosis in chickens. In: Taylor AER, Muller R (eds) Vaccination against parasites, vol 18. Symposia of the British society for Parasitology. Blackwell Scientific, Oxford

Rose ME, Mockett APA (1983) Antibodies to coccidia: detection by the enzyme-linked immunosorbent assay (ELISA). Parasite Immunol 5:479-490

Rose ME, Lawn AM, Millard BJ (1984) The effect of immunity on the early events in the life cycle of *Eimeria tenella* in the caecal mucosa of the chicken. Parasitology 88:190-210

Ryley JF (1982) Treatment of veterinary protozoan infections. In: Mettrick DF, Desser SS (eds) Parasites - their world and ours. Elsevier Biomedical, Amsterdam, pp 319-326

Shirley MW, Hoyle SR (1981) The antigenicity of *Eimeria maxima* populations obtained from commercial farms. J Parasitol 67:587-588

Shirley MW, Rollinson D (1979) Coccidia: the recognition and characterization of populations of *Eimeria*. In: Taylor AER, Muller R (eds) Problems in the identification of parasites and their vectors. Blackwell Scientific, Oxford, pp 7-30

Shirley MW, Jeffers TK, Long PL (1983) Studies to determine the taxonomic status of *Eimeria mitis*, Tyzzer 1929, and *E. mivati*, Edgar and Seibold, 1964. Parasitology

Sokolić A, Movsesijan M, Abu Ali N, Tanielian Z (1974) Immunofluorescence in studies on *Eimeria tenella*. Acta Vet Yugoslav 24:55-60

Tadros W, Laarman JJ (1976) *Sarcocystis* and related coccidian parasites: a brief general review, together with a discussion on some biological aspects of their life cycles and a new proposal for their classification. Acta Leiden 44:1-107

Tanielian Z, Abu Ali N, Gannoum BJ, Sokolić A, Borojević D (1976) Circulating antibody response in chickens to homologous and heterologous antigens of *Eimeria tenella*, *E. necatrix* and *E. brunetti*. Acta Parasitol Jugoslav 7:79-84

Tilahun G, Stockdale PHG (1982) Sensitivity and specificity of the indirect fluorescent antibody test in the study of four murine coccidia. J Protozool 29:129-132

Wisher MH (1983) Sporozoite antigens of coccidia (Abstr). J Cell Biochem [Suppl] 7A:25

Entamoeba Histolytica, Antigens and Amoebiasis

A. CHAYEN and B. AVRON

1 Introduction

Amoebiasis, the condition of harbouring the protozoan parasite *Entamoeba histolytica*, is a major health problem throughout the world (WHO EXPERT COMMITTEE REPORT 1969). The prevalence of the parasite is estimated to be a minimum of 5% in the United States, reaching a maximum in developing countries, where the disease is sometimes endemic (PITTMAN 1980).

Department of Biophysics and Unit for Molecular Biology of Parasitic Diseases, The Weizmann Institute of Science, 76 100 Rehovot, Israel

Current Topics in Microbiology and Immunology, Vol. 120
© Springer-Verlag Berlin · Heidelberg 1985

1.1 Problems in the Study of Amoebiasis

In a biochemical or immunological approach to parasites, interest centres around the immunological status of the parasite inside the host, the antigens giving rise to an immune response at different stages of the parasite's life cycle, and the identification of antigens that could give rise to a protective response. *E. histolytica* has a relatively simple life cycle with only two stages; the infective cyst form and the vegetative motile trophozoite. Despite this, there are a number of considerations that make amoebiasis a less than trivial problem:

1. The vast majority of people who harbour *E. histolytica* display no clinical symptoms, with the parasite apparently living harmlessly inside the host. The question therefore arises as to whether all strains of *E. histolytica* have the potential to cause disease or whether two distinct substrains exist, one of which cannot under any conditions attack the cells of the host. In the former case, the problem becomes that of indentifying the stimuli that result in activation to a harmful parasite. In the latter situation it becomes important to identify and to distinguish between the virulent and avirulent varieties of the organism.

2. There is a distinct lack of evidence that a humoral response *in vivo* can be effective in harming the parasite or in conferring immunity against reinfection.

A study of the antigens present is probably the best way to clarify the problems. Information about the antigens on the surface of *E. histolytica* is not very extensive. Although there is evidence for antigenic variation between strains, much of the serological work has been done using a crude extract from a single strain. Moreover, the immune response has generally been classified by immunodiffusion rather than by the identification of the antigens involved.

Problems in the study of antigens important to the host-parasite relationship stem, at least in part, from:

1. Lack of an axenic *in vitro* culture system in which the complete life cycle of the parasite can be maintained.

2. Lack of an effective animal model.

3. The probability that during the lengthy process of adaptation to axenic culture conditions selection pressures may operate to alter the surface properties.

1.2 *Entamoebae*

This review will be concerned mainly with *E. histolytica*, which is the main species pathogenic to man (TRISSL et al. 1978). However, before a discussion of the parasite at a molecular level can be undertaken, it is useful to consider a general description of the organism and the disease.

The genus *Entamoeba* includes the human parasites *E. histolytica, E. coli, E. polecki* (CHACIN-BONILLA 1983) and *E. gingivalis* (BELDING 1965; SUOZA 1982). *E. hartmanni*, a non-pathogenic amoeba, has been distinguished from *E. histolytica* morphologically only on the grounds of size, the latter being the larger (GLEASON et al. 1963; BELDING 1965; DESPOMMIER 1981). *Entamoebae* of non-human origin include the primate parasite *E. chattoni* (SARGEAUNT et al. 1982a), as well as the free-living *E. moshkovskii* (SCAGLIA et al. 1982) and the reptilian parasite *E. invadens* (GEIMAN and RATCLIFFE 1936).

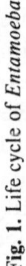

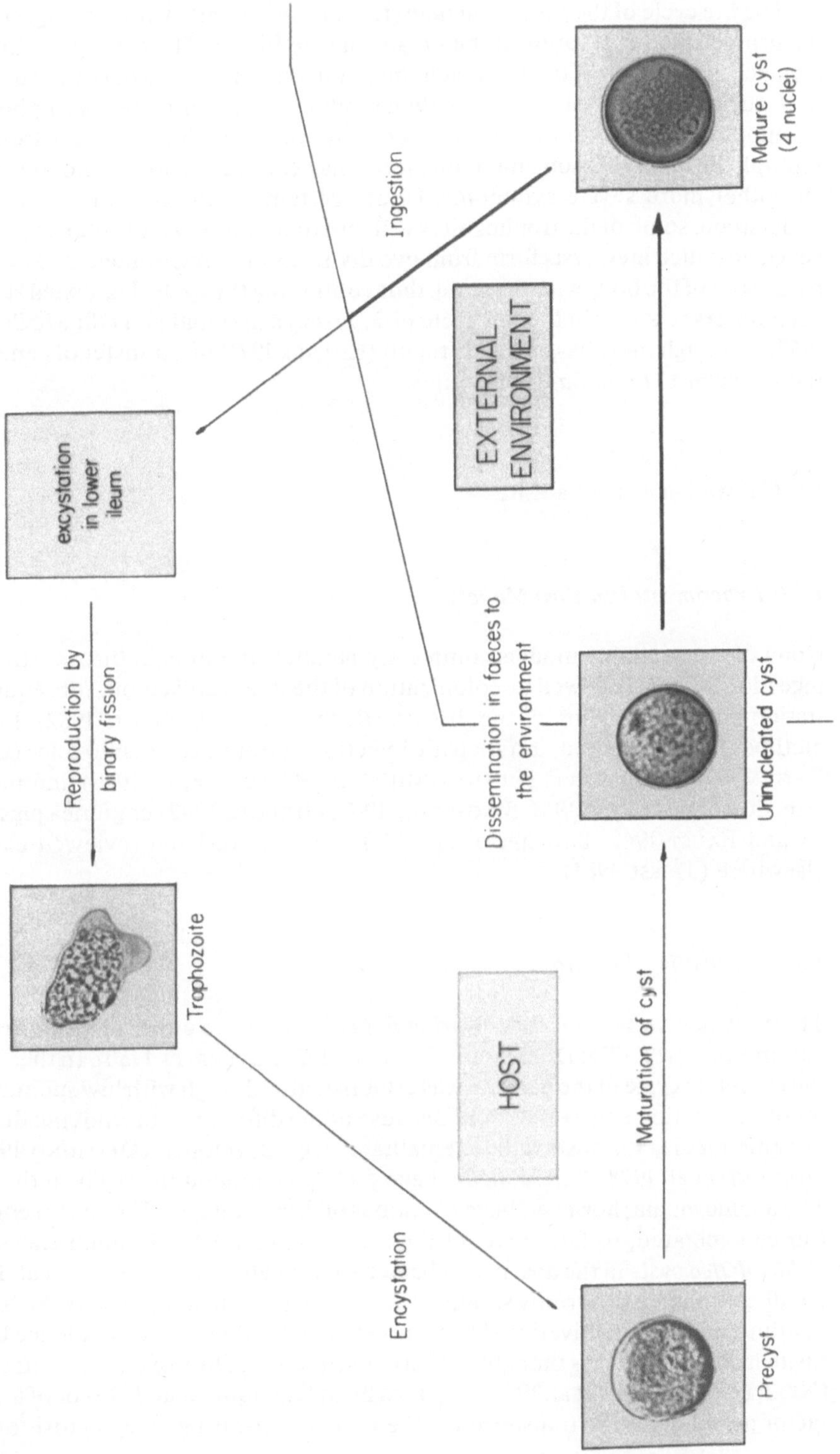

Fig. 1. Life cycle of *Entamoeba*

The life cycle of the parasite in man (DOBELL 1928) starts with oral ingestion of the quadrinucleated cyst forms of the organism (see Fig. 1). The cysts pass through the stomach and intestinal tract to reach the ileum and caecum, where they excyst. The four-nucleate (metacystic) amoeba then produces eight uninucleate trophozoites by a complicated series of nuclear and cytoplasmic divisions. These motile trophozoites multiply by binary fission and in this form may cause diarrhoea, amoebic dysentery and other more severe symptoms. Under certain conditions, which are not well understood, some of the trophozoites will lose their motility and enter a precystic stage. Quadrinucleated cysts form from two divisions of a single-nucleated cyst and are passed out of the body with the stool, thus continuing the cycle. No sexual stages have been observed so far in the life cycle of *E. histolytica* (DOBELL 1928; McCONNACHIE 1969), although there has been one report (ENTNER 1971) of a transfer of genetic material between *Entamoeba* trophozoites.

1.3   Cultivation of the Parasite

1.3.1   Experimental In Vivo Models

None of the available models completely parallels the human disease, that is, oral ingestion of cysts followed by colonization of the ileum and colon, subsequent encystation and transmission of the disease (RAVDIN and GUERRANT 1982). Four main methods have been used; intrahepatic injection of trophozoites into mice (GOLD and KAGAN 1978) or hamsters (LUSHBAUGH et al. 1978; RAI et al. 1982) and intracaecal infection of rats (NEAL 1951; SINGH et al. 1963; GUIRGES 1982) or guinea pigs (CARRERA and FAUST 1949; PHILLIPS et al. 1972). These have been reviewed extensively elsewhere (TRISSL 1982).

1.3.2   Cultivation In Vitro

The early *in vitro* systems for culturing *E. histolytica* were either xenic or monoxenic cultures (CULTER 1918; DOBELL and LAIDLAW 1926; EVERRITT 1950). In these cultures the complete cycle of the parasite was established, although with low spontaneous rates of encystation and excystation. Because of the difficulties of studying the parasite in xenic media, various axenic media have been developed (DIAMOND 1968, 1980; DIAMOND et al. 1978). These media have greatly facilitated the study of the parasite. The axenic media, however, have a number of disadvantages. Firstly, no encystation can be generated, so far, in any of them (DIAMOND 1980), although excystation of *E. histolytica* cysts in the absence of bacteria has been reported (REES et al. 1950). Secondly, virulence of various strains of *E. histolytica* is lost, or greatly declines, with continuous axenic cultivation (DIAMOND et al. 1973). The loss of virulence by the parasite in axenic media is thought to be a consequence of loss of encystment capability (NEAL 1965; PHILLIPS et al. 1972; PHILLIPS 1973; DIAMOND et al. 1973) or of a virulence factor postulated to be transferred to the *Entamoeba* during phagocytosis of bacteria

(WITTNER and ROSENBAUM 1970). An alternative suggestion was that viruses may be involved in the virulence of the parasite (DIAMOND et al. 1972; KNIGHT et al. 1975) but no correlation could be demonstrated between the virulence potential of an *E. histolytica* strain and an amoebal virus (MATTERN et al. 1978a).

Entamoeba invadens has sometimes been used, in culture, as a model for *E. histolytica* because of the morphological and biochemical similarities between them (GEIMAN and RATCLIFFE 1936; McCONNACHIE 1955; CHAVEZ et al. 1978; CARABEZ-TREJO and DE LA TORRE 1980; CERVANTES-MAMOA and MARTINEZ-PALOMO 1980; ARROYO BEGOVICH et al. 1980a, b; SIRIJINTAKARN and BAILEY 1980; CERBON and FLORES 1981). The advantage of using the *E. invadens* model lies in the possibility of inducing trophozoites to form cysts under specific axenic culture conditions (RENGPIEN and BAILEY 1975). From these media the various stages in the encystation process can be separated to homogeneity (AVRON et al. 1983).

2 Pathology

A consideration of the pathology of amoebiasis may help towards the diagnosis and also towards an understanding of the method of operation of the parasite. It has been suggested that all the clinical symptoms seen in amoebiasis are "mistakes" on the part of the parasite (ELSDON-DEW 1964), which would otherwise exist in a harmless relationship with the host. Certainly, in cases of disease, particularly in extraintestinal infection, cysts are not formed or passed (HARRIES 1982). In acute amoebic dysentery mainly trophozoites are passed in the faeces. Since these probably cannot infect a new host, the infection could be said to be non-productive for the parasite.

2.1 Intestinal Amoebiasis

Invasion of the mucosal layer, when it does occur, is seen most often in the caecum, then (in descending order of frequency) in the ascending colon, rectum, sigmoid and appendix. Lesions formed by the parasite as it moves through the layers of mucosa have the appearance of discrete flask- or diamond-shaped ulcers and can be identified as slightly raised areas with central yellow pits, often surrounded by petechial haemorrhage (BRANDT and TAMAYO 1970; HARRIES 1982). The intervening mucosa seems unaffected at this stage (HARRIES 1982). After treatment, most of the intestinal amoebic ulcerations appear to heal without scarring (BRANDT and TAMAYO 1970).

Intestinal amoebiasis may become more extensive resulting in confluent ulceration. Advanced intestinal amoebiasis results in loss of mucosa and submucosa over large areas. Occasionally there is thickening of the intestinal wall accompanied by loss of muscular layers, and the contents of the intestine are prevented from entering the peritoneal cavity only by an oedematous peritoneal membrane which may eventually rupture (BRANDT and TAMAYO 1970). Repeated invasion of the colon, followed by secondary infection, may result in the formation of an amoeboma or amoebic granuloma (HARRIES 1982), which can be mistaken for a malignant tumor (PATTERSON and SCHOPPE 1982).

2.2 Extraintestinal Amoebiasis

Extraintestinal complications may follow the intestinal phase of the disease but can also occur in patients who have not previously displayed symptoms (BRANDT and TA-MAYO 1970). One such complication is direct extension to the skin, which results in cutaneous amoebiasis (PATTERSON and SCHOPPE 1982). It is characterized by the appearance, in the perianal and perineal regions, of a very painful, rapidly growing ulcer with irregular margins (BRANDT and TAMAYO 1970). In severe infections, intestinal amoebiasis may perforate into the peritoneal cavity, resulting in peritonitis (HARRIES 1982), a form considered to be the most dangerous complication of intestinal amoebiasis (ADAMS and MacLEOD 1977a).

Both cutaneous amoebiasis and peritonitis are very rare complications compared with dissemination of *E. histolytica* through the portal venous system (RAVDIN and GUERRANT 1982) to the liver (BRANDT and TAMAYO 1970; ADAMS and MacLEOD 1977b; HARRIES 1982; PATTERSON and SCHOPPE 1982), where an abscess is formed. Single abscesses are found in about two-thirds of the cases, the right lobe of the liver being affected about eight times more often than the left one (HARRIES 1982). The wall of the abscess consists of fibroblasts, macrophages and lymphocytes; leucocytes are scant (PATTERSON and SCHOPPE 1982). Although trophozoites can be found in the wall of the abscess, cysts are never present (HARRIES 1982). Bacteria are not found and the "anchovy" pus in the vicinity of the necrotic tissue is sterile (PRICE 1981).

3 Virulence

As already mentioned, the majority of people harbouring *E. histolytica* do not display clinical signs of the disease and are therefore described as asymptomatic. Asymptomatic individuals pass only cysts, in contrast to symptomatic individuals, who pass trophozoites as well. The reasons for the variable susceptibility to the disease are not well understood. Two main schools of thought exist concerning the question of virulence within the *E. histolytica* strains. The first assumes that all strains of *E. histolytica* are potentially pathogenic and that some unknown stimulus converts the harmless commensal to a truly histolytic parasite (ELSDON-DEW 1964; TRISSL 1982). The factors responsible for this transformation have not been identified. The nutritional state of the host (ADAMS and MacLEOD 1977a; DIAMOND 1982), the cytotoxicity and surface properties of the parasite (BOS and VAN DE GRIEND 1977; TRISSL et al. 1977, 1978; MEEROVITCH and GHADIRIAN 1978; CALDERON et al. 1980; RAVDIN et al. 1980, 1982; PRASAD et al. 1981; MUNOZ et al. 1982; GADASI and KESSLER 1983) as well as rapid faecal transmission (POWELL et al. 1966) have been implicated. Alternatively, it has been suggested that virulence may be an intrinsic property of the parasite.

The virulence of amoebae has been designated according to their behaviour in a number of assay systems. For axenic strains it is usually determined as the ability of the organisms to give rise to abscesses when introduced into animals, notably into the liver (NEAL 1960; LUSHBAUGH et al. 1978). Another common criterion for virulence of *Entamoebae* is their ability to phagocytose red blood cells. Virulent amoebae phagocytosed more red cells than non-virulent strains (TRISSL et al. 1978). This led to the design of a selection scheme in which amoebae that phagocytosed bacteria loaded with bromodeoxyuridine were destroyed with UV light. The resultant strain phagocytosed less and was less virulent, as compared with the original strain (OROZCO et al. 1982). A useful assay system *in vitro* (MATTERN et al. 1978b) involves a monolayer of adherent cells which are incubated either with whole trophozoites or with cell extracts. The effect of the amoebae (reviewed by RAVDIN and GUERRANT 1982) can be divided into two main stages: one in which the cells are rounded up and detached from the substrate and a second stage in which the cells are killed. The effect of amoebae or amoebal extracts on the permeability of cell layers or model membranes has been measured by variations in the electrical resistance (LYNCH et al. 1982; YOUNG et al. 1982).

Differences in the surface charge of amoebae were investigated by the electrophoretic mobility and the binding of cationic ferritin at neutral pH. By these criteria, the virulent strains were not charged relative to a negative charge on non-virulent amoebae (TRISSL et al. 1977). Amoebae from monoxenic cultures more nearly resembled the virulent amoebae. The cause of the surface charge is not known, but it cannot be attributable to sialic acid because the trophozoites are reported to lack this sugar (FERIA-VELASCO et al. 1973).

3.1 Interaction with Bacteria

It has been observed that the interaction of amoebae with certain bacteria increased the frequency of liver abscesses in hamsters injected intrahepatically with the amoebae (WITTNER and ROSENBAUM 1970). The organisms used were strains grown axenically which had lost this virulent tendency during the axenic culture (DIAMOND et al. 1973). BRACHA AND MIRELMAN et al. 1984 extended this to an *in vitro* assay system and found that the interaction of amoebae with certain bacteria was through bacterial receptors for mannose (BRACHA et al. 1982) and increased the ability of the parasites to cause cells to detach from a substrate. For this it was necessary for the bacteria to be ingested, but irradiated bacteria could also be effective. The mechanisms involved in this interaction are not clear. Very short (1-h) incubation times appear to be sufficient to see an effect (MIRELMAN et al. 1983).

3.2 Enzymatic Activities

A number of proteins and enzymes which could conceivably be involved in the cytopathic behaviour of the amoebae have been described and semipurified. NEAL (1960) used assays for gelatinase and caseinase to investigate the "proteolytic index" of different strains of *E. histolytica*. Strains from asymptomatic carriers gave a lower proteolytic index than those from patients with the disease. When the virulent amoebae were cultured *in vitro*, the proteolytic index dropped and the virulence decreased, only to increase again when the amoebae were passaged through hamster livers. However, the level of enzymatic activity dropped before the decrease in invasiveness. Therefore it was suggested that the proteolytic activity was not directly needed for invasion, but was accompanying another factor. Amoebae with or without hyaluronidase could be equally virulent (JARUMILINTA and MAEGRAITH 1969). In contrast, the amount of collagenase activity associated with the plasma membrane was found to be greater in a more virulent strain as compared with that in a number of less virulent ones (GADASI and KESSLER 1983).

3.3 Amoebal Toxins

Much attention has been focussed on the identification of an amoebal toxin which would be responsible for the cytotoxic effects (reviewed by RAVDIN and GUERRANT 1982; UDEZULU et al. 1982). A number of groups have semipurified what may be a family of molecules that includes a 25- to 35-k species (LUSHBAUGH et al. 1979) and a group of four proteins with molecular weights of 25, 58, 65 and 68 k (McGOWAN et al. 1982). These last four were present in much higher amounts in the more virulent strains. Toxin activity on cells in tissue culture was inhibited by both immune and non-immune serum. An acid and a neutral proteinase with molecular weights of 41 k and 27 k respectively, semipurified by McLAUGHLIN and FAUBERT (1977), may be related to the family of toxins. The acid proteinase gave a precipitin line against immune serum, as did the toxin fraction of McGOWAN et al. (1982).

Amoebae have also been reported to contain ionophore activity which could be used directly as a cytotoxin or could be employed to deliver another toxin to cells or bacteria. The amoebapore (LYNCH et al. 1982) was reported to have a subunit molecular weight of 13 k and to be present in higher amounts in the more virulent strains. A molecule with similar properties, but with a molecular weight of 30 k, was found to be released from amoebae upon stimulation with concanavalin A (YOUNG et al. 1982). The presence of lectin activity in trophozoites (KOBILER and MIRELMAN 1980, 1981) may allow binding of the amoebae to epithelial cells as a prerequisite for cytotoxity.

3.4 *Entamoeba histolytica* Zymodemes

Isoenzymes are a further series of molecules which may be used to distinguish virulent and non-virulent strains of amoebae. REEVES and BISCHOFF (1968) demonstrated that different species of *Entamoeba* could be distinguished by their enzyme patterns and that, within *E. histolytica*, there were different isoenzymes. These observations have been extended by Sargeaunt and colleagues to differentiate between invasive and non-invasive strains. Following a number of studies in different areas, 18 zymodeme patterns with respect to phosphoglucomutase, hexokinase, glucose phosphate isomerase and malate dehydrogenase were identified (SARGEAUNT et al. 1978; SARGEAUNT and WILLIAMS 1980). The marker for pathogenicity was defined in this system as the presence of a band in the phosphoglucomutase pattern coupled, except in one zymodeme, with an advanced band in the hexokinase pattern. Six of the zymodemes could be associated with a clinical condition in the host, whereas the remaining zymodemes were isolated from asymptomatic carriers (SARGEAUNT et al. 1982b). In a study in South Africa, all the isolates from cases of liver abscess or amoebic dysentery showed one of two zymodemes (SARGEAUNT et al. 1982c). Surprisingly, 3 out of a total of 67 asymptomatic cyst passers also had isolates with a pathogenic zymodeme which could represent subclinical disease. It is difficult to see how the zymodeme pattern could be directly related to the pathogenicity. Axenic strains that have lost their virulence maintain the original zymodeme characteristic of a pathogenic organism (SARGEAUNT et al. 1982b).

The observation of different zymodemes has been used to support the second hypothesis concerning the virulence and non-virulence within *E. histolytica*. That is, that there are two separate species within *E. histolytica*, the so-called *Entamoeba dispar* (non-virulent) and *Entamoeba dysenteria* (the cause of the disease). The substrains would be distinguished on the basis of their zymodemes and would be immutable (SARGEAUNT 1982). All the axenic strains currently under study are derived from patients with the disease and have the pathogenic zymodeme II.

The identification of the zymodeme pattern requires preculture of the amoebae with the concurrent danger that some selection force may be operating. Also, it requires skilled personnel. The study of the zymodemes present in different regions could be useful in settling the question of the species within *E. histolytica*. For this purpose it would also be very useful to have studies on axenized strains which were derived from asymptomatic individuals and which had non-pathogenic zymodemes.

4 Clinical Diagnosis

4.1 Morphological Identification

Traditionally the diagnosis of amoebiasis has been made by identifying the parasite in the stool. In the case of acute amoebic dysentery the presence of hematophagous trophozoites in the faeces, in material aspirated from lesions at protoscopy, or in rectal biopsies, is considered as good evidence (ADAMS and MACLEOD 1977a; PATTERSON et al. 1980; PATTERSON and SCHOPPE 1982). For identification of non-hematophagous trophozoites, histological stains, based on iodine or periodic acid, have been used. The specificity of the iodine stain was found to be comparable, in rectal biopsies, to direct and indirect immunofluorescence techniques while the periodic acid stain was only a little less sensitive (GILMAN et al. 1980). The presence of cysts in the faeces has been detected morphologically and the cysts distinguished as *E. histolytica* by the size and number of the nuclei (MCCONNACHIE 1969).

The methods mentioned have a number of disadvantages. First of all they require that the operator has a fair degree of skill and experience. A number of treatments and drugs which the patient may have been given can complicate diagnosis. KROGSTAD et al. (1978) recommended that patients should not have received any of a list that included antibiotics and barium, for 10–14 days before examination of the stools. Another complication of morphological identification arises in many patients with liver abscesses, and even some with amoebic colitis, who do not have any rectal involvement and may not pass cysts or trophozoites (ADAMS and MACLEOD 1977b).

4.2 Detection of Anti-Amoeba Antibodies

In cases of invasive amoebiasis, a high titer of anti-amoeba antibodies is often found in the serum. PATTERSON et al. (1980) reported that more than 96% of their patients gave a positive reaction with an indirect hemagglutination assay (IHA). However, in another study (AIKAT et al. 1979) it was shown that only one-third of 79 patients who died of liver abscesses had shown positive serology.

In order to support the morphological means of diagnosis, a number of serological assays have been developed to detect antiamoebal antibodies. These include complement fixation tests, IHA (KESSEL et al. 1965), gel diffusion precipitin tests (LUNDE and DIAMOND 1969); PATTERSON et al. 1980), latex agglutination assays (MORRIS et al. 1970; STAMM et al. 1973) and, more recently, enzyme-linked immunosorbent assay (ELISA; YANG and KENNEDY 1979; BOS et al. 1980; LIN et al. 1981) and a thin-layer immunoassay (NILSSON et al. 1980). The IHA and latex agglutination assays seem to recognize the same antibodies, which are different from those in the precipitin tests (MADDISON et al. 1965a). Other methods involving immunoelectrophoresis (ALPER et al. 1976; SHEEHAN et al. 1979) and immunofluorescence (GOLDMAN et al. 1962) have also been used. These assays, with the exception of the fluorescence, have generally been done on a crude antigen preparation made (as in LUNDE and DIAMOND 1969) by homogenizing amoebae and centrifugation to obtain a soluble extract. YAP et al. (1970), who used instead antigens from cysts of E. invadens, got results which were reportedly very similar to those observed when using E. histolytica trophozoite antigen. Although cross-reactivity between strains of E. histolytica is relied upon, it has been shown (CHANG et al. 1979; ISHAQ and PADMA 1980) that different isolates have distinct, as well as shared, antigens. Cross-reaction with other Entamoeba strains and with related parasites does not seem, empirically, to be a big problem. Low cross-reactivity has been reported with Entamoeba moshkovskii, Entamoeba coli, the Laredo strain and Giardia lamblia (GOLDMAN et al. 1962; PARKHOUSE et al. 1978; ROOT et al. 1978).

4.3 Detection of Amoebic Antigens

Tests to detect amoebal antigens rather than antiamoeba antibodies are less widely used. An ELISA assay for use on faeces has been reported (ROOT et al. 1978). A limited selection of antigens as antigen-antibody complexes may be detected in the serum by radio-binding assay (PILLAI and MOHIMEN 1982; PILLAI et al. 1982) although none was identified in a previous study (MEEROVITCH at al. 1978).

4.4 Mistaken Diagnosis

The existing serological assays have been compared with each other. A major disadvantage of the assays, and a reason why serology cannot be relied upon as the sole means of diagnosis, is the persistence of antibody in the apparent absence of disease. Positive reactions in IHA have been seen as much as 20 years later (PATTERSON et al. 1980). The ELISA and IHA assays appear to be the most sensitive, which may make them useful methods for epidemiological studies. However, the persistence of a positive result makes them difficult to interpret (SPENCER et al. 1981). The precipitin test has been reported to be less sensitive but may showed a greater degree of correlation to the state of disease. In one study a positive precipitin test was found to persist for a maximum of 6 months after "successful" treatment (PATTERSON et al. 1980), but persistence up to 6 years has also been noted (STAMM et al. 1976). It is generally agreed that there is no clear correlation between the severity of the disease and the titer of the antibody, although it has been suggested that patients with liver abscesses tend to have the higher titers of serum antibody (HEALY 1971; ABIOYE et al. 1972). Asymptomatic carriers of the disease, who nevertheless may pass quantities of cysts, may have negative serology (MADDISON et al. 1965b). It seems clear that serological assays, while they may be useful, should always be used in conjunction with other methods.

4.5 Antigens

Antigens present in different amounts in virulent and non-virulent strains are of particular interest. Electron microscopy, coupled with the use of cytochemical reagents to detect carbohydrates, identifies a surface coat, or glycocalyx, over the plasma membrane of the amoeba (LUSHBAUGH and MILLER 1974; PINTO DA SILVA et al. 1975; MARTINEZ-PALOMO 1982). The glycocalyx, which appears to be formed of short 6- to 10-nm) filaments, has been reported to be more prominent in invasive forms of the amoeba obtained from colonic lesions (EL-HASHIMI and PITTMAN 1970) or liver abscesses (PROCTOR 1976). Variation in the thickness of the surface coat between amoebae cultured under axenic or monoxenic conditions has also been noted (LUSHBAUGH and MILLER 1974; MARTINEZ-PALOMO et al. 1980).

ALEY et al. (1980) used concanavalin A to cross-link the membrane of the amoebae and isolated pure fractions of plasma membrane. They recovered 18 peptides, several of which bound concanavalin A. A lipidopeptidophosphoglycan, extractable with phenol-water, may also be present on the surface. Rats injected intraperitoneally with trophozoites produced an IgA response, detected in the bile, to this molecule (ACOSTA et al. 1982). Pathogenic amoebae were shown to be selectively agglutinated by concanavalin A (MARTINEZ-PALOMO et al. 1973), which suggests a difference in the exposed mannose or glucose residues on virulent and non-virulent strains. It is not known whether the differences in the sugars reside in the glycoproteins, the glycocalyx, the glycolipids or elsewhere.

Some attempts have been made to identify the antigens detected in the diagnostic assays. PARKHOUSE et al. (1978) identified a number of glycoproteins that could be immunoprecipitated with immune sera. In a recent study, the reactivity of immune sera to various, semipurified fractions of amoebae was examined (AUST-KETTIS et al. 1983). It was found that a very variable response was given by different patients. The majority of the response appeared to be directed against the cell surface, because it could be absorbed out on whole amoebae. Ten polypeptides that stained for carbohydrate also bound human antibodies. According to sodium dodecyl sulphate polyacrylamide gel electrophoresis (SDS-PAGE), the most prominent antigenic bands had molecular weights of 40 k and 88 k, which compares to the 81-k protein immunoprecipitated by PARKHOUSE et al. (1978). No correlation was noted between the antigens recognized and the clinical symptoms, duration of the illness, or origin of the strain of amoeba. The authors concluded that purification of amoebal preparations was unlikely to yield an antigen that would increase the accuracy of diagnosis.

In some studies (BOONPUCKNAVIG et al. 1967; SAWHNEY et al. 1980), part of the immune response was found to be directed against intracellular antigens, notably in a high-speed sedimentable fraction. This could represent misdirection of the immune response, in which irrelevant antibodies are produced that cannot recognize the living parasite. Considering the rapid rate of turnover of the membrane of the amoeba, which is increased in the presence of antibody to surface components (AUST-KETTIS and SUNDQVIST 1978, 1981; MARTINEZ-PALOMO 1982), it seems likely that antigens may be present both in and on the amoeba at different stages. This may confuse studies to localize the antigens.

5 Protection and Escape

The existence of protection of the host against invasive amoebiasis is indicated by negative, rather than positive, observation. There is no good evidence in humans for any protection against reinfection (KRUPP 1970; OYERINDE et al. 1977). There is, however, some evidence for immunity in animal models (SEPULVEDA 1982). A high molecular weight fraction of *E. histolytica* was found to be the best immunogen for protection against caecal abscesses in guïnea pigs (VINAYAK et al. 1980). In contrast, extracts of whole amoebae and a lower molecular weight fraction were found to be more effective in protection against liver abscesses in hamsters (GHADIRIAN et al. 1980).

The involvement of the immune system in protection from the parasite was also implicated by other clinical observations. For example, corticosteroid treatment of patients with amoebic dysentery often results in exacerbation of the disease (KANANI and KNIGHT 1969; STUIVER and GOUD 1978). These reports cite a number of cases in which patients with a history of mild, intestinal disturbances developed liver abscesses, following treatment with corticosteroids, without any evidence for reinfection with amoebae.

5.1 The Role of Complement

A general method of controlling invasion by amoebae could be the complement system. Amoebae were reported to be killed by human sera from individuals with no clinical history or laboratory evidence of amoebiasis. ORTIZ-ORTIZ et al. (1978) demonstrated that this effect was not due to natural anti-amoeba antibodies. Moreover, treatment with ethylenediamine-tetraacetic acid or cobra venom factor, destruction of factor B of the alternate complement pathway or heat inactivation at 56 °C indicated that the cytopathogenic effect was complement dependent. It was also demonstrated that when *E. histolytica* trophozoites were incubated with normal human serum, a correlation existed between the consumption of C3, but not of C1, C4 or C2, thereby demonstrating that trophozoites of *E. histolytica* activate the alternative pathway of the complement system. This was confirmed by other groups, who suggested that activation of the classical pathway can also occur since incubation of amoebae with the IgG fraction of an antiserum followed by unheated serum accelerated the lysis observed (HULDT et al. 1979). It could be demonstrated that, in the presence of 5%-10% unheated serum, there was very little lysis but an increase in the concentration and release of an acid hydrolase (*N*-acetyl-ß-glucosaminidase). Huldt and co-workers therefore suggested that complement cleavage products might stimulate the amoebae to increased production and release of hydrolytic enzymes which could contribute to the pathogenicity of the parasite. That is, complement activation may be contributing on the one hand to host protection, and on the other hand to the pathogenesis of *E. histolytica.* Complement is present in the intestine in lower amounts than in the serum (TOMASI and BIENENSTOCK 1968) and the intestinal contents are anticomplementary (discussed by BEFUS and BIENENSTOCK 1982), so that the role of complement as a defence against intestinal parasites remains under debate. Even in the blood, the amoebae may be resistant. REED et al. (1983) compared the susceptibility to lysis by human sera of pathogenic and non-pathogenic strains of *E. histolytica,* as characterized by their zymodeme patterns. They demonstrated that pathogenic strains were significantly less susceptible to lysis than non-pathogenic ones. A similar suggestion had been made by CALDERON and TOVAR-GALLEGOS (1980).

5.2 Cellular Response

In the past decade a major effort has been made to elucidate the cellular response to invasion of the host by *E. histolytica.* Immunity following infection or vaccination of golden hamsters has been described. Animals injected intradermally with trophozoites developed lesions at the site of injection. Once these had healed, the animals were injected intrahepatically with more trophozoites. The vaccinated animals were protected, although the titers of antibody in this group were no higher than in the control animals which did develop liver abscesses (MEEROVITCH et al. 1978). HARRIS and BRAY (1976) tested the cellular sensitivity of patients in areas where amoebiasis is highly endemic. They were able to demonstrate that symptomless carriers and patients with intestinal disease did have circulating antibodies to *E. histolytica* but did not possess a detectable population of specifically sensitized lymphocytes. In contrast, patients suffering from hepatic disease possessed a circulating population of lymphocytes capable of responding *in vivo* to amoebic antigen and also had substantially raised specific anti-amoebic IgM and IgG.

5.2.1 T Cells

The involvement of T cells was suggested from a number of observations. SAVANAT et al. (1973a) demonstrated that an aqueous extract of *E. histolytica* could stimulate *in vitro* blast transformation of peripheral blood lymphocytes. Out of 23 patients, 21 showed positive transformation, with the cells producing 23%-31% of the response stimulated by phytohaemagglutinin. The response was specific since lymphocytes from healthy individuals (or individuals with other diseases) did not give a blastogenic response. However, there was no correlation between the immunoelectrophoretic pattern of the serum and the degree of transformation, indicating that the blast transformation did not reflect the humoral antibody response. The non-lipid fraction of a crude *E. histolytica* extract was shown to be mitogenic for purified peripheral T-lymphocytes but not for non-T-lymphocytes *in vivo* (DIAMANTSTEIN et al. 1981). This mitogenic response was shown to be of a different nature than that observed for concanavalin A. The authors concluded that, either there were different binding sites in the crude *E. histolytica* extract on the same T cells or that the mitogenic agent and the concanavalin A were acting on different subsets of T cells.

GHADIRIAN and MEEROVITCH (1981) investigated the effect of T-cell depletion on experimental hepatic amoebiasis in hamsters. The mean weight of the liver abscesses in neonatally thymectomized hamsters, 10 days postinfection, was significantly higher than in sham-operated controls. A combination of neonatal thymectomy and anti-T-cell serum treatment enhanced both the size of the primary liver abscess and the size of metastases to other sites, suggesting that T cells may control the metastatic spread of the parasites from the initial site of inoculation to other organs. Moreover, treatment of hamsters with anti-T-cell serum shortly before or after intrahepatic challenge increased significantly the size of the abscesses, but a similar treatment 2–3 weeks prior to challenge had no effect.

5.2.2 Polymorphonucleocytes and Macrophages

Polymorphonucleocytes (PMNs) and macrophages were also shown to affect amoebae, at least *in vitro*. The peripheral blood lymphocytes and macrophages of vaccinated or protected hamsters were not phagocytosed by the amoebae but rather caused their death *in vivo*. In contrast, spleen cells from infected animals showed no significant cytotoxic effect on the parasite (GHADIRIAN and MEEROVITCH 1982a). Animals in which macrophage activity had been inhibited (by treatment with silica) developed more abscesses and metastases, when injected intrahepatically with trophozoites, than did control animals. Moreover, animals treated with *bacille* Calmette-Guérin (BCG) (which have enhanced macrophage activity) developed significantly smaller abscesses in the liver and had fewer metastatic foci (GHADIRIAN and MEEROVITCH 1982b).

Polymorphonucleocytes were shown to be able to kill non-pathogenic *E. histolytica* trophozoites (GUERRANT et al. 1981). Amoebae of a low-virulence strain which were surrounded by PMNs (more than 300/amoeba) were fragmented and ingested by the PMNs by means that appeared to be independent of oxygen or of complement. In contrast, contact with more virulent amoebae (defined by their ability to cause hepatic abscesses in hamsters and destroy cell monolayers in tissue culture) caused loss of PMN motility, PMN degranulation, death and occasional phagocytosis by the amoebae. PMNs which were in the vicinity of virulent amoebae, but not in contact with them, remained alive. The mechanism of the killing of the amoebae by the PMNs is not known. Amoebae, although generally considered anaerobes, can withstand 5% O_2 and even have an affinity for oxygen (WEINBACH and DIAMOND 1974). A possible role for toxic oxygen intermediates released by PMNs or mononuclear leucocytes was suggested by MURRAY et al. (1981). These authors employed a glucose-glucose oxidase system, which generates H_2O_2, to study the effect of peroxide on the parasite. Amoebicidal activity was demonstrated which depended upon the number of parasites present. A linear relationship was observed between the inoculum size and the flux of H_2O_2 required to kill 50% of the trophozoites. In a xanthine-xanthine oxidase system, which generated O_2^- in addition to H_2O_2, inoculum size-dependent killing of the parasite *in vivo* could only be observed when catalase (a scavenger of peroxide) was omitted from the system.

5.2.3 Immunosuppression

There has been a suggestion that the amoeba may cause a degree of immune suppression. MADDISON et al. (1968) reported that 83% of patients with amoebic liver abscesses could be demonstrated to have a hypersensitive skin reaction to a crude extract of *E. histolytica*; 75% of these patients had an immediate-type hypersensitivity reaction and 25% had an delayed-type reaction.

Migration inhibition factor (MIF) was shown to appear in hamsters with amoebic liver abscesses 5 days after inoculation, to disappear between the 10th and 20th days, and to reappear 25 days postinoculation. Similarly, guinea pigs, injected intrahepatically with amoebae, showed blast transformation after 4 days but antibody and MIF response were only apparent after 8 days (BRAY and HARRIS 1977). Healing of the lesions, when it did occur, correspondingly began after 8 days. Patients with amoebic liver abscesses also revealed diminished cell-mediated immunity to *E. histolytica* antigen when tested for MIF production (ORTIZ-ORTIZ et al. 1975), although control skin reactions to streptokinase-streptodornase were normal. When the same patients were tested 10 days after discharge from hospital, they all had normal delayed-type skin reaction and MIF production to *E. histolytica* antigen.

Allergens capable of eliciting an IgE response to *E. histolytica* were demonstrated in an extract of axenically grown *E. histolytica* (SAVANAT et al. 1973b; USAWATTANA-KUL et al. 1982). In general, the question of whether individuals with amoebiasis have raised levels of IgE has proved difficult to resolve because of the high incidence, in the areas investigated, of coinfection with parasitic helminths, which are known to cause an increase in IgE (discussed by TRISSL 1982). Levels of IgA were not found to be significantly raised in patients with intestinal amoebiasis (ABIOYE et al. 1972); however, this work was done on sera, which is not the most probable place to detect increases in IgA.

5.3 Escape from the Immune Response

It has been suggested that the best form of defence for the amoeba is to attack. That is, cells that approach close to the amoeba are engulfed before the amoeba can be damaged. In electron microscope studies, plasma cells, lymphocytes, neutrophils and some eosinophils could be seen in the vicinity of amoebic infection (PITTMAN et al. 1973) but the trophozoites were separated from the viable tissue by a zone of necrotic material. Epithelial penetration by *E. histolytica* elicited a series of degenerative changes in the PMNs migrating in the epithelium, resulting in the lysis of the PMNs. This occurred in PMNs in the immediate vicinity of the amoeba both with and without direct contact (TAKEUCHI and PHILLIPS 1975).

Rapid turnover of the membrane (SERRANO et al. 1977) may allow the organism to escape the effects of antibody or complement components (HULDT et al. 1979). Rapid redistribution and internalization of antibodies and concanavalin A from the surface of the parasite was observed (AUST-KETTIS and SUNDQVIST 1978), with a distinct difference in the behaviour of single and double antibody layers. Single antibody layers were mainly removed by release or shedding while double-layer antibodies, similarly to concanavalin A, could cross-link and redistribute antigens, finally being interna-

lized and digested (Aust-Kettis and Sundqvist 1980, 1981). This may explain the resistance of pathogenic amoebae to attack by PMNs and complement (Guerrant et
al. 1981) and resistance to a humoral response.

6 Conclusion

Entamoeba histolytica invasion into tissues has been shown to provoke both humoral
and cellular immune responses. People with antibody titers to specific amoebic antigens may still be reinfected, suggesting that a humoral response alone is not sufficient to protect the host from invasion by the parasite. Sensitivity of the parasite to
complement *in vitro* has been shown. A cellular response to the parasite by lymphocytes, PMNs, macrophages and eosinophils has also been demonstrated both *in vitro*
and *in vivo*, so that the reasons for the inefficiency of the host defence may lie in the
protective mechanims of the parasite rather than in the immune response of the host.
Rapid turnover of membrane components and the cytotoxic potential of *E. histolytica*
allow it to evade the consequence of an encounter with an immune response, which
may otherwise have been effective.

One of the central questions which is being addressed today with relation to
amoebiasis is "when do the *Entamoeba* parasites invade the host?" Several theoretical
possiblities exist:

1. All *Entamoeba* parasites are capable of invading host tissues but do so only if the
microenvironmental conditions are right. Invasion of host tissues by the parasite will
lead to the onset clinical symptoms.

2. All parasites invade host tissues but are dealt with successfully by the immune
response (causing a subclinical infection) while the symptomatic patients are selective individuals in which the immune response has failed in successful protection.

3. There are two types of *Entamoeba* parasites: those which are pathogenic and will
invade host tissues under the appropriate conditions and those which are non-pathogenic, or true commensals, and will therefore never be causative agents of clinical
symptoms.

To date, the true possibility has not been identified. If it is found that all parasites
are potential causes of disease, probably all those that harbour the parasite should be
treated as a preventive measure. On the other hand, if there are pathogenic and non-
pathogenic subspecies then studies on specific antigenic components of the parasites
should help in the development of a simple and rapid diagnostic technique for identification of the two strains. The most desirable outcome would be the development of
an assay to detect "antigens" specific to the pathological organisms. In any event a
simplification of the crude extract used as the basis for most of the serological assays
in general use would be advantageous. This may be difficult in view of the heterogeneous responses given by different individuals. In most of the studies of this problem
a single strain of *Entamoeba histolytica* was used as the source of antigen and a consideration of a number of strains could be useful. An alternative approach to solve the
problems of a diagnostic assay might be to detect characteristic sequences of DNA in
extracts of amoebae by hybridization to prepared probes.

It should be pointed out that good clinical results are usually obtained if the disease has been correctly diagnosed. A particular danger lies in mistaking intestinal amoebiasis for ulcerative colitis or related diseases for which the treatment is steroids. This may lead to disastrous results for the amoebiasis sufferer (KANANI and KNIGHT 1969). In a study in the United States, inaccuracy in diagnosis was concluded to be a result of scarcity of experienced technicians, a lack of awareness of the condition and difficulty in obtaining good samples (KROGSTAD et al. 1978). There are a number of drugs that are used to treat patients with amoebiasis (reviewed extensively by KNIGHT 1980; WOLFE 1982). Classically, emetine, an alkaloid of the *ipecacuanha* plant, was the major drug for treatment (IMPERATO 1981). However, today the drug of choice seems to be metronidazole, trade name Flagyl (ADAMS 1977; GILLIN and DIA-MOND 1981). New drugs [(tinidazole (WELCH et al. 1978) and fexinidazole (RAETHER and SEIDENATH 1983)] with higher efficacy and fewer side effects have now been introduced onto the market and these may be the drugs of the future.

A central question to which little effort has been given so far is identification of cyst-specific antigens. Identification of cyst antigens is of prime importance since understanding of the encystation and excystation processes may lead to another avenue of attack on the parasite – the eradication, rather then prevention, of amoebiasis.

References

Abioye AA, Lewis EA, McFarlane H (1972) Clinical evaluation of serum immunoglobulins in amoeiasis. Immunology 23:937–946

Acosta G, Barranco C, Isibasi A, Campos R, Kumate J (1982) Excrecion de anticuerpos de clase IgA especificos anti-amiba enbilis de ratas inmunizadas con trofozoitos de *Entamoeba histolytica* cultivados en medio axenico. Arch Invest Med (Mex) [Suppl 3] 13:255–259

Adams EB (1977) Metronidazole in invasive amebiasis. In: Finegold SM (ed) Metronidazole: proceedings of the international metronidazole conference. Montreal, Quebec, May 26–28, 1976. Excerpta Medica, Amsterdam, pp 88–93

Adams EB, MacLeod IN (1977a) Invasive amebiasis – I. amebic dysentery and its complications. Medicine 56:315–323

Adams EB, MacLeod IN (1977b) Invasive amebiasis – II. amebic liver abscess and its complications. Medicine 56:325–334

Aikat BK, Bhusnurmath SR, Pal AK, Chhuttani PN, Datta DV (1979) The pathology and pathogenesis of fatal hepatic amoebiasis – a study based on 79 autopsy cases. Trans R Soc Trop Med Hyg 73:188–192

Aley SB, Scott WA, Cohn ZA (1980) Plasma membrane of *Entamoeba histolytica*. J Exp Med 152:391–404

Alper EI, Littler C, Monroe LS (1976) Counterelectrophoresis in the diagnosis of amebiasis. Am J Gastroenterol 65:63–67

Arroyo-Begovich A, Carabez-Trejo A, de la Torre M (1980a) Tinsion de quistes de *Entamoeba invadens*, *Entamoeba histolytica* y *Entamoeba coli* con aglutinina de germen de trigo marcada con particulas de oro coloidal. Arch Invest Med (Mex) [Suppl 1] 11:25–30

Arroyo-Begovich A, Carabez-Trejo A, Ruiz-Herrera J (1980b) Identification of the structural component in the cyst wall of *Entamoeba invadens*. J Parasitol 66:735–741

Aust-Kettis A, Sundqvist KG (1978) Dynamics of the interaction between *Entamoeba histolytica* and components of the immune response I. Capping and endocytosis: influence of inhibiting and accelerating factors; variation of the expression of surface antigens. Scand J Immunol 7:35–44

Aust-Kettis A, Sundqvist KG (1980) Dynamics of the interaction between *Entamoeba histolytica* and components of the immune response II. On the distinction of surface bound and internalized anti-amoeba antibodies. Scand J Immunol 12:443–451

Aust-Kettis A, Sundqvist KG (1981) Dynamics of the interaction between *Entamoeba histolytica* and components of the immune response III. Fate of antibodies after binding to the cell surface. Scand J Immunol 13:473–481

Aust-Kettis A, Thorstensson R, Utter G (1983) Antigenicity of *Entamoeba histolytica* strain NIH-200: a survey of clinically relevant antigenic components. Am J Trop Med Hyg 32:512–522

Avron B, Bracha R, Deutsch MR, Mirelman D (1983) *Entamoeba invadens* and *E. hystolytica*: separation and purification of precysts and cysts by centrifugation on discontinuous density gradients of Percoll. Exp Parasitol 55:265–269

Befus D, Bienenstock J (1982) Factors involved in symbiosis and host resistance at the mucosal interface. Prog Allergy 31:76–177

Belding DL (1965) The parasitic amoebae of man. In: Textbook of parasitology, 3rd edn. Appleton-Century-Crofts, New York, pp 31–108

Boonpucknavig S, Lynraven GS, Nairn RC, Ward HA (1967) Subcellular localization of *Entamoeba histolytica* antigen. Nature 216:1232–1233

Bos HJ, Van de Griend RJ (1977) Virulence and toxicity of axenic *Entamoeba histolytica*. Nature 265:341–343

Bos HJ, Schouten WJ, Noordpool H, Makbin M, Oostburg BFJ (1980) A seroepidemiological study of amebiasis in Surinam by the enzyme linked immunosorbent assay (ELISA). Am J Trop Med Hyg 29:358–363

Bracha R, Kobiler D, Mirelman D (1982) Attachment and ingestion of bacteria by trophozoites of *Entamoeba histolytica*. Infect Immun 36:396–406

Bracha R, Mirelman D (1984) Virulence of Entamoeba histolytica trophozoites: effects of bacteria, microaerobic conditions and metronidazole. J Exp Med 160:353–368

Brandt H, Tamayo RP (1970) Pathology of human amebiasis. Human Pathol 1:351–385

Bray RS, Harris WG (1977) Cellular immune responses to amoebic liver abscess in the guinea-pig. Clin Exp Immunol 29:147–151

Calderon J, Tovar-Gallegos GR (1980) Resistance to immune lysis induced by antibodies in *Entamoeba histolytica*. In: Van den Bossche H (ed) The host invader interplay. Elsevier North Holland, Amsterdam, pp 227–230

Calderon J, Munoz Ma-deL, Acosta HM (1980) Surface redistribution and release of antibody-induced caps in *Entamoebae*. J Exp Med 151:184–193

Carabez-Trejo A, de la Torre M (1980) Staining of cyst wall of *Entamoeba invadens, E. histolytica* and *E. coli* with colloidal gold labeled wheat germ agglutinin. Eur J Cell Biol 22:617

Carrera GM, Faust EC (1949) Susceptibility of the guinea pig to *Endamoeba histolytica* of human origin. Am J Trop Med Hyg 29:647–667

Cerbon J, Flores J (1981) Phospholipid composition and turnover of pathogenic amoebas. Comp Biochem Physiol [B] 69:487–492

Cervantes-Mamoa A, Martinez-Palomo A (1980) Estudio del ciclo vital de *Entamoeba invadens* mediante cinematografía espaciada. Arch Invest Med (Mex) [Suppl 1] 11:31–40

Chacin-Bonilla L (1983) *Entamoeba polecki* infection in Venezuela. Report of a new case. Trans R Soc Trop Med Hyg 77:137

Chang S-M, Lin C-M, Dusanic DG, Cross JH (1979) Antigenic analysis of two axenized strains of *Entamoeba histolytica* by two dimensional immunoelectrophoresis. Am J Trop Med Hyg 28:845–853

Chavez B, Martinez-Palomo A, De la Torre M (1978) Estructura ultramicroscopica de la pared de quistes de *Entamoeba invadens, Entamoeba histolytica* y *Entamoeba coli*. Arch Invest Med (Mex) [Suppl 1] 9:113–117

Culter DW (1918) A method for the cultivation of *Entamoeba histolytica*. J Pathol Bacteriol 22:22–27

Despommier DD (1981) The laboratory diagnosis of *Entamoeba histolytica*. Bull NY Acad Med 57:212–216

Diamantstein T, Klos M, Gold D, Hahn H (1981) Interaction between *Entamoeba histolytica* and the immune system: I. mitogenicity of *Entamoeba histolytica* extracts for human peripheral T lymphocytes. J Immunol 126:2084–2086

Diamond LS (1968) Techniques of axenic cultivation of *Entamoeba histolytica* Schaudinn, 1903 and *E. histolytica*-like amebae. J Parasitol 54:1047–1056

Diamond LS (1980) Axenic cultivation of *Entamoeba histolytica*: progress and problems. Arch Invest Med (Mex) [Suppl 1] 11:47–54

Diamond LS (1982) Amebiasis: nutritional implications. Rev Infect Dis 4:843–850

Diamond LS, Mattern CFT, Bartgis IL (1972) Viruses of *Entamoeba histolytica*. I. Identification of transmissible virus-like agents. J Virol 9:326–341

Diamond LS, Phillips BP, Bartgis IL (1973) Virulence of axenically cultivated *Entamoeba histolytica*. Arch Invest Med (Mex) [Suppl 1] 4:99–104

Diamond LS, Harlow DR, Cunnick CC (1978) A new medium for the axenic cultivation of *Entamoeba histolytica* and other *Entamoeba.* Trans R Soc Trop Med Hyg 72:431–432

Dobell C (1928) Researches on the intestinal protozoa of monkeys and man. II. Description of the whole life history of *Entamoeba histolytica* in cultures. Parasitology 20:365–412

Dobell C, Laidlaw PP (1926) On the cultivation of *Entamoeba histolytica* and some other entozoic amoebae. Parasitology 18:283–318

El-Hashimi W, Pittman F (1970) Ultrastructure of *Entamoeba histolytica* trophozoites obtained from the colon and from in vitro cultures. Am J Trop Med Hyg 19:215–226

Elsdon-Dew R (1964) Amoebiasis. Exp Parasitol 15:87–96

Entner N (1971) "Mating" in *Entamoeba histolytica*? Nature (New Biol) 232:256

Everritt MG (1950) The relationship of population growth to *in vitro* encystation of *Endamoeba histolytica.* J Parasitol 36:586–594

Feria-Velasco A, Martinez-Zedillo G, Trevino-Garcia Manzo N, Gutierrez-Pastrana MD (1973) Investigacion del acido sialico en la cubierta exterior de trofozoitos de *Entamoeba histolytica.* Estudio bioquimico y citoquimico de alta resolucion. Arch Invest Med (Mex) [Suppl 1] 4:33–38

Gadasi H, Kessler E (1983) Correlation of virulence and collagenolytic activity in *Entamoeba histolytica.* Infect Immun 39:528–531

Geiman QM, Ratcliffe HL (1936) Morphology and life-cycle of an amoeba producing amoebiasis in reptiles. Parasitology 28:208–228

Ghadirian E, Meerovitch E (1978) Behavior of axenic IP-106 strain of *Entamoeba histolytica* in the golden hamster. Am J Trop Med Hyg 27:241–247

Ghadirian E, Meerovitch E (1981) Effect of immunosuppression on the size and metastasis of amoebic liver abscesses in hamsters. Parasite Immunol 3:329–338

Ghadirian E, Meerovitch E (1982a) *In vitro* amoebicidal activity of immune cells. Infect Immun 36:243–246

Ghadirian E, Meerovitch E (1982b) Macrophage requirement for host defence against experimental hepatic amoebiasis in the hamster. Parasite Immunol 4:219–225

Ghadirian E, Meerovitch E, Hartmann D-P (1980) Protection against amebic liver abscess in hamsters by means of immunization with amebic antigen and some of its fractions. Am J Trop Med Hyg 29:779–784

Gillin FD, Diamond LS (1981) Inhibition of clonal growth of *Giardia lamblia* and *Entamoeba histolytica* by metronidazole, quinacrine, and other antimicrobial agents. J Antimicrobial Chemother 8:305–316

Gilman R, Islam M, Paschi S, Goleburn J, Ahmad F (1980) Comparison of conventional and immunofluorescent techniques for the detection of *Entamoeba histolytica* in rectal biopsies. Gastroenterology 78:435–439

Gleason NN, Goldman M, Carver RDK (1963) Size and nuclear morphology of *Entamoeba histolytica* and *Entamoeba hartmanni* trophozoites in cultures and in man. Am J Hyg 77:1–14

Gold D, Kagan IG (1978) Susceptibility of various strains of mice to *Entamoeba histolytica.* J Parasitol 64:937–938

Goldman M, Gleason NN, Carver RK (1962) Antigenic analysis of *Entamoeba histolytica* by means of fluorescent antibody III. Reactions of the Laredo strain with five anti-hystolytica sera. Am J Trop Med Hyg 11:341–346

Guerrant RL, Brush J, Ravdin JI, Sullivan JA, Mandell GL (1981) Interaction between *Entamoeba histolytica* and human polymorphonuclear neutrophils. J Infect Dis 143:83–93

Guirges SY (1982) Virulence of *Entamoeba histolytica* strains from Bombay (India) to rats. J Hyg Epidemiol Microbiol Immunol 26:141–151

Harries J (1982) Amoebiasis: a review. J R Soc Med 75:190–197

Harris WG, Bray RS (1976) Cellular sensitivity in amoebiasis – preliminary results of lymphocytic transformation in response to specific antigen and to mitogen in carrier and disease states. Trans R Soc Trop Med Hyg 70:340–343

Healy GR (1971) Laboratory diagnosis of amebiasis. Bull N Y Acad Sci 47:478–493

Huldt G, Davies P, Allison AC, Schorlemmer HU (1979) Interaction between *Entamoeba histolytica* and complement. Nature 277:214–216

Imperato PJ (1981) A historical overview of amebiasis. Bull NY Acad Med 57:175–187

Ishaq M, Padma MC (1980) Antigenic variations among strains of *Entamoeba histolytica.* Ann Trop Med Parasitol 74:373–375

Jarumilinta R, Maegraith BG (1969) Enzymes of *Entamoeba histolytica.* Bull WHO 41:269–273

Kanani SR, Knight R (1969) Relapsing amoebic colitis of 12 years standing exacerbated by cortico-steroids. Br Med J 2:613–614

Kessel JF, Lewis WP, Pasquel CM, Turner JA (1965) Indirect hemagglutination and complement fixation tests in amebiasis. Am J Trop Med Hyg 14:540–550

Knight R (1980) The chemotherapy of amoebiasis. J Antimicrob Chemother 6:577–593

Knight R, Bird RG, McCaul TF (1975) Fine structural changes at Entamoeba histolytica rabbit kidney cell (RK13) interface. Ann Trop Med Parasitol 69:197–202

Kobiler D, Mirelman D (1980) Lectin activity in Entamoeba histolytica trophozoites. Infect Immun 29:221–225

Kobiler D, Mirelman D (1981) Adhesion of Entamoeba histolytica trophozoites to monolayers of human cells. J Infect Dis 144:539–546

Krogstad DG, Spencer HC, Healy GR, Gleason MN, Sexton J, Herron CA (1978) Amebiasis: epidemiologic studies in the United States, 1971–1974. Ann Intern Med 88:89–97

Krupp IM (1970) Antibody response in intestinal and extraintestinal amebiasis. Am J Trop Med Hyg 19:57–62

Lin TM, Halbert SP, Chiu CT, Zarco R (1981) Simple standardized enzyme-linked immunosorbent assay for human antibodies to Entamoeba histolytica. J Clin Microbiol 13:646–651

Lunde MN, Diamond LS (1969) Studies on antigens from axenically cultivated Entamoeba histolytica and Entamoeba histolytica-like amoebae. Am J Trop Med Hyg 18:1–6

Lushbaugh WB, Miller JH (1974) Fine structural topochemistry of Entamoeba histolytica Schaudinn 1903. J Parasitol 60:421–433

Lushbaugh WB, Kairalla AB, Loadholt CB, Pittman FE (1978) Effect of hamster liver passage on the virulence of axenically cultivated Entamoeba histolytica. Am J Trop Med Hyg 27:248–254

Lushbaugh WB, Kairalla AB, Cantey JR, Hofbauer AF, Pittman FE (1979) Isolation of a cytotoxin-enterotoxin from Entamoeba histolytica. J Infect Dis 139:9–17

Lynch EC, Rosenberg IM, Gitler C (1982) An ion channel forming protein produced by Entamoeba histolytica. EMBO J 1:801–804

Maddison SE, Powell SJ, Elsdon-Dew R (1965a) Comparison of hemagglutinins and precipitins in amebiasis. Am J Trop Med Hyg 14:551–553

Maddison SE, Powell SJ, Elsdon-Dew R (1965b) Application of serology to the epidemiology of amebiasis. Am J Trop Med Hyg 14:554–557

Maddison SE, Kagan IG, Elsdon-Dew R (1968) Comparison of intradermal and serologic tests for the diagnosis of amebiasis. Am J Trop Med Hyg 17:540–547

Martinez-Palomo A (1982) The biology of Entamoeba histolytica. In: Brown KN (ed) Tropical medicine research studies series 2. Research Studies Press, Chichester

Martinez-Palomo A, Gonzales-Robles A, de la Torre M (1973) Selective agglutination of pathogenic strains of Entamoeba histolytica induced conA. Nature (New Biol) 245:186–187

Martinez-Palomo A, Orozco E, Gonzalez-Robles A (1980) Entamoeba histolytica: topochemistry and dynamics of the cell surface. In: Van den Bossche H (ed) The host invader interplay. Elsevier North Holland, Amsterdam, pp 55–68

Mattern CFT, Diamond LS, Keister DB (1978a) Amebal viruses and the virulence of Entamoeba histolytica. Arch Invest Med (Mex) [Suppl 1] 9:165–166

Mattern CFT, Keister DB, Caspar PA (1978b) Experimental amebiasis. III A rapid in vitro assay for virulence of Entamoeba histolytica. Am J Trop Med Hyg 27:882–887

McConnachie EW (1955) Studies on Entamoeba invadens Rodhain, 1934, in vitro and its relationship to some other species of Entamoeba. Parasitol 45:452–481

McConnachie EW (1969) The morphology, formation, and development of cysts of Entamoeba. Parasitology 59:41–53

McGowan K, Deneke CF, Thorne GM, Gorbach SL (1982) Entamoeba histolytica cytotoxin: purification, characterization, strain virulence and protease activity. J Infect Dis 146:616–625

McLaughlin J, Faubert G (1977) Partial purification and some properties of a neutral sulfhydryl and an acid proteinase from Entamoeba histolytica. Can J Microbiol 23:420–425

Meerovitch E, Ghadirian E (1978) Restoration of virulence of axenically cultivated Entamoeba histolytica by cholesterol. Can J Microbiol 24:63–65

Meerovitch E, Hartman DP, Ghadirian E (1978) Protective immunity and possible autoimmune regulation in amebiasis. Arch Invest Med (Mex) [Suppl 1] 9:247–252

Mirelman D, Feingold C, Wexler A, Bracha R (1983) Interactions between Entamoeba histolytica, bacteria and intestinal cells. In: Cytopathology of parasitic disease. CIBA foundation 99. Pitman, London, pp 2-30

Morris MN, Powell SJ, Elsdon-Dew R (1970) Latex agglutination test for invasive amoebiasis. Lancet 1:1362-1363

Munoz MdeL, Calderon J, Rojkind M (1982) The collagenase of *Entamoeba histolytica.* J Exp Med 155:42-51

Murray HW, Aley SB, Scott WA (1981) Susceptibility of *Entamoeba histolytica* to oxygen intermediates. Mol Biochem Parasitol 3:381-391

Neal RA (1951) The duration and epidemiological significance of *Entamoeba histolytica* infections in rats. Trans R Soc Trop Med Hyg 45:363-370

Neal RA (1960) Enzymic proteolysis by *Entamoeba histolytica;* biochemical characteristics and relationship with invasiveness. Parasitology 50:531-550

Neal RA (1965) Influence of encystation on invasiveness of *Entamoeba histolytica.* Exp Parasitol 16:369-372

Nilsson L, Petchclai B, Elwig H (1980) Application of thin layer immunoassay (TIA) for demonstration of antibodies against *Entamoeba histolytica.* Am J Trop Med Hyg 29:524-529

Orozco ME, Guarneros G, Martinez-Palomo A (1982) Clonas de *Entamoeba histolytica* deficientes en fagocitosis presentan deficiencia en virulencia. Arch Invest Med (Mex) [Suppl 3] 13:137-143

Ortiz-Ortiz L, Zamacona G, Sepulveda B, Capin NR (1975) Cell-mediated immunity in patients with amebic abscess of the liver. Clin Immunol Immunopathol 4:127-134

Ortiz-Ortiz L, Capin R, Capin NR, Sepulveda B, Zamacona G (1978) Activation of the alternative pathway of complement by *Entamoeba histolytica.* Clin Exp Immunol 34:10-18

Oyerinde JPO, Ogunbi O, Alonge AA (1977) Age and sex distribution of infections with *Entamoeba histolytica* and *Giardia intestinalis* in the Lagos population. Int J Epidemiol 6:231-234

Parkhouse M, Cid ME, Calderon J (1978) Identificacion de antigenos de membrana de *Entamoeba histolytica* con anticuerpos de pacientes de amibiasis. Arch Invest Med (Mex) [Suppl 1] 9:211-218

Patterson M, Schoppe LE (1982) The presentation of amoebiasis. Med Clin North Am 66:689-705

Patterson M, Healy GR, Shabot JM (1980) Serologic testing for amoebiasis. Gastroenterology 78:136-141

Phillips BP (1973) *Entamoeba histolytica:* concurrent irreversible loss of infectivity/pathogenicity and encystment potential after prolonged maintenance in axenic culture *in vitro.* Exp Parasitol 34:163-167

Phillips BP, Diamond LS, Bartgis IL, Stuppler SA (1972) Results of intracecal inoculation of germfree and conventional guinea pigs and germfree rats with axenically cultivated *Entamoeba histolytica.* J Protozool 19:498-499

Pillai S, Mohimen A (1982) A solid-phase sandwich radioimmunoassay for *Entamoeba histolytica* proteins and the detection of circulating antigens in amoebiasis. Gastroenterology 83:1210-1216

Pillai S, Mohimen A, Mehra S (1982) A radioimmunoprecipitation polyethylene glycol assay for circulating *Entamoeba histolytica* antigens. J Immunol Methods 55:273-276

Pinto da Silva P, Martinez-Palomo A, Gonzalez-Robles A (1975) Membrane structure and surface coat of *Entamoeba histolytica* topochemistry and dynamics of the cell surface: cap formation and microexudate. J Cell Biol 64:538-550

Pittman FE (1980) Intestinal amebiasis. Pract Gastroenterol (March): 33-39

Pittman FE, El-Hashimi WK, Pittman JC (1973) Studies of human amebiasis II. Light and electron-microscopic observations of colonic mucosa and exudate in acute amebic colitis. Gastroenterology 65:558-603

Powell SJ, Maddison SE, Elsdon-Dew R (1966) Rapid faecal transmission and invasive amoebiasis in Durban. S Afr Med J 40:646-649

Prasad AK, Das SR, Sagar P (1981) Cholesterol induced changes in Concanavalin A agglutinability of *Entamoeba histolytica.* Indian J Exp Biol 19:1172-1174

Price ME (1981) Amoebic liver abscess in a Norfolk factory worker. Br Med J 283:1175

Proctor EM (1976) Ultrastructure of trophozoites of *Entamoeba histolytica* from human amoebic liver abscess. Trans R Soc Trop Med Hyg 70:256-257

Raether W, Seidenath H (1983) The activity of Fexinidazole (HOE 329) against experimental infections with *Trypanosoma cruzi,* trichomonads and *Entamoeba histolytica.* Ann Trop Med Parasitol 77:13-26

Rai GP, Gupta AK, Das SR (1982) A preliminary study on the pathogenicity of axenically grown *Entamoeba histolytica* for the golden hamster. Aust J Exp Biol Med Sci 60:97-99

Ravdin JI, Guerrant RL (1982) A review of the parasite cellular mechanisms involved in the pathogenesis of amebiasis. Rev Infect Dis 4:1185-1207

Ravdin JI, Croft BY, Guerrant RL (1980) Cytopathogenic mechanisms of *Entamoeba histolytica*. J Exp Med 152:377–390

Ravdin JI, Sperelakis N, Guerrant RL (1982) Effect of ion channel inhibitors on the cytopathogenicity of *Entamoeba histolytica*. J Infect Dis 146:335–340

Reed SL, Sargeaunt PG, Braude AI (1983) Resistance to lysis by human serum of pathogenic *Entamoeba histolytica*. Trans R Soc Trop Med Hyg 77:248–253

Rees CW, Reardon LV, Bartgis IL (1950) The excystation of *Entamoeba histolytica* without bacteria in microcultures. Parasitology 40:338–342

Reeves RE, Bischoff JM (1968) Classification of *Entamoeba* species by means of electrophoretic properties of amebal enzymes. J Parasitol 54:594–600

Rengpien S, Bailey GB (1975) Differentiation of *Entamoeba*: a new medium and optimal conditions for axenic encystation of *E. invadens*. J Parasitol 61:24–30

Root DM, Cole FX, Williamson JA (1978) The development and standardization of an ELISA method for the detection of *Entamoeba histolytica* antigens in fecal samples. Arch Invest Med (Mex) [Suppl 1] 9:203–210

Sargeaunt PG (1982) Amoebiasis. J R Soc Med 75:920–921

Sargeaunt PG, Williams JE (1980) The epidemiology of *Entamoeba histolytica* in Mexico City. Trans R Soc Trop Med Hyg 74:653–656

Sargeaunt PG, Williams JE, Grene JD (1978) The differentiation of invasive and noninvasive *Entamoeba histolytica* by isoenzyme electrophoresis. Trans R Soc Trop Med Hyg 72:519–521

Sargeaunt PG, Williams JE, Jones DM (1982a) Electrophoretic isoenzyme patterns of *Entamoeba histolytica* and *Entamoeba chattoni* in a primate survey. J Parasitol 29:136–139

Sargeaunt PG, Williams JE, Bhojnani R, Kumate J, Jimenez E (1982b) A review of isoenzyme characterization of *Entamoeba histolytica* with particular reference to pathogenic and nonpathogenic stocks isolated in Mexico. Arch Invest Med (Mex) [Suppl 3] 13:89–94

Sargeaunt PG, Jackson TFHG, Simjee A (1982c) Biochemical homogeneity of *Entamoeba histolytica* isolates especially those from liver abscess. Lancet 1:1386–1388

Savanat T, Viriyanond P, Nimitmongkol N (1973a) Blast transformation of lymphocytes in amebiasis. Am J Trop Med Hyg 22:705–710

Savanat T, Bunnag D, Chongsuphajaisiddhi T, Viriyanond P (1973b) Skin test for amebiasis: an appraisal. Am J Trop Med Hyg 22:168–173

Sawhney S, Chakravarti RN, Jain P, Vinayak VK (1980) Immunogenicity of axenic *Entamoeba histolytica* antigen and its fractions. Trans R Soc Trop Med Hyg 74:26–29

Scaglia M, Villa M, Gatti S, Strosselli M, Grazioli V (1982) *Entamoeba moshkovskii*: a new isolate from sewage sludges in Italy. Trans R Soc Trop Med Hyg 76:703

Sepulveda B (1982) Amebiasis: host-pathogen biology. Rev Infect Dis 4:1247–1253

Serrano R, Deas JE, Warren LG (1977) *Entamoeba histolytica*: membrane fractions. Exp Parasitol 41:370–384

Sheehan DJ, Bottone EJ, Pavletich K, Clare-Heath SR (1979) *Entamoeba histolytica*: efficacy of microscopic, cultural, and serological techniques for laboratory diagnosis. J Clin Microbiol 10:128–133

Singh N, Das SR, Saxena U (1963) Virulence of strains of *Entamoeba histolytica* from India, with an account of a method for obtaining cent percent infection in rat. Ann Biochem Exp Med 23:237–242

Sirijintakarn P, Bailey GB (1980) The relationship of DNA synthesis and cell cycle events to encystation by *Entamoeba invadens*. Arch Invest Med (Mex) [Suppl 1] 11:3–10

Spencer HC, Sullivan JJ, Mathews HM, Sauerbrey M, Bloch M, Chin W, Healy GR (1981) Serologic and parasitologic studies of *Entamoeba histolytica* in El Salvador 1974–1978. Am J Trop Med Hyg 30:63–68

Stamm WP, Ashley MJ, Parelkar SN (1973) Evaluation of a latex agglutination test for amoebiasis. Trans R Soc Trop Med Hyg 67:211–213

Stamm WP, Ashley MJ, Bell K (1976) The value of amoebic serology in an area of low endemicity. Trans R Soc Trop Med Hyg 70:49–53

Stuiver PC, Goud TJLM (1978) Corticosteroids and liver amoebiasis. Br Med J 2:394–395

Suoza EMD (1982) *Entamoeba gingivalis* on children's mouths. Rev Microbiol 13:173–179

Takeuchi A, Phillips BP (1975) Electron microscope studies of experimental *Entamoeba histolytica* infection in the guinea pig. I. Penetration of the intestinal epithelium by trophozoites. Am J Trop Med Hyg 24:34–48

Tomasi TB, Bienenstock J (1968) Secretory immunoglobulins. Adv Immunol 9:1–96

Trissl D (1982) Immunology of *Entamoeba histolytica* in human and animal hosts. Rev Infect Dis 4:1154–1184

Trissl D, Martinez-Palomo A, Arguello C, De la Torre M, De la Hoz R (1977) Surface properties related to concanavalin A-induced agglutination. A comparative study of several *Entamoeba* strains. J Exp Med 145:652–655

Trissl D, Martinez-Palomo A, De la Torre M, De la Hoz R, Perez de Suarez E (1978) Surface properties of *Entamoeba*: increased rates of human erythrocyte phagocytosis in pathogenic strains. J Exp Med 148:1137–1145

Udezulu IA, Leitch GJ, Bailey GB (1982) Use of indomethacin to demonstrate enterotoxic activity in extracts of *Entamoeba histolytica* trophozoites. Infect Immun 36:795–801

Usawattanakul W, Tapchaisri P, Thammapalerd N, Tharavanij S, Kojima S (1982) Mouse IgE response to an allergen from *Entamoeba histolytica*. J Parasitol 68:398–401

Vinayak VK, Sawhney S, Jain P, Chakravarti RN (1980) Protective effects of crude and chromatographic fractions of axenic *Entamoeba histolytica* in guinea-pigs. Trans R Soc Trop Med Hyg 74:483–487

Weinbach EC, Diamond LS (1974) *Entamoeba histolytica*: aerobic metabolism. Exp Parasitol 35:232–243

Welch JS, Rowsell RJ, Freeman C (1978) Treatment of intestinal amoebiasis and giardiasis – efficacy of Metronidazole and Tinidazole compared. Med J Aust 1:469–471

WHO Expert Committee Report (1969) Amoebiasis. WHO Tech Rep Ser 421:5–52

Wittner M, Rosenbaum RM (1970) Role of bacteria in modifying virulence of *Entamoeba histolytica*. Am J Trop Med Hyg 19:755–761

Wolfe MS (1982) The treatment of intestinal protozoan infections. Med Clin North Am 66:707–720

Yang J, Kennedy MT (1979) Evaluation of enzyme linked immunosorbent assay for the serodiagnosis of amoebiasis. J Clin Microbiol 10:778–785

Yap EH, Zaman V, Aw SE (1970) The use of cyst antigen in the serodiagnosis of amoebiasis. Bull WHO 42:553–561

Young JD-E, Young TM, Lu LP, Unkeless JC, Cohn ZA (1982) Characterization of a membrane pore-forming protein from *Entamoeba histolytica*. J Exp Med 156:1677–1690

Parasite Antigens, Their Role in Protection, Diagnosis and Escape: The Leishmaniases

J. ALEXANDER[1] and D.G. RUSSELL[2]

1 Introduction

The leishmaniases comprise a group of diseases of major and increasing public health importance (WHO Technical Report 701, 1984). The genus consists of many species which are epidemiologically diverse and complex. Nevertheless, all leishmanias are transmitted by sandfly vectors, either of the genus *Phlebotomus* (Old World) or *Lutzomyia* (New World), and they are all obligate intracellular parasites in their vertebrate hosts where they are found within macrophage phagolysosomes (ALEXANDER and VICKERMAN 1975; CHANG and DWYER 1976). Although the forms found infecting macrophages, amastigotes, are for the most part morphologically identical, the diseases produced in man display widely different clinical manifestations. This depends primarily on the species of parasite initiating the infection, but also in part on the general state of health or age, and more particularly,

1 Department of Bioscience and Biotechnology, University of Strathclyde, The Todd Centre, Glasgow G4 0NR, Scotland, United Kingdom
2 Max-Planck-Institut für Biologie, Corrensstraße 38, D-7400 Tübingen

Current Topics in Microbiology and Immunology, Vol. 120
© Springer-Verlag Berlin · Heidelberg 1985

Table 1. Major species of *Leishmania* causing disease in man

Species	Subspecies	Distribution	Disease	Tendency to self-cure
L.donovani	*L.d.donovani*	Old World	VL	No
	L.d.infantum	Old World	PKDL	No
	L.d.chagasi	New World		
L.major		Old World	CL	Rapid
L.tropica		Old World	CL	Slow
			LR	No
L.aethiopica		Old World	CL	Slow
			DCL	No
L.braziliensis	*L.b.braziliensis*			
	L.b.guyanensis	New World	MCL	No
	L.b.panamensis		CL	Yes
L.mexicana	*l..m.mexicana*			
	L.m.amazonensis	New World	CL	Variable
	L.m.pifanoi		DCL	No
L.peruviania		New World	CL	Mostly

VL, visceral leishmaniasis; PKDL, post-kala-azar dermal leishmaniasis; CL, cutaneous leishmaniasis; LR, leishmaniasis recidivans; DCL, diffuse cutaneous leishmaniasis; MCL, mucocutaneous leishmaniasis

the genetic make-up of the host (BLACKWELL and ALEXANDER 1985). The major disease-causing species of *Leishmania* in man are summarised in Table 1.

This group of parasites pose intriguing problems for clinicians and researchers. In many cases, the insect vector and/or the reservoir hosts – for the disease is often a zoonosis – remain to be identified. Diagnosis in man is often not straightforward, not only because of the spectral nature of the disease (TURK and BRYCESON 1971) but also because parasites are often difficult to detect, especially in some of the more chronic infections. Positive delayed-type hypersensitivity (DTH) following cutaneous inoculation with "leishmanin" (phenol-killed parasites), although often a useful indication of a protective immune response, is not species specific, nor does it indicate either heterologous immunity or the species of immunising or sensitising parasite (reviewed in ZUCKERMAN 1975). This has necessitated the development of more reliable techniques for taxonomically identifying disease-causing organisms. The interaction of leishmanias with their host cells is also of considerable interest, as macrophages not only have a formidable armoury of microbicidal mechanisms, but also in their accessory cell function, they play a crucial role in the control of the immune response.

In this review, we hope to update the progress that has been made in recent years in the biochemical, immunological and functional characterisation of leishmanial antigens. In addition, we shall outline the contribution these studies have made to our understanding of parasite biology, parasite interaction with host cells and with the immunological system in general. We shall also indicate those areas where further research would be useful and discuss progress towards developing successful vaccines.

2 Biology of the Parasite

2.1 Sandfly Stages

For convenience, many workers rather oversimplify the life cycle of *Leishmania*, suggesting that the parasite exists as two distinct forms, the amastigote found within the macrophage of the vertebrate host and the motile, flagellated promastigote in the sandfly gut or in cell-free culture medium. However, amastigotes, three promastigote types and morphologically distinct paramastigotes have been described as the parasites migrate forward in the sandfly gut (reviewed in KILLICK-KENDRICK 1979). Within the midgut, two types of promastigotes have been described; the long, thin nectomonad (body length, $>12\,\mu m$), which is free swimming or attached by interdigitation of the flagellum to the ciliated epithelium of the midgut, and the short, broad haptomonad (body length, $<12\,\mu m$) bound by hemidesmosome attachments of the flagellum to the cuticular intema of the stomodeal valve (KILLICK-KENDRICK et al. 1974). As the parasites migrate into the pharynx, they become round to oval, and the kinetoplast lies posterior to the nucleus. These forms, paramastigotes, give rise in turn to the characteristic free-swimming promastigotes found in the proboscis, which have a short body length ($6\,\mu m$), but a comparatively long flagellum. The mechanisms by which the parasites are induced to attach or detach and migrate through the sandly gut are unknown. However, promastigotes have been shown to migrate along chemical gradients (BRAY 1983a) in a manner reminiscent of the "running and tumbling" response of flagellated bacteria (ALEXANDER and BURNS 1983). They are induced to do this by sugars normally stored in high concentrations in the sandfly crop and released into the alimentary canal just posterior to the pharynx (BRAY 1983a). It would seem likely that promastigotes have as yet uncharacterised lectin-like receptors which modulate this movement.

The forms within the fly are particularly important as these, or at least a subpopulation, initiate natural infection and ultimately, it is against these forms that man must be protected. For this reason, the number of parasites deposited in the skin, their morphological form and antigenic composition are of great interest. Although the true metacyclic has not been identified, the proboscis forms would seem the most likely candidates owing to their proximity to wounds. ADLER and THEODOR (1931) have reported that there maybe as many as several hundred proboscis forms, and we have recently confirmed this in the laboratory in *L. major*-infected *Phlebotomus papatasii* (KILLICK-KENDRICK and ALEXANDER, unpublished results). The infectivity of paramastigotes from the pharynx for the vertebrate host remains to be examined. By contrast, studies on the infectivity of promastigotes from the midgut have been well documented. Although it was invariably considered that midgut promastigotes were infective to susceptible vertebrates (ADLER 1964), some workers suggested that, during certain stages of their life cycle, the parasites were not necessarily infective (PARROT and DONATIEN 1927; FENG and CHUNG 1941). These observations have recently been clarified by SACKS and PERKINS (1984, 1985), who found that infectivity for macrophages was growth cycle dependent both in the fly and in cell-free culture. Rapidly dividing log phase promastigotes had low infectivity for host cells and animals, while stationary phase promastigotes had a correspondingly high infectivity (SACKS and PERKINS 1984; SACKS et al. 1985). Changes in infectivity were associated with changes in cell surface characteristics (see Sect. 3). The infective forms had a long flagellum and comparatively short body, somewhat similar to proboscis forms. On entering the macrophage, infective forms transform to the small (3 µm), non-motile amastigote, which can also be distinguished from promastigotes by monoclonal antibodies (MCMAHON-PRATT and DAVID 1982; HANDMAN and HOCKING 1982) and both qualitative and quantitative differences in metabolic pathways (COOMBS 1982).

2.2 Entry into the Vertebrate Host

On entering the vertebrate host, it would appear that promastigotes activate complement via the alternative pathway (BRAY 1983a). This not only creates a C5A gradient that chemotactically attracts macrophages, but also produces the third component of complement, which, when cleaved, binds to promastigotes and facilitates phagocytosis via the macrophage iC3b receptor CR3 (BLACKWELL et al. 1985). Promastigotes of L. m. mexicana, L. enriettii and L. major (BRAY 1983a; MOSSER and EDELSON 1984) are rapidly lysed by complement, but L. m. mexicana amastigotes (BRAY 1983a) and up to 30% of L. b. panamensis and 10% of L. donovani stationary phase "infective" promastigotes were found to be resistant to human complement (FRANKE et al. 1985). We would suspect that "infective" sandfly stages would also be resistant to complement lysis, but this remains to be tested.

2.3 Entry into the Macrophage

The entry of Leishmania into macrophages occurs in two main stages: attachment of the parasite to the host cell membrane via some form of receptor/ligand binding, followed by its enclosure within a macrophage phagosome.

Receptor/ligand binding may involve the Fc receptor, which needs the presence of specific antisera to opsonise the parasite and is therefore of little relevance to the naive host (BRAY 1983b). Alternatively, in the immune host, cytophilic antibody may bind to the host cell via the Fc receptor, thus presenting a Fab site for parasite attachment (HERMAN 1980). Attachment may also occur via the C3b receptors (see above), which are sensitive to trypsin but not chymotrypsin. Even in the absence of serum, this method of entry may take place, as macrophages can secrete complement (EZEKOWITZ et al. 1984).

An alternative form of attachment may be by way of lectin-like glycoprotein receptors and glycolipid ligands via terminal saccharides or sialic acid residues. These can be demonstrated in two ways: first, by competitive binding of the ligands with lectins or of the lectin-like receptors with sugars or, secondly, by attacking the ligand with specific glycosidases or the sialic acid residue with sialidase or neuraminidase.

Studies using these techniques have so far identified glycoprotein or glycolipid ligands having either terminal mannose, glucose or N-acetyl-glucosamine saccharides or sialic acid residues on L. donovani promastigotes (CHANG 1981). On the other hand, BRAY (1983b) demonstrated a reverse wheat germ agglutinin-like receptor on the surface of *L. m. mexicana* promastigotes which binds to a terminal *N*-acetyl-glucosamine glycoprotein ligand on the macrophage. More recent work comparing *L. donovani* amastigote and promastigote attachment has suggested a role for the well-characterised macrophage mannose/fucose receptor in promastigote, but not amastigote attachment (CHANNON et al. 1984). These results correspond with studies demonstrating mannose-rich glycoconjugates on the surface of *L. donovani* promastigotes (DWYER and GOTTLIEB 1983). They may also relate to the findings of HERNANDEZ (1982) for *L. braziliensis* where the number of receptor sites for concanavalin A was reduced during promastigote to amastigote transformation. Alternatively, ZEHAVI et al. (1983) have provided evidence for a macrophage galactose-specific receptor in the attachment of *L. tropica* promastigotes. For a more detailed discussion of parasite attachment to the host cell, see Sect. 5.

Following attachment, perturbation of the macrophage membrane may activate the membrane-bound enzyme NADPH oxidase and trigger a respiratory burst resulting in the production of potentially toxic oxygen radicals. However, promastigotes of *L. major, L. donovani* and *L. m. mexicana* trigger the respiratory burst to a markedly greater degree than amastigotes [(MURRAY 1981, 1982; CHANNON et al. 1984; ALEXANDER, unpublished observations) (Fig. 1a, b)]. This and the fact that promastigotes are more susceptible to killing by hydrogen peroxide than amastigotes correlates with their low survival rate in peritoneal macrophages as compared with amastigotes (MURRAY 1982). CHANNON and co-workers (1984) have suggested that differences in respiratory burst activity may arise because amastigotes possess an azide-sensitive mechanism which either competes for superoxide anions O_2^- or causes localized inactivation of respiratory burst activity. However, they suggest that the more likely explanation is that there are qualitative and quantitative differences in the distribution of surface ligands involved in binding the parasite to the macrophage membrane. That binding and phagocytosis can occur without a respiratory burst has been demonstrated for human macrophage C3b receptors (WRIGHT and SILVERSTEIN 1983). It is, perhaps, of particular significance that our preliminary observations suggest that proboscis promastigotes of *L. major*, unlike midgut forms, fail to trigger the respiratory burst of mouse peritoneal macrophages (ALEXANDER and KILLICK-KENDRICK, unpublished observations). This would suggest that proboscis forms are already pre-adapted for a life in the vertebrate host.

The entry of *Leishmania* parasites into macrophages would seem to be predominantly by means of the so-called "zipper" process of phagocytosis (SILVERSTEIN 1977; BLACKWELL and ALEXANDER 1983), although an active role for the promastigote (AIKAWA et al. 1982) and the amastigote (WYLER 1982) in uptake cannot be ruled out.

2.4 Intracellular Stages

Both in vitro (ALEXANDER and VICKERMAN 1975; CHANG and DWYER 1976) and in vivo (BERMAN et al. 1981) studies have demonstrated that, once inside the macrophage, leishmanias reside within the potentially hostile environment of the phago-

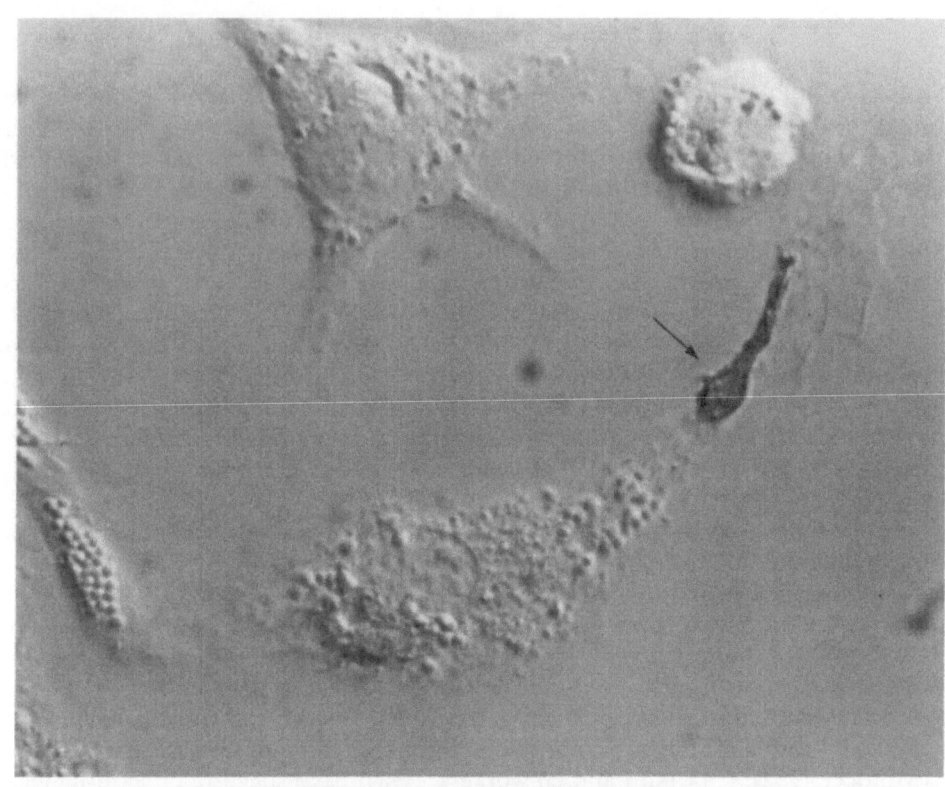

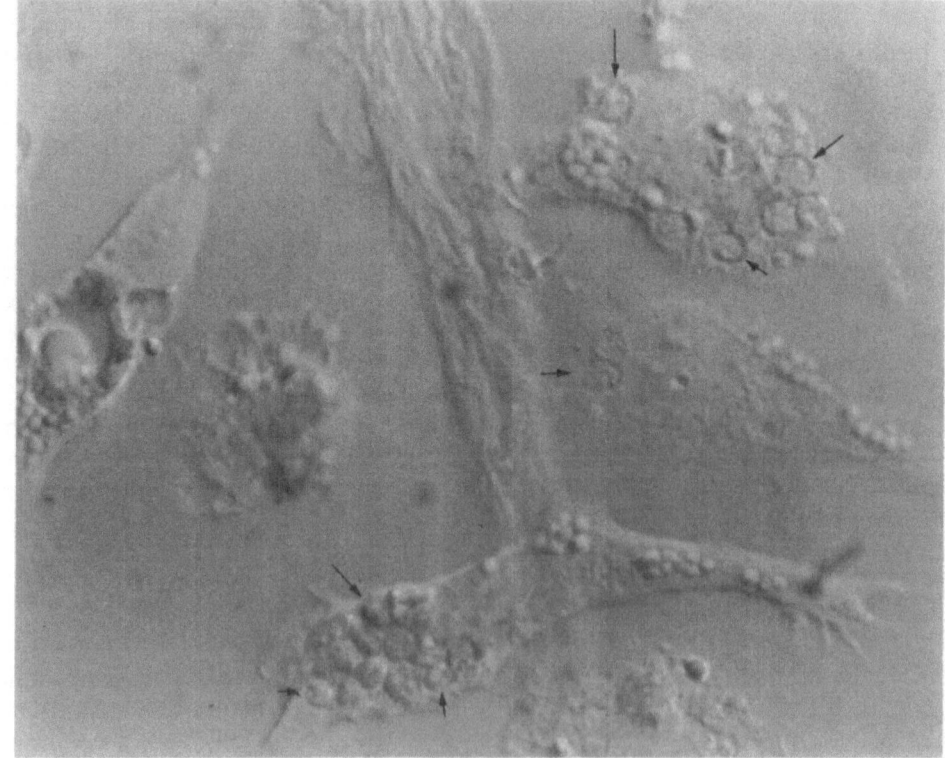

lysosome. Numerous theories have been proposed to account for parasite survival within the parasitophorous vacuole (ALEXANDER and VICKERMAN 1975; LEWIS and PETERS 1977; CHANG and DWYER 1978; COOMBS 1982; EL-ON et al. 1980; DWYER and GOTTLIEB 1983; HANDMAN et al. 1981). Not surprisingly, perhaps, amastigotes would seem much better adapted to this environment than cultured promastigotes. For example, treating macrophages with inhibitors of phagosome-lysosome fusion only moderately enhances *L. m. mexicana* amastigote survival but dramatically increases the survival of promastigotes (ALEXANDER 1981a). Parasite survival has been related either to the inactivation of lysosomal hydrolytic enzymes by the parasite through the release of metabolic waste and secretory factors (EL-ON et al. 1980; COOMBS 1982; KUTISH and JANOVY 1981) or to the refactory nature of the parasite surface membrane (CHANG and DWYER 1978; HANDMAN et al. 1981). This, in turn, has been speculatively associated with 68- and 43-kd glycoproteins on the promastigote surface (DWYER and GOTTLIEB 1983). Nevertheless, the location of parasites within the macrophage phagolysosome has the beneficial side effect of leaving them open to attack with lipophilic or charged drugs which can be rapidly endocytosed by macrophages and concentrated within parasitophorous vacuoles. Thus, the efficacy of anti-leishmanial compounds has been enhanced several hundredfold by encapsulating drugs in liposomes (ALVING and STECK 1979). Recently, SHEPHERD et al. (1983) have demonstrated the mannose/fucose receptor-mediated, mannaninhibitable accumulation of B-glucoronidase into the parasitophorous vacuole of macrophages infected with *Leishmania amazonensis*. This would indicate that mannosylated molecules might also be used effectively for targeting leishmanicidal drugs.

2.5 Host Responses and Infection-Limiting Factors

After the initial interaction, leishmanias may have to withstand changes in the macrophage response not only as a result of T cell-mediated lymphokine activation, but also T cell-independent activation. The latter may not only control innate host susceptibility, but also the subsequent development of the T cell-mediated response (reviewed in BLACKWELL and ALEXANDER 1983). Thus, innate resistance to *L. donovani* in mice is under the control of a single gene (*Lsh*) which maps to chromosome 1 (BRADLEY et al. 1979) and is expressed in Kupffer's cells 2 days after contact with parasites (CROCKER et al. 1984). Subsequent to macrophage involvement, the growth of *L. donovani* is influenced by the H-2 complex (BLACKWELL et al. 1980). Similarly, a primary macrophage defect (HOWARD et al. 1980)

←――

Fig. 1a, b. Interference phase micrographs of resident peritoneal macrophages infected with a promastigotes or **b** amastigotes of *L. m. mexicana*. Macrophages were incubated in 0.1% nitroblue tetrazolium, and the macrophage respiratory burst was visualised locally by the production of insoluble formazan (reduced nitroblue tetrazolium). The respiratory burst was triggered by 85%–96% of promastigotes but only 10%–22% of amastigotes. **a** Promastigote (*arrow*) in a macrophage surrounded by insoluble formazan. × 1500. **b** Amastigotes (*arrows*) in macrophages with no evidence of formazan production. × 1500 which has been associated with a skin macrophage subpopulation (GORCZINSKI and MacRAE 1981) plays a pivotal role in the induction of specific Lyt1$^+$2$^-$IJ$^-$T suppressor cells in mice susceptible to *L. major*. Susceptibility is not under *Lsh* or H-2 gene control, but rather, appears to map to chromosome 8 and is designated *Scl* (BLACKWELL et al. 1984). These differences in genetic-controlled responses at the level of the tissue macrophage may reflect and explain the predilection of different species of *Leishmania* for different host tissues and thus different macrophage subpopulations. *L. donovani* and *L. major* (HOCHMEYER et al. 1984) and *L. major* and *L. mexicana* (SCOTT et al. 1983) show marked differences in their relative abilities to survive in the same macrophage populations under similar conditions of activation. These results not only indicate differing host susceptiblity, but may reflect the particular macrophage subpopulation used in the study (ALEXANDER 1981b). Analysing and characterising the *Lsh* and *Scl* gene products should greatly increase our understanding of the differential interaction of *L. donovani* and *L. major* with host cells.

2.6 Antigen Presentation

In addition to their microbicidal and degradative functions, those macrophage subpopulations bearing Ia antigens play a pivotal role in the initiation and subsequent regulation of lymphocyte proliferation (reviewed by UNANUE 1981, 1984). This makes the interaction of the leishmanias with their host cells all the more intriguing. The requirement for presenting cells in stimulating the proliferation of *L. major* specific T cells has been demonstrated (LOUIS et al. 1981). The proliferation of T cell blasts, generated in vitro in response to the parasite, was found to depend on the presence of syngenic spleen cells. The response was inhibited by treating spleen cells with anti-Ia antibodies and complement. In an elegant piece of work, GORCZYNSKI and MACRAE (1981) were able to demonstrate subpopulations of skin macrophages from *L. major*-susceptible BALB/c mice which not only allowed rapid parasite growth but induced the proliferation of suppressor T cells from naive lymphocyte populations. The same macrophage population incubated with soluble *L. major* excretion factor did, however, promote the proliferation of sensitised T-lymphocytes. These results suggest that the immunological dysfunction resulting in non-healing lesions in BALB/c mice infected with *L. major* lies somewhere in processing the correct antigen from live, intact parasites. However, by selectively inoculating characterised antigens, even into those animals normally susceptible to leishmaniasis, a protective immune response may be generated. Although a number of workers have demonstrated antigen on the surface of infected macrophages (FARAH et al. 1975; BERMAN and DWYER 1981; HANDMAN and HOCKING 1982), this has not as yet been associated with the concomitant expression of class II molecules (Ia antigens). The search for and characterisation of defined immunogenic molecules in antigen presentation is essential to our understanding of immunoregulatory mechanisms and important in the development of successful vaccines.

3 Biochemical Characterisation of Parasite Antigens

The study of exposed surface antigens of *Leishmania* has expanded considerably since the application of ^{125}I surface labelling to these cells. Early studies did, however, produce some conflicting results. HANDMAN and CURTIS (1982) reported finding a fairly simple pattern with six major polypeptides on the surface of *L. major* promastigotes, whereas a year later, GARDINER and DWYER (1983) labelled over 20 polypeptides by the same method on the same species. More recent studies in other laboratories on a variety of species agree with HANDMAN and CURTIS that promastigotes have a fairly limited number of surface proteins (LEPAY et al. 1983; RAMASAMY et al. 1983), and the autoradiograph from a two-dimensional gel of surface-iodinated *L.m.mexicana* promastigotes (Fig. 2) is probably now the accepted picture. The most abundant polypeptide in Fig. 2 has a molecular weight of 63 kd. ETGES and co-workers (1985) have isolated a 63-kd polypeptide from *L. major* promastigotes. This polypeptide accounted for 70% of the incorporated label on the parasite surface, indicating that, unless it has an inordinate number of accessible tyrosine residues, it is a major constituent of the parasite's plasmalemma. Analysis of the 63-kd polypeptide by peptide mapping and immunoprecipitation demonstrated that similar, abundant membrane proteins are also present in *L. donovani* and *L. tropica*. Immunological studies on cross-reactive antigens of different *Leishmania* species suggest that this 63-kd protein is a highly conserved protein represented in both New and Old World *Leishmania* (LEPAY et al. 1983; RAMASAMY et al. 1983; GARDINER et al. 1984).

The carbohydrate nature of the promastigote surface was first indicated by the parasite agglutinating ability of various lectins (DWYER 1974; DAWIDOCWIZ et al. 1975). Since then, it has been shown that the majority of iodinated polypeptides on the promastigote surface possess saccharide residues (ETGES et al. 1985; LEPAY et al. 1983; RAMASAMY et al. 1983). Mapping of saccharide heterogeneity using immobilized lectins to precipitate iodinated polypeptides has shown that the 63-kd glycoprotein (Gp 63) from *L.m.mexicana* promastigotes contains mannose, *N*-acetyl-glucosamine and *N*-acetyl-galactosamine residues (RUSSELL and WILHELM 1985). An interesting consequence of these lectin interaction experiments has been the identification of a highly infective subpopulation of *Leishmania* promastigotes. SACKS and PERKINS (1984, 1985) demonstrated that the promastigote stage does not, in fact, represent a single homogenous cell population but one undergoing differentiation, both in culture and in the fly, from a non-infective to an infective stage. The development of the infective stage coincides with the loss of the promastigote's ability to bind the *d*-galactose-reactive lectins, peanut agglutinin and *Ricinus communis* agglutinin (SACKS et al. 1985). This indicates that the carbohydrate content, and therefore the antigenic make-up, of the parasite's surface changes as it attains peak infectivity. On the basis of Western blot analysis and immunoprecipitation of surface-labelled organisms, an antigen of roughly 116 kd was reported on the surface of infective but not of non-infective forms.

The surface antigens of the amastigote form of the parasite have not been so extensively studied. HANDMAN and CURTIS (1983) have, however, published a comparative study of *L. major* amastigote and promastigote surface proteins. They found that the pattern generated by the amastigote has greater complexity, but that anti-promastigote antisera immunoprecipitates most of the major amastigote surface

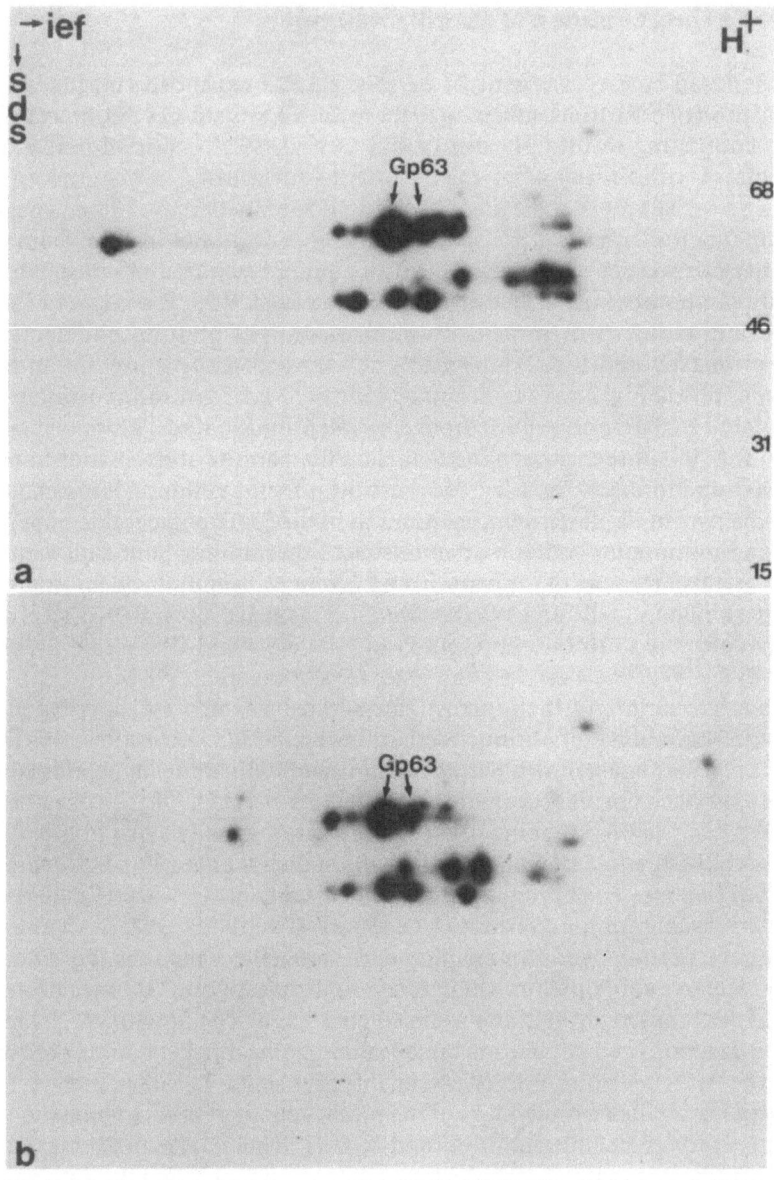

Fig. 2. a An autoradiograph from a two-dimensional gel of surface-iodinated promastigotes of *L. m. mexicana* processed immediately after labelling. If classified solely by molecular weight (*vertical axis*) these cells appear to possess only a few surface polypeptides; however, when also separated by iso-electric focussing (*horizontal axis*), a slightly more complex pattern emerges. The greatest amount of radiolabel is associated with a glycoprotein of molecular weight 63 kd. This polypeptide, designated Gp 63, exists in at least two isoforms (Russell and Wilhelm 1985) and is reported to be present in several *Leishmania* spp. (Etges et al. 1985) **b** An autoradiograph from a two-dimensional gel of a detergent extract obtained from a sample identical to that in **a**, which was precipitated with concanavalin A-Sepharose. This gel demonstrates that the majority of iodinable elements on the surface of *L. m. mexicana* promastigotes are mannose-containing glycoproteins, a fact previously discussed in Dwyer and Gottlieb (1983). Both two-dimensional gels were prepared and run under the conditions described in Russell et al. (1984)

proteins, indicating that the majority of antigens are common to both parasite stages. Two abundant polypeptides of 43 and 94 kd respectively appear to be amastigote specific, whilst a 50-kd polypeptide is confined to the promastigote form. As regards comparison of the saccharide content of the surface of the two stages, a recent study by WILSON and PEARSON (1984) on *L. donovani* showed that, whilst Con A (mannose-specific) and *Ricinus communis* (galactose-specific) lectins reacted with both forms, peanut agglutinin (galactose-specific) bound only to promastigotes and wheat germ agglutinin (*N*-acetyl-glucosamine-specific) only to amastigotes. Whether this alteration of surface sugars is achieved by a change in the saccharide constituents of glycoproteins common to both forms or in a change of complete glycoproteins has yet to be investigated.

In addition to the protein and glycoprotein antigens discussed above, *Leishmania* promastigotes also possess a surface-bound glycoconjugate, the carbohydrate-rich portion of which is shed into the medium, known as the excreted factor (EF) (SCHNUR et al. 1972; SLUTZKY and GREENBLATT 1979; SEMPREVIVO and MACLEOD 1981; KANE-SHINO et al. 1982). Initial experiments by Slutzky and Greenblatt and co-workers (SLUTZKY and GREENBLATT 1977, 1979) identified a soluble, carbohydrate-rich antigen in the medium of growing *L. major* promastigote cultures. Subsequently, they isolated a 75-kd glycoconjugate from promastigote medium and found that antisera against this compound reacted with both a promastigote membrane constituent of similar size and with a molecule present in macrophages containing amastigotes (SLUTZKY and GREENBLATT 1979; SLUTZKY et al. 1979). The authors suggested that the EF is in fact a glycoprotein which is cleaved to release its carbohydrate protein into the medium. SEMPREVIVO and MACLEOD (1981) published a study on the EF of *L. donovani* promastigotes, indicating that it contains carbohydrate, lipid, peptide, phosphate ester and sphingosine moieties that gave it amphipathic properties facilitating micelle formation. More revealing, however, is the recent study of HANDMAN et al. (1984) on *L. major* EF showing it contains glucose, galactose, phosphate and sulphate, but no detectable amino acids, although it does incorporate radiolabelled palmitic acid. As the excreted factor was purified by immunoaffinity using a monoclonal antibody directed against the EF, instead of the rather harsh solvent conditions described previously, this study does suggest that this lipid is an intrinsic part of the molecule. A comparison of soluble and membrane bound forms of the EF by charge-shift electrophoresis indicated that the membrane form possesses a hydrophobic region, possibly a lipid or a small hydrophobic protein. It was possible to mimic the conversion from membrane to soluble form using pronase. The in *vivo* mechanism by which EF is shed from the promastigote surface in unknown, two possibilities are either that it may be released in vesicular form – the pinching off of vesicles from the parasite surface has been observed by HERNANDEZ (1983) – or it may be cleaved off by the activity of a protease of phospholipase associated with the promastigote plasmalemma. HANDMAN and co-workers (1984) also demonstratead similar, though immunologically distinct, molecules released from promastigotes of *L.mexicana* and *L.donovani*.

The existence of a variety of plasmalemma-associated enzymes has recently been reported by several laboratories. GOTTLIEB and DWYER (1981) demonstrated acid phosphatase activity on the external face of *L. donovani* promastigotes. The following year, GLEW and co-workers (1982) isolated the enzyme, also from *L. donovani*, and found it to be a mannose-containing glycoprotein with an apparent molecular weight of 170 kd. It is an acid hydrolase with a broad substrate specificity. In a recent review, DWYER and GOTTLIEB (1983) reported a soluble acid phosphatase with different substrate specificity that may be excreted by the amastigote form whilst inside the parasitophorous vacuole. DWYER and GOTTLIEB (1981, 1983) have also demonstrated both 3' and 5' nucleotidase activity associated with *L. donovani* promastigote plasmalemmas. These enzymes are capable of hydrolysing both 3' and 5' AMP. Cytochemical staining of isolated promastigote membranes demonstrated that their activity is confined to the outer surface of the plasmalemma. As these organisms possess a purine salvage pathway, it is likely that these externally orientated enzymes represent the initial dephosphorylation step facilitating transmembrane transport of nucleotides.

4 Applications of Monoclonal Antibody Research

One of the major problems of a general biochemical description of *Leishmania* antigens is that, because techniques differ between laboratories, it is virtually impossible to cross-reference results. This problem could be overcome if the antigen were to be defined by a specific or unique property, such as the determination of its biological function or its interaction with a defined reagent. Monoclonal antibodies constitute one particularly useful range of reagents. Such probes of known specificity have many potential applications in, for example, the examination of stage- and species-specific antigens and the characterisation of antigens important in macrophage/lymphocyte interaction and acquired immunity.

4.1 Diagnosis

Disease diagnosis was initially made on clinical and geographical criteria and thereafter on epidemiological and parasitological grounds. More recently, *Leishmania* species have been distinguished by such methods as the buoyant density of nuclear and kinetoplast DNA (CHANCE et al. 1974), by analysing sequence homologies in kinetoplast DNA (ARNOT and BARKER 1981), by the use of DNA probes (BARKER and BUTCHER 1983) and, most commonly, by the electrophoretic mobility of isoenzymes (GARDENER et al. 1974; MILES et al. 1979; AL-TAQI and EVANS 1978). However, the recent successful application of *Leishmania* species and subspecies-specific monoclonal antibodies to identify parasites may prove the most effective and convenient method of rapid diagnosis (McMAHON-PRATT and DAVID 1981; McMAHON-PRATT et al. 1982; HANDMAN and HOCKING 1982; DE IBARRA et al. 1982; JAFFE and McMAHON-PRATT 1983; JAFFE et al. 1984; McMAHON-PRATT and DAVID 1982; McMAHON-PRATT et al. 1985). McMAHON-PRATT and colleagues (1985) have generated an extensive catalogue of monoclonal antibodies capable of differentiating not only Old and New World species, but also species and subspecies of the *mexicana* and *braziliensis* "complexes". The resultant library has been used, with impressive accuracy, to screen over 100 isolated stocks of *Leishmania*. Testing with monoclonal antibodies can be done microscopically using immunofluorescence or immunoperoxidase techniques, which have the advantage of using comparatively few intact parasites and which can be rapidly adapted to testing parasites from the sandfly or from infected tissue. However, monoclonal antibodies are often stage specific (HANDMAN and HOCKING 1982; MCMAHON-PRATT and DAVID 1982), and ideally, cultured parasites, i.e. promastigotes, should be used in diagnosis.

It remains to be seen how dependably these reagents operate in the field. However, if used with the awareness that some antigenic drift will doubtless occur, a multiple monoclonal antibody assay has great potential for fast, accurate diagnosis.

If, as is often the case with some chronic leishmanial infections, parasites cannot be isolated from individuals suspected of having leishmaniasis, diagnosis is dependent either on the presence of anti-leishmanial antibody or a positive "leishmanin" skin test. Neither technique, however, indicates the species causing the disease. This problem could be circumvented by using monoclonal antibodies to characterise species-specific parasite antigen. Following purification, these antigen preparations could be used in a skin test or in an immunoassay against serum obtained from suspected infections.

4.2 Stage-Specific Antigens

With respect to the study of stage-specific antigens, HANDMAN and HOCKING (1982) isolated several monoclonal antibodies that recognise the surface antigens of *L. major* promatigotes and amastigotes. Most antibodies reacted with antigens common to both parasite stages, an unsurprising result considering their earlier study on surface-iodinated cells (HANDMAN and CURTIS 1982). They found only one promastigote-specific antibody and noted that most antibodies were directed against carbohydrate epitopes. Four of the antibodies recognized antigens displayed on the surface of amastigote-containing macrophages, one of which has subsequently been show to react with the promastigote's EF (HANDMAN et al. 1984). As regards species-specificity, some antibodies bound only to *L. major*, whilst others cross-reacted with *L. mexicana*, *L. donovani*, *L. enrietti* and even *Crithidia fasciculata*. In addition, in 1982, MCMAHON-PRATT and DAVID reported two monoclonal antibodies specific for the *L. mexicana* promastigote form. These antigens, polypeptides of 92 kd and 43 kd, were absent from amastigotes and amastigote-infected macrophages, appearing during transformation to promastigotes. The use of monoclonal antibodies as transformation markers was extended by FONG and CHANG (1982), who isolated four monoclonal antibodies, all recognizing epitope(s) on the same surface antigen of *L. mexicana* promastigotes. These antibodies all precipitated a triplet of polypeptides of approximately 68 kd. Thus, these laboratories have demonstrated that there are stage-specific antigens within the same species of *Leishmania*. The monoclonal antibodies isolated may have potential for monitoring gene expression during cell transformation.

4.3 Antigen Presentation

Another area where monoclonal antibodies may be particularly useful is in the study of antigen presentation by macrophages. Although it is known that parasite antigens are displayed on the surface of infected macrophages (see Sect. 2.6), the mechanism by which the parasite protein reaches the host cell surface is unknown. Current suggestions are that the antigen may either be incorporated into the phagolysosomal membrane, which subsequently fuses with the macrophages plasmalemma (DWYER 1978) or, alternatively, that antigens may be excreted into extracellular fluid before being absorbed, either alone or as an immune complex, onto the host cell surface (HANDMAN et al. 1979). As at least two laboratories have isolated monoclonal antibodies that recognize parasite antigens on the macrophage surface (DEIBARRA et al. 1982; HANDMAN and HOCKING 1982), this makes possible a study on the mechanism of antigen presentation using immuno-electron microscopy to trace the pathway and immunoprecipitation from cell fractions to monitor any processing of the antigen during its passage to the host cell surface.

4.4 Protective Antibodies

From a medical viewpoint, the most exciting application of monoclonal antibody technology is in the identification of possible candidates for vaccine development. HANDMAN and HOCKING (1982) reported that two of their monoclonal antibodies, directed against *L.major* surface determinants, were able to promote increased killing of parasites by macrophages. The parasites suffered no effect if grown as promastigotes in the presence of antibody. The authors suggested that the antibodies had blocked some mechanism important for the promastigote-amastigote transformation process, although demonstration of this supposition would require further experimentation. ANDERSON et al. (1983) examined the in vivo effects of pre-incubating *L.mexicana* promastigotes in various monoclonal antibodies prior to injection into mice. They found that a cocktail of three antibodies could confer complete protection against an inoculation of up to 10^4 parasites. The antibodies used recognized antigens of 43 kd and 46 kd and an undisclosed antigen. The authors felt that the monoclonal antibody 1X-5H9-C9 against the 46-kd antigen was the crucial component of the treatment. The protection conferred by this antibody was observed even in C3-depleted mice pretreated with cobra venom factor, indicating that it was not a function of complement activity.

5   Functional Characterisation of Parasite Surface Antigens

Although biochemical and serological characterisation of parasite antigens generates a considerable amount of useful information, the problems experienced in reproducing techniques or in the availability of a monoclonal antibody make it considerably difficult to cross-reference antigens. Therefore, the most desirable way of defining an antigen is by a property specific to that molecule, the most significant being its biological function. The characterisation of *Leishmania* antigens by their biological activity has, to date, only been possible in three cases, all of which are concerned with the interaction between the parasite and its host cell, the macrophage. It is not surprising that this is the point in the parasite's life cycle that has attracted the most attention because the successful attainment of an intracellular existence is vital to both the promastigote during initial infection and the amastigote during amplification of infection.

The microbicidal activity of macrophages usually associated with *Leishmania* killing is the production of free oxygen radicals (MAUEL 1984a; MAUEL 1984b; MAUEL et al. 1984). During the initial uptake stage of infection, the promastigote stimulates respiratory burst activity to a much greater degree than the amastigote (CHANNON et al. 1984). Controlling the potential killing activity of the prospective host cell would obviously be of great benefit to the parasite. Recently, REMALEY et al. (1984) reported that the promastigote acid phosphatase described by GOTTLIEB and DWYER (1981) and isolated by GLEW et al (1982) is capable of suppressing the oxygen radical production of human neutrophils. Using a fluorometric assay system, they demonstrated that, upon stimulation with a chemoattractant peptide, the manufacture of oxygen radicals by neutrophils is severely depressed in the presence of the parasite acid phosphatase. They calculated that each promastigote carries enough phosphatase to reduce superoxide anion production of three neutrophils by 50%. Obviously, an enzyme with this biological activity would be extremely useful in establishing an infection by inhibiting the macrophage-killing responses.

The phagocytosis of *Leishmania* parasites by macrophages involves two distinct steps: firstly, adhesion or attachment of the parasite to the macrophage surface and, secondly, its subsequent internalisation. The latter is probably achieved by the "zipper" mechanism proposed by GRIFFIN et al. (1975) and SILVERSTEIN (1977) involving recruitment of additional macrophage receptors around the circumference of the parasite. This uptake procedure is likely to be a direct consequence of the initial attachment of the *Leishmania* to the phagocyte. The attachment of the promastigote to the macrophage has been examined in several laboratories. It is a receptor-mediated event that is saturable (CHANG 1981; KLEMPNER et al. 1983) and inhibitable by simple and complex sugars (CHANG 1981; ZENIAN 1981; BRAY 1983b and CHANNON et al. 1984). It has been suggested that the success of this initial infection may be influenced by the specificity of the macrophage surface receptor involved in uptake (BLACKWELL and ALEXANDER 1983; ADAMS and HAMILTON 1984). It is therefore of interest to identify the parasite surface ligands and their complementary macrophages surface receptors.

HANDMAN and GODING (1985) reported that the EF from *L.major* promastigotes is capable of binding specifically to the macrophage-like cell line J774. The soluble EF was purified by immunoaffinity chromatography, using a monoclonal antibody W1C-79.3 against EF (DEIBARRA et al. 1982) and co-incubated with macrophages prior to labelling by immunofluorescence. When warmed to 37° C, the labelled macrophages were able to cap the bound EF, further evidence that the macrophage receptor complexes with the cell's contractile elements. Using F(ab) fragments from the monoclonal antibody, the authors showed that they were able to inhibit significantly the attachment of both EF and living promastigotes to the macrophages. This suggests that the macrophage-binding moiety on *L.major* promastigotes is the membrane-bound form of their EF (see Fig. 3).

Another study of *Leishmania* promastigote/macrophage interaction suggests that other macrophage-binding moieties exist on the surface of this parasite stage (RUSSELL and WILHELM 1985). In this study, the predominant glycoprotein of *L.m.mexicana* promastigotes (Gp 63) was isolated from a crude membrane fraction by deter-

* **EXCRETED FACTOR**

⁂ **MEMBRANE BOUND FORM**

Y **MACROPHAGE RECEPTOR**

Fig. 3. Diagrammatic representation of EF-mediated binding of leishmanial promastigotes to macrophages. The carbohydrate-rich EF occurs in two forms: a soluble form shed from the parasite surface and an alternate form with a hydrophobic region, probably lipid, that is inserted into the parasitie plasmalemma (Handman et al. 1984). Both the soluble and membrane-bound forms associate with receptors that are situated on the macrophage surface, indicating that it is a means by which *L. major* promastigotes attach to their prospective host cells (Handman and Goding 1985)

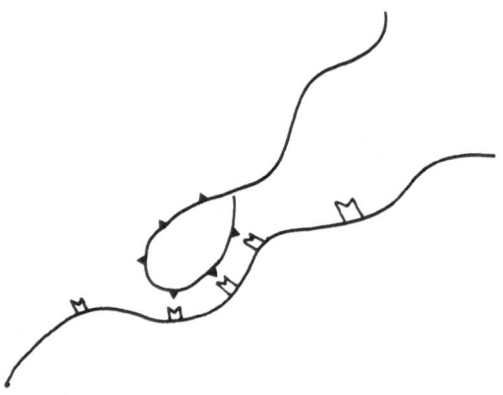

▲ **Gp 63**

M **MACROPHAGE RECEPTOR**

Fig. 4. Diagrammatic representation of an alternative macrophage-binding receptor demonstrated on the surface of *L. m. mexicana* (Russell and Wilhelm 1985). A polyclonal antiserum raised against the predominant glycoprotein, Gp 63 (see Fig. 2), produced F(ab) fragments capable of blocking macrophage-specific attachment. This suggested that these promastigotes bind via Gp 63 and a complementary receptor on the macrophage surface

gent extraction, lectin affinity chromatography and sodium dodecyl sulphate polyacrylamide gel electrophoresis. The Gp 63 was excised from the gels and used to generate a polyclonal antiserum. The specificity of this antiserum was tested by Western blotting and by a two-site immunoradiometric assay developed especially to detect EF (HANDMAN and GODING 1984). The antiserum identified a single 63-kd

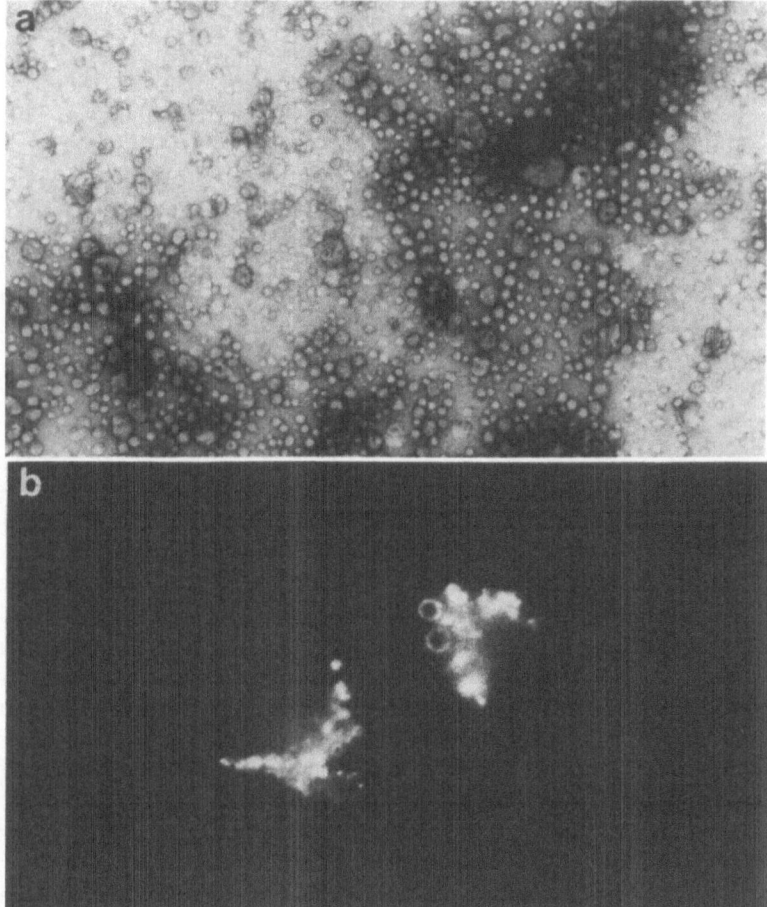

Fig. 5. a Electronmicrograph of a negatively stained preparation of proteoliposomes containing Gp 63 (see Fig. 2) from *L.m.mexicana*. The liposomes were produced by incubating phospholipids isolated from *Crithidia fasciculata* with Gp 63 in the presence of octyl glucoside, which was subsequently dialysed out of the preparation. × 45 000. **b** Immunofluorescence micrograph of J774 macrophage-like cell line following co-incubation with the Gp 63-containing proteoliposomes. The preparation was stained with anti-Gp 63 IgG and an FITC-conjugated second antibody. The uptake of these proteoliposomes by macrophages was inhibitable by the mannose-rich oligosaccharide mannan (from *Saccharomyces cerevisiae*), indicating that it is receptor mediated (Russell, unpublished observations). This supports the proposal that these promastigotes employ the surface-exposed glycoprotein Gp 63 as at least one of the mechanisms involved in host cell entry

polypeptide and showed no cross-reactivity with the promastigote's EF (Russell, unpublished results). Immunoelectromicroscopy demonstrated that the antiserum recognized epitopes exposed on the outer surface of the promastigote and that these epitopes were present all over the cell plasmalemma. F(ab) fragments made from the anti-Gp 63 IgG caused a 68% inhibition of macrophage-specific binding by *L.m.mexicana* promastigotes. This depression of attachment was observed for both J774, the macrophage-like cell line, and peritoneal macrophages. The background, or non-specific binding, was estimated using a fibroblast cell line and found to be unaffected by the anti-Gp 63F(ab) fragments, indicating that Gp 63-mediated binding (see Fig.4) is indeed a macrophage-specific attachment mechanism. Additional evidence of Gp 63-medicated attachment and uptake has been gained from examining the interaction of Gp 63-containing proteoliposomes with macrophages (Fig. 5). These vesicles are readily phagocytosed by the macrophages in a manner that is inhibitable by the Oligosaccharide mannan (Russell, unpublished results), which is known to depress promastigote/macrophage binding (CHANNON et al. 1984). These studies indicate that more than one macrophage-binding moiety exists on the surface of *Leishmania* promastigotes. However, whether the Gp 63 and the EF, both of which are known to occur on most species of *Leishmania,* operate cooperatively or independently in the different species remains to be examined.

6 Protective Immunity and Vaccination

A protective immune response against *Leishmania*, like that against other obligate intracellular parasites, is thought to be largely cell mediated and, as a multitude of experimental and clinical studies suggest, tends to coincide with the development of DTH (reviewed by TURK 1975). Although some studies have suggested a potential protective role for the humoral response in leishmaniasis, other studies have suggested that it is largely irrelevant to the outcome of disease. Still further studies suggest that it may in fact be detrimental and could exacerbate infection. A recent series of publications have, however, indicated the active involvement of humoral immunity during both clinical and experimental infections. These observations have important implications in developing effective vaccine protocols and merit further consideration.

Non-healing visceral and cutaneous leishmaniasis have long been associated with high immunoglobulin levels and negative DTH reactions, while healing or cured individuals develop strong cutaneous DTH as immunoglobulin levels fall (reviewed by TURK and BRYCESON 1971). This has encouraged workers to suggest a possible disease-enhancing or DTH-blocking effect of antibody (GARNHAM and HUMPHREY 1969). The role of antibody in an anergic model of cutaneous leishmaniasis has therefore recently been studied (SACKS et al. 1984). BALB/c mice develop non-healing lesions when infected subcutaneously with *L.major*, and a protective cell-mediated response is inhibited by the generation of Lyt1^+2^- suppressor T cells (LIEW et al. 1982). By examining mice rendered B cell deficient by treatment with antimouse IgM from birth, SACKS et al. (1984) demonstrated that, whereas immunity in innately resistant mice is independent of antibody, suppressed DTH in susceptible mice is B cell dependent, and thus, B cell-depleted BALB/c mice are resistant to *L.major*. A non-healing response in B cell-depleted mice can be adoptively restored with suppressor T cells from infected normal BALB/c mice. In addition, because a non-healing response can be restored to prophylactically whole-body, sublethally irradiated BALB/c mice with normal BALB/c spleen cells but not spleen cells from B cell-depleted mice, the induction of suppressor T cells would indeed appear to be B cell dependent.

Conversely, a number of recent studies have suggested a protective role for antibody, at least in cutaneous leishmaniasis. Thus, in immunohistopathology studies,

RIDLEY and RIDLEY (1984) have shown parasite elimination to be associated with immunologically induced host cell lysis, concomitant with and perhaps the result of rising levels of IgG, IgM and IgE, the complement components of the classical pathway and the formation of immune complexes. Further evidence for a protective role for antibodies has been provided by using *Leishmania* species-specific monoclonal antibodies in both *in vitro* (HANDMAN and HOCKING 1982) and in vivo (ANDERSON et al. 1984) studies. Thus, two *L. major*-specific monoclonal antibodies have been shown to inhibit the growth of *L. major* within macrophages *in vitro* (HANDMAN and HOCKING 1982), while three *L. mexicana*-specific monoclonal antibodies passively protect BALB/c mice if inoculated together with the parasite into the footpad (ANDERSON et al. 1984). It would be interesting to know the mechanisms by which these antibodies are protective, characterise the surface antigens to which they are directed (Sec. 4.4) and test the immunogenicity and vaccine potential of the purified antigen.

Further research using radioattenuated *L.major* vaccines (HOWARD et al. 1982) or those based on *L.major* EF (STEINBERGER et al. 1984) have also indicated a positive contribution for antibody in protection. HOWARD et al. (1982) found that intravenous and intraperitoneal but not subcutaneous inoculations with radioattenuated *L.major* promastigotes protected BALB/c mice not only against homologous challenge but also against *L.m.mexicana*, *L.m.amazonensis* and *L.braziliensis guyanensis.* Similarly, STEINBERGER et al. (1984) protected C3H mice against *L.major* by intraperitoneal but not subcutaneous inoculations with galactose-rich (SLUTZKY and GREENBLATT 1982) EF cross-linked with muramyl dipeptide. In both studies, protection was associated with raised antileishmanial antibodies and negative DTH.

While these studies emphasise that the contribution of the humoral response to the outcome of leishmanial infection cannot be ignored, the vast bulk of current evidence continues to implicate cell-mediated immunity as the controlling factor (SHEPPARD et al. 1983; BRYCESON et al. 1972; PRESTON et al. 1972; SKOV and TWOHY 1974; REZAI et al. 1980; LIEW et al. 1982). While this does not preclude the possibility of synergy between T cell subpopulations (SKOV and TWOHY 1974) and T and B cells (ALEXANDER and PHILLIPS 1978) throughout the course of infection, it does emphasise the complexity of immunoregulatory mechanisms operating in both cutaneous and visceral leishmaniasis.

Recently, more sophisticated approaches towards analysing the role of the components of cellular immunity in the protective response have used *Leishmania*-specific T cell clones (LOUIS et al. 1982a, 1982b; LIEW 1983; SHEPPARD et al. 1983; LIMA et al. 1984). LIEW (1983) has cloned a Lyt1⁺2⁻IJ⁻ suppressor T cell from *L.major*-susceptible BALB/c mice which inhibits lymphocyte proliferation and DTH responses *in vivo*. Other workers have isolated effector T cells with a similar phenotype Lyt1⁺2⁻ which specifically activate *L.major*-infected macrophages resulting in parasite destruction (LOUIS et al. 1982b) and induce specific DTH reactions in syngeneic mice (LIMA et al. 1984). SHEPPARD et al. (1983) isolated T cell lines and clones that proliferated in response to antigen present in the membrane of intact *Leishmania donovani* promastigotes. While about 40% of the clones produced had highly restricted specificity, 60% were more extensively cross-reactive with other *Leishmania* species and even with trypanosomes. The majority of clones appeared to recognise carbohydrate-containing antigens, and absorption with solid substrate-bound lectins indicated that these antigens contained both mannose and galactose ligands. As some of these T cell clones passively transferred DTH when injected locally, while others secreted lympho-kine that activated macrophages, their further characterisation would be useful. More particularly, characterisation of cross-species antigens that activate the T cell clones may be useful in producing a standardised multipurpose vaccine.

It is not wholly surprising that antigen-specific T cell clones raised against species of *Leishmania* are activated by one or more different species. Examples of cross-reactivity and cross-immunity are well documented in the literature (reviewed by ALEXANDER and BLACKWELL 1985), and an understanding of the mechanisms involved would be of extreme importance in developing effective vaccines.

Successful vaccination at least for cutaneous leishmaniasis has long been practiced in the Middle East using live organisms. Two such current vaccination programmes are well underway in Israel (GREENBLATT 1980) and in the USSR (SERGIEV 1977). Recently, workers have successfully induced homologous and heterologous protection with radioattenuated organisms in laboratory models of cutaneous leishmaniasis (ALEXANDER and BLACKWELL 1985; HOWARD et al. 1982). Similar studies in man are also promising (MARYINK et al. 1979). Some of these results are particularly encouraging, as they show that vaccination has proved effective even in normally non-curing forms of leishmaniasis.

7 Summary

In recent years, the application of molecular and modern immunological techniques have added greatly to our understanding of the leishmaniases. Diagnostic procedures which had already been substantially improved by the biochemical analysis of parasites have been rationalised with the advent of species- and subspecies-specific monoclonal antibodies. Analysis of parasite stage-specific antigens has increased our understanding of infectivity and, more particularly, of the molecular basis underlying the interaction of the parasites with their phagocytic host cells. This knowledge should prove invaluable in the future design of highly specific chemotherapeutic immunotherapeutic agents. Modern molecular techniques should also allow the manufacture of the species-specific antigens that have already been characterised. The identification of antigens that induce species-specific cutaneous DTH reactions would not only be useful as diagnostic reagents, but also in monitoring the degree of host resistance.

At the same time, immunologists have elucidated in the mouse model both the immunoregulatory mechanisms and the underlying genetic controls operating in visceral and cutaneous leishmaniasis. These findings have enabled us to speculate both on how the host response could affect and on how best to apply potential vaccine protocols. So far, some vaccination studies using both live and attenuated whole parasites as antigen have proved successful against both heterologous and homologous challenge infection. The use of *Leishmania*-specific T cell clones of known function in a proliferative assay should eventually lead to the isolation and characterisation of cross-species-immunoprotective antigens. Thus, in the next few years, we anticipate exciting and rapid progress in the understanding and treatment of leishmaniasis.

References

Adams DO, Hamilton TA (1984) The cell biology of macrophage activation. Ann Rev Immunol 2:238–318

Adler S (1964) *Leishmania.* Adv Parasit 2:35–96

Adler S, Theodor O (1931) Investigations on Mediterranean kala-azar I-V. Proc Roy Soc Lond 125 B : 491–519

Aikawa M, Hendricks LD, Ito Y, Jagusiak M (1982) Interactions between macrophage-like cells and *Leishmania braziliensis in vitro.* Am J Pathol 108:50–59

Alexander J (1981a) *Leishmania mexicana:* inhibition and stimulation of phagosome-lysosome fusion in infected macrophages. Exp Parasit 52:261–270

Alexander J (1981b) Interaction of *Leishmania mexicana mexicana* with mouse macrophages in vitro. In: Forster O, Landry M (eds) Heterogeneity of Mononuclear Phagocytes. Academic, New York, pp 447–454

Alexander J (1982) A radio-attenuated *Leishmania major* vaccine markedly increases the resistance of CBA mice to subsequent infection with *Leishmania mexicana mexicana.* Trans Roy Soc Trop Med Hyg 76:646–649

Alexander J, Blackwell JM (1985) The immunological significance of genetically determined cross-reactivity between taxonomically distinct *Leishmania* species. Annales de Parasitologie (in press)

Alexander J, Burns RG (1983) Differential inhibition by erythro-9-[3-(2-hydroxynonyl)] adenine of flagella-like and cilia-like movement of *Leishmania* promastigotes. Nature 305:313–315

Alexander J, Phillips RS (1978) *Leishmania tropica* and *Leishmania mexicana mexicana*: cross-immunity in mice. Exp Parasit 45:93–100

Alexander J, Vickerman K (1975) Fusion of host cell secondary lysosomes with the parasitophorous vacuoles of *Leishmania mexicana*-infected macrophages. J. Protozool. 22:502–508

Alving CR, Steck EA (1979) The use of liposome-encapsulated drugs in leishmaniasis. Trends Biochem Sci 4:175–177

Al-Taqi M, Evans DA (1978) Characterisation of *Leishmania* spp. from Kuwait by isoenzyme electro-phoresis. Trans Roy Soc Trop Med Hyg 72:56–65

Anderson S, David JR, McMahon-Pratt D (1983) In vivo protection against *Leishmania mexicana* mediated by monoclonal antibodies. J Immunol 131:1616–1618

Arnot DE, Barker DC (1981) Biochemical identification of cutaneous leishmanias is by analysis of kineto-plast DNA: II. Sequence homologies in *Leishmania* k DNA. Mol Biochem Parasit 3:47–56

Barker BC, Butcher J (1983) The use of DNA probes in the identification of leishmanias: descrimation between isolates of the *mexicana* and *braziliensis* complexes. Trans Roy Soc Trop Med Hyg 77:285–298

Berman JD, Fioretti TB, Dwyer DM (1981) In vivo and in vitro localization of *Leishmania* within macro-phage phago-lysomes: use of colloidel gold as a lysosomal label. J Protozool 28:239–242

Blackwell JM, Alexander J (1983) The macrophage and parasitic protozoa. Trans Roy Soc Trop Med Hyg 77:636–645

Blackwell JM, Alexander J (1985) Different host genes recognise and control infection with taxonomically distinct *Leishmania* species. Annales de Parasitologie (in press)

Blackwell JM, Ezekowitz RAB, Roberts MB, Channon JY, Sim RB, Gordon S (1985) Macrophage comple-ment and lectin-like receptors blind *Leishmania* in the absence of serum. J Exp Med 162:324–331

Blackwell JM, Freeman JC, Bradley DJ (1980) Influence of H-2 complex on acquired resistance to *Leish-mania donovani* infection in mice. Nature 283:72–74

Bradley DJ, Taylor BA, Blackwell JM, Evans EP, Freeman J (1979) Regulation of *Leishmania* populations within the host: III. Mapping of the locus controlling susceptibility to visceral leishmaniasis in the mouse. Clin Exp Immunol 37:7–14

Bray RS (1983a) *Leishmania*: chemotactic responses of promastigotes and macrophages in vitro. J Proto-zool 30:322–329

Bray RS (1983b) *Leishmania mexicana mexicana*: attachment and uptake of promastigotes to and by macrophages in vitro. J Protozool 30:314–322

Bryceson ADM, Preston PM, Bray RS, Dumonde DC (1972) Experimental cutaneous leishmaniasis: II. Effects of immunosuppression and antigenic competition on the course of infection with *Leishmania enriettii* in the guinea pig. Clin Exp Immunol 10:305–335

Chance HL, Peters W, Schchory L (1974) Biochemical taxonomy of *Leishmania*: I. Observations of DNA. Ann Trop Med Parasit 68:307–316

Chang KP (1981) *Leishmania donovani* – macrophage binding mediated by surface glycoproteins/antigens. Mol Biochem Parasitol 4:67–76

Chang KP, Dwyer DM (1976) Multiplication of a human parasite *(Leishmania donovani)* in phagolysosomes of hamster macrophages in vitro. Science 193:678–680

Chang KP, Dwyer DM (1978) *Leishmania donovani* – hamster macrophage interaction in vitro: cell entry, intracellular survival and multiplication of amastigotes. J Exp Med 147:515–530

Channon JY, Roberts MB, Blackwell JM (1984) A study of the differential respiratory burst activity elicited by promastigotes and amastigotes of *Leishmania donovani* in murine resident peritoneal macrophages. Immunol 53:345–355

Coombs GH (1982) Proteinases of *Leishmania mexicana* and other flagellate protozoa. Parasit 84:149–155

Crocker PR, Blackwell JM, Bradley DJ (1984) Expression of the natural resistance gene *Lsh* in resident liver macrophages. Infect Immun 43:1033–1040

Dawidcowiz K, Kernandez AG, Infante RB (1975) The surface membrane of *Leishmania*: I. Effects of lectins on different stages of *Leishmania braziliensis*. J Parasitol 61:950–953

DeIbarra AAL, Howard JG, Snary D (1982) Monoclonal antibodies to *Leishmania tropica major:* specificities and antigen location. Parasitol 85:523–531

Dwyer DM (1974) Lectin-binding saccharides on a parasitic protozoan. Science 184:471–473

Dwyer DM (1978) *Leishmania* – host cell membrane cycling *in vitro*. J Protozool 25:28A

Dwyer DM, Gottlieb M (1983) The surface membrane chemistry of *Leishmania*: its possible role in parasite sequestration and survival. J Cell Biochem 23:35–45

El-On J, Bradley DJ, Freeman JC (1980) *Leishmania donovani:* action of excreted factor on hydrolytic enzyme activity of macrophages from mice with genetically different resistance to infection. Exp Parasit 49:167–174

Etges RJ, Bouvier J, Hoffman R, Bordier C (1985) Evidence that the major surface proteins of three *Leishmania* species are structurally related. Mol Biochem Parasitol 14:141–149

Ezekowitz RAB, Sim RB, Hill M, Gordon S (1984) Local opsonization by secreted macrophage complement components: role of receptors for complement in uptake of zymosan. J Exp Med 159:244–260

Farah FS, Samra SA, Nuwayri-Salti (1975) The role of macrophage in cutaneous leishmaniasis. Immunol 29:755–764

Feng LC, Chung HL (1941) Experiments on the transmission of kala-azar from dogs to hamsters by Chinese sandflies. Chinese Med J 60:489–496

Fong D, Chang KP (1982) Surface antigenic change during differentiation of a parasitic protozoan, *Leishmania mexicana:* identification of monoclonal antibodies. Proc Nat Acad Sci USA 79:7366–7370

Franke ED, McGreevy PB, Katz SP, Sacks DL (1985) Growth cycle-dependent generation of complement resistant *Leishmania* promastigotes. J Immunol 134:2713–2718

Gardener PJ, Chance ML, Peters W (1974) Biochemical taxonomy of *Leishmania*: II. Electrophoretic variation of malate dehydrogenase. Ann Trop Med Parasit 68:317–328

Gardiner PR, Dwyer DM (1983) Radioiodination and identification of externally disposed membrane components of *Leishmania tropica*. Mol Biochem Parasitol 8:283–295

Gardiner PR, Jaffe CL, Dwyer DM (1984) Identification of cross-reactive promastigote cell surface antigens of some leishmanial stocks by ^{125}I-labelling and immunoprecipitation. Inf Immun 43:637–643

Garnham PCC, Humphrey JH (1969) Problems in Leishmaniasis related to immunology. Current Topics Microbiol Immun 48:29–42

Glew RH, Czuczman MS, Diven WF, Berens RL, Pope MT, Katsoulis DE (1982) Partial purification and characterization of particulate acid phosphatase of *Leishmania donovani* promastigotes. Comp Biochem Physiol 72B:581–590

Gorczinski RM, MacRae S (1981) Analysis of subpopulations of glass-adherent mouse skin cells controlling resistance/susceptibility of infection with *Leishmania tropica* and correlation with the development of independent proliferative signals to Lyt-1[+]/Lyt-2[+] T lymphocytes. Cell Immun 67:74–89

Gottlieb M, Dwyer DM (1981) *Leishmania donovani* surface membrane acid phosphatase activity of promastigotes. Exp Parasitol 52:117–128

Greenblatt CL (1980) The present and future of vaccination for cutaneous leishmaniasis. In: Mizrahi A, Hertman I, Klingberg MA, Kohn A (eds) New Developments with Human Vaccines. Alan R. Liss, New York, pp 259–285

Griffin FM, Griffin JA, Leider JW, Silverstein SD (1975) Studies on the mechanism of phagocytosis: I. Requirements for circumferential attachment of particle-bound ligands to specific receptors on the macrophage plasma membrane. J Exp Med 142:1263–1282

Handman E, Ceredig R, Mitchell GG (1979) Murine cutaneous leishmaniasis: Disease patterns in intact and nude mice of various genotypes and examination of some differences between normal and infected macrophages. Aus J Exp Biol Med Sci 57:9-29

Handman E, Curtis JM (1982) *Leishmania tropica*: surface antigens of intracellular and flagellate forms. Exp Parasit 54:243-249

Handman E, Greenblatt CL, Goding JW (1984) An amphipathic sulphated glycoconjugate of *Leishmania* characterization with monoclonal antibodies. EMBO J 3:1206-2301

Handman E, Goding JW (1985) The *Leishmania* receptor for macrophages is a lipid-containing glycoconjugate. EMBO J 4:329-336

Handman E, Hocking RE (1982) Stage-specific, strain-specific and cross-reactive antigens of *Leishmania* species identified by monoclonal antibodies. Inf Immun 37:28-37

Handman E, Mitchell GF, Goding JW (1981) Identification and characterization of protein antigens of *Leishmania tropica* isolates. J Immunol 126:508-512

Herman R (1980) Cytophilic and opsonic antibodies in visceral leishmaniasis. Infect Immun 28:585-593

Hernandez AG (1982) Lectins as a tool in parasite research. In: Chance ML, Walton BC (eds) Biochemial Characterization of *Leishmania*. UNDP/WORLD BANK/WHO, Geneva pp 181-196

Hernandez AG (1983) In: Cytopathology of Parasitic Disease, UBA Foundation Symposium 99. Pitman, London, pp 138-156

Hochmeyer WT, Walters D, Gore RW, Williams JS, Fortier AH, Nacy CA (1984) Intracellular destruction of *Leishmania donovani* and *Leishmania tropica* amastigotes by activated macrophages: dissociation of these microbicidal effector activities in vitro. J Immunol 132:3120-3125

Howard JG, Hale C, Chan Liew WL (1980) Immunological regulation of experimental cutaneous leishmaniasis: I. Immunogenetic aspects of susceptibility to *Leishmania tropica* in mice. Parasite Immunol 2:303-314

Howard JG, Nicklin S, Hale C, Liew FY (1982) Prophylactic immunization against experimental leishmaniasis: I. Protection induced in mice genetically vulnerable to fatal *Leishmania tropica* infection. J Immunol 129:2206-2210

Jaffe CL, Bennet E, Grimaldi G, McMahon-Pratt D (1984) Production and characterization of species-specific monoclonal antibodies against *Leishmania donovani* for immunodiagnosis. J Immunol 133:440-447

Jaffe CL, McMahon-Pratt D (1983) Monoclonal antibodies specific for *L. tropica*: I. Characterization of antigens associated with stage- and species-specific determinants. J Immunol 131:1987-1993

Kaneshiro ES, Gottlieb M, Dwyer DM (1982) Cell surface origin of antigens shed by *Leishmania donovani* during growth in axenic culture. Inf Immun 37:558-567

Killick-Kendrick R (1979) Biology of *Leishmania* in phlebotomine sandflies. In: Lumsden WHR, Evans DA (eds) Biology of the Kinetoplastida, vol 2. Academic, London, pp 395-460

Killick-Kendrick R, Molyneux DH, Ashford RW (1974) *Leishmania* in phlebotomid sandflies: I. Modifications of the flagellum associated with attachment to the midgut and oesophageal valve of the sandfly. Proc Roy Soc Lond Series B 187:409-419

Klempner MS, Cendron M, Wyler DJ (1983) Attachment of plasma membrane vesicles of human macrophages to *Leishmania tropica* promastigotes. J Inf Dis 148:377-384

Kutish GF, Janovy J (1981) Inhibition of *in vitro* macrophage digestion capacity by infection with *Leishmania donovani* (Protozoa: Kinetoplastida). J Parasit 67:457-462

Lepay DA, Nogueira N, Cohn Z (1983) Surface antigens of *Leishmania donovani* promastigotes. J Exp Med 157:1562-1572

Lewis DH, Peters W (1977) The resistance of intracellular *Leishmania* parasites to digestion by lysosomal enyzmes. Ann Trop Med Parasit 71:295-312

Liew FY (1983) A cloned T cell line expressing specific suppressive activity against *in vitro* and *in vivo* responses to *Leishmania tropica*. Nature 305:630-632

Liew FY, Hale C, Howard JG (1982) Immunological regulation of experimental cutaneous leishmaniasis: V. Characterization of effector and specific suppressor T cells. J Immunol 128:1917-1922

Lima GC, Engers HD, Louis JA (1984) Adoptive transfer of delayed-type hypersensitivity reactions specific for *Leishmania major* antigens to normal mice using murine T cell populations and clones generated *in vitro*. Clin exp Immunol 57:130-135

Louis JA, Lima GMC, Engers HD (1982a) Murine T lymphocyte responses specific for the protozoan parasites *Leishmania tropica* and *Trypanosoma brucei*. Clin Immunol Allergy 2:597

Louis JA, Moedder E, MacDonald HR, Engers HD (1981) Recognition of protozoan parasites by murine T lymphocytes: II. Role of the H-2 gene complex in interactions between antigen-presenting macrophages and *Leishmania* immune T lymphocytes. J Immunol 126:1661-1666

Louis JA, Zubler RH, Coutinho SG, Lima GMC, Behir R, Mauel J, Engers HD (1982b) The *in vitro* generation and functional analysis of murine T cell populations and clones specific for a protozoan parasite, *Leishmania tropica*. Immunol Rev 61:215–243

Maryink W, du Costa CA, Magalkaes PA, Melo MN, Dias M, Oliviera Lima A, Michalick MS, Williams P (1979) A field trial of a vaccine against American dermal leishmaniasis. Trans Roy Soc Trop Med Hyp 73:385–387

Mauel J (1984a) Mechanisms of survival of protozoan parasites in mononuclear phagocytes. Parasitol 88:579–592

Mauel J (1984b) Intracellular parasite killing induced by electron carriers: I. Effects of electron carriers on intracellular *Leishmania* spp. in macrophages from different genetic backgrounds. Mol Biochem Parasitol 13:83–96

Mauel J, Schnyder J, Baggiolini M (1984) Intracellular parasite killing induced by electron carriers: II. Correlation between parasite killing and the induction of oxidative events in macrophages. Mol Biochem Parasitol 13:97–110

McMahon-Pratt D, Bennet E, David JR (1982) Monoclonal antibodies that distinguish subspecies of *Leishmania braziliensis*. J Immunol 129:926–927

McMahon-Pratt D, Bennet E, Grimaldi G, Jaffe CL (1985) Subspecies- and species-specific antigens of *Leishmania mexicana* characterized by monoclonal antibodies. J Immunol 134:1935–1940

McMahon-Pratt D, David JR (1981) Monoclonal antibodies that distinguish between New World species of *Leishmania*. Nature 291:581–583

McMahon-Pratt D, David JR (1982) Monoclonal antibodies recognizing determinants specific for the promastigote stage of *Leishmania mexicana*. Mol Biochem Parasitol 6:317–327

Miles MA, Povoa MM, de Souza AA, Lainson R, Shaw JJ (1979) Some methods for the enzymatic characterization of Latin-American *Leishmania* with particular reference to *Leishmania mexicana amazonensis* and subspecies of *Leishmania hertigi*. Trans Roy Soc Trop Med Hyg 74:243–252

Mosser DM, Edelson PJ (1984) Activation of the alternative complement pathway by *Leishmania* promastigotes: parasite lysis and attachment to macrophages. J Immunol 132:1501–1506

Murray HW (1981) Susceptibility of *Leishmania* to oxygen intermediates and killing by normal macrophages. J Exp Med 153:1302–1315

Murray HW (1982) Cell-mediated immune response to visceral leishmaniasis: II. Oxygen-dependent killing of intracellular *Leishmania donovani* amastigotes. J Exp Med 153:1302–1315

Parrot L, Donatien A (1927) Le parasite du bouton d'Orient chez le phlébotome. Infection naturelle et infection expérimentale de *Phlebotomus papatasi* (Scop). Arch Inst Past d'Algerie 5:9–21

Preston PM, Carter RL, Leuchars E, Davies AJS, Dumonde DC (1972) Experimental cutaneous leishmaniasis: III. Effects of thymectomy on the course of infection of CBA mice with *Leishmania tropica*. Clin exp Immunol 10:337–357

Ramasamy R, Kar SK, Jamnada H (1984) Cross-reacting surface antigens on *Leishmania* promastigotes. Int J Parasitol 13:337–341

Remaley AT, Kuhns DB, Basford RE, Glew RH, Kaplan SS (1984) Leishmanial phosphatase blocks neutrophil O_2^--production. J Biol Chem 259:1173–1175

Rezai HR, Farrell J, Soulsby EL (1980) Immunological responses of *L. donovani* infection in mice and significance of T cells in resistance to experimental leishmaniasis. Clin exp Immunol 40:508–514

Ridley MJ, Ridley DS (1984) Cutaneous leishmaniasis: immune complex formation and necrosis in the acute phase. Br J exp Path 65:327–336

Russell DG, Miller D, Gull K (1984) Tubulin heterogeneity in the trypanosome *Crithidia fasciculata*. Mol Cell Biol 4:779–790

Russell DG, Wilhelm H (1985) The involvement of gp 63, the major surface glycoprotein, in the attachment of *Leishmania* promastigotes to macrophages. J Immunol (submitted)

Sacks DL, Hieny S, Sher A (1985) Identification of cell surface carbohydrate and antigenic changes between non-infective and infective developmental stages of *Leishmania major* promastigotes. J Immunol 135:564–569

Sacks DL, Perkins PV (1984) Identification of an infective stage of *Leishmania* promastigotes. Science 223:1417–1419

Sacks DL, Perkins PV (1985) Development of infective stage *Leishmania* promastigotes within phlebotomine sandflies. Am J Trop Med Hyg (in press)

Sacks DL, Scott PA, Asofsky R, Sher AF (1984) Cutaneous leishmaniasis in anti-IgM-treated mice: enhanced resistance due to functional depletion of a B cell-dependent T cell involved in the suppressor pathway. J Immunol 132:2072–2077

Schnur LF, Zuckerman A, Greenblatt CL (1972) Leishmanial serotypes as distinguished by the gel diffusion of factors excreted *in vitro* and *in vivo*. Isr J Med Sci 8:932–942

Scott P, Sacks D, Sher A (1983) Resistance to macrophage-mediated killing as a factor influencing the pathogenesis of chronic cutaneous leishmaniasis. J Immunol 131:966–971

Semprevivo LH, MacLeod LH (1981) Characterization of the exemetabolite of *Leishmania donovani* as a moved glycopeptidophosphosphingolipid. Biochem Biophys Res Comm 193:1179–1185

Semprevivo LH, De Tolla LJ, Passmore HC, Palczuk NC (1981) Spectral model of leishmaniasis in congenic strains of mice. J Parasit 67:8–14

Sergiev VP (1977) Control measures against cutaneous leishmaniasis. In: Ecologie des Leishmaniasis. Colloque International du CNRS, no 239, Montpellier. pp 321–323

Shepherd VL, Stahl PD, Bernd P, Rabinovitch M (1983) Receptor-mediated entry of B-glucoronidase into the parasitophorous vacuoles of macrophages infected with *Leishmania mexicana amazonensis*. J exp Med 157:1471–1482

Sheppard HW, Scott PA, Dwyer DM (1983) Recognition of *Leishmania donovani* antigens by murine T lymphocyte lines and clones. J Immunol 131:1496–1503

Silverstein SC (1977) Endocytic uptake of particles of mononuclear phagocytes and the penetration of obligate intracellular parasites. Am J Trop Med Hyg 26:161–168

Skov CB, Twohy DW (1974) Cellular immunity to *Leishmania donovani*: II. Evidence for synergy between thymocytes and lymph node cells in reconstitution of acquired resistance to *L. donovani* in mice. J Immunol 113:2012–2019

Slutzky GM, El-On J, Greenblatt CL (1979) Leishmanial excreted factor: protein-bound and free forms from promastigote cultures of *Leishmania tropica* and *Leishmania donovani*. Inf Immun 26:916–924

Slutzky GM, Greenblatt CL (1977) Isolation of a carbohydrate rich immunologically active factor from cultures of *Leishmania tropica*. FEBS Lett 80:401–404

Slutzky GM, Greenblatt CL (1979) Analysis by SDS-polyacrylamide gel electrophoresis of an immunologically active factor of *Leishmania tropica* from growth media, promastigotes and infected macrophages. Biochem Med 21:70–77

Slutzky GM, Greenblatt CL (1982) Identification of galactose as the immunodominant sugar of leishmanial excreted factor and subsequent labelling with galactose oxidase and sodium boro [^{3}H] hydride. Inf Immun 37:10–14

Steinberger A, Slutzky GM, El-On J, Greenblatt CL (1984) *Leishmania tropica:* protective response in C3H mice vaccinated with excreted factor cross-linked with synthetic adjuvant muramyl dipeptide. Exp Parasit 58:223–229

Turk JL (1975) Interaction between B and T lymphocytes in delayed hypersensitivity. In: Van Furth (ed) Mononuclear Phagocytes in Immunity Infection and Pathology. Blackwell, Oxford, pp 533–537

Turk JL, Bryceson ADM (1971) Immunological phenomena in leprosy and related diseases. Advan Immunol 13:209–266

Unanue ER (1981) The regulatory role of macrophages in antigenic stimulation: II: symbiotic relationship between lymphocytes and macrophages. Adv Immun 31:1–136

Unanue ER (1984) Antigen-presenting function of the macrophage. Ann Rev Immun II:395–428

WHO Technical Report No. 701 (1984) The Leishmaniases

Wilson ME, Pearson RD (1984) Stage-specific variations in lectin binding to *Leishmania donovani*. Inf Immun 46:128–134

Wright SD, Silverstein SC (1983) Receptors for C3b and iC3b promote phagocytosis but not the release of toxic oxygen from human phagocytes. J exp Med 158:2016–2023

Wyler DJ (1982) In vitro parasite-monocyte interactions in human leishmaniasis: evidence for an active role for the parasite in attachment. J Clin Invest 70:82–88

Zehavi V, El-On J, Pearlman K, Abrahams JC, Greenblatt CL (1983) Binding of *Leishmania* promastigotes to macrophages. *Z. parasitenkd.* 69:405–414

Zenian A, (19817 *Leishmania tropica*: biochemical aspects of promastigotes attachment to macrophages in vitro. Exp Parasitol 51:175–187

Zuckerman A (19757 Current status of the immunology of blood and tissue protozoa: I. *Leishmania*. Exp Parasit 38:370–400

Parasite Antigens in Protection, Diagnosis and Escape: *Plasmodium*

C. I. NEWBOLD

D'ou viennent les elements parasitaires qui se trouvent dans le sang des malades attients de fievre palustre? Par quelle voie s'introduisent-ils dans l'economie? Comment provoquentils la fievre intermittente et les autres manifestations de l'impaludisme? On ne peut encore poser ces importantes questions.[1]

LAVERAN 1880

Nuffield Department of Clinical Medicine, University of Oxford, John Radcliffe Hospital, Headington, Oxford OX3 9DU, United Kingdom

1 Where do these parasitic elements found in the blood of malaria patients come from? How do they get into the human system? How do they cause intermittent fever and other signs of malaria? Only now is one able to pose these important questions.

Current Topics in Microbiology and Immunology, Vol. 120
© Springer-Verlag Berlin · Heidelberg 1985

1 Introduction

1.1 Scope of This Chapter

In the hundred years since Laveran first described the malaria parasite as the cause of intermittent fevers, the basic biological questions posed at the end of his paper have been answered. The pathology of acute or chronic malaria infection is however still far from being fully understood and despite the accumulated knowledge of a century, the disease persists as one of the world's major health problems. Optimism about worldwide eradication by organised vector control and chemotherapy was maintained up until the early 1960s. Since this time there has been a resurgence of malaria in many areas of the world due to a number of factors many of which are outside the scope of this discussion. Deterioration in control of malaria has however been associated with a rapid spread of both insecticide resistance among mosquitos and drug resistance of the parasites. Such chemical weapons remain as the major means of controlling malaria so that for the reasons outlined above, research in recent years has begun to focus on possible means of immunoprophylaxis. This chapter will be devoted mainly to work which has attempted to define and characterise parasite antigens important in the induction of protective immunity, and evasion of the host's immune response.

It will be impossible here to discuss the immunology of the disease since this would form a chapter in itself and has been extensively reviewed elsewhere (COHEN 1979; JAYWARDENA 1981; BROWN 1982). The relevance of parasite antigens to diagnosis will also not be discussed in detail since malaria diagnosis is not usually a problem (except perhaps in relationship to blood banks). Immunodiagnosis is probably more relevant to the study of epidemiology and will therefore only be mentioned briefly in this context.

In order to place the work on malarial antigens in perspective it will first be necessary to outline the major features of both the parasite and the disease as it occurs in endemic areas and in the laboratory.

1.2 The Parasite

1.2.1 Life Cycle

A detailed description of the life cycle can be found in any textbook of parasitology. For reference, a summary is presented in Fig. 1.

In common with many other parasitic diseases, malaria is transmitted via a complex life cycle involving both vertebrate and invertebrate hosts. The invasive stage (sporozoite) is inoculated into the vertebrate host during a blood meal by a female *Anopheles* mosquito. From the site of entry the sporozoites travel via the blood stream rapidly to the liver. Here invasion of parenchymal cells occurs followed by several rounds of intracellular asexual division. The time taken for this stage of the infection varies with parasite species but ends with the rupture of the liver cell and the liberation of large numbers of merozoites. Circulating erythrocytes are invaded very

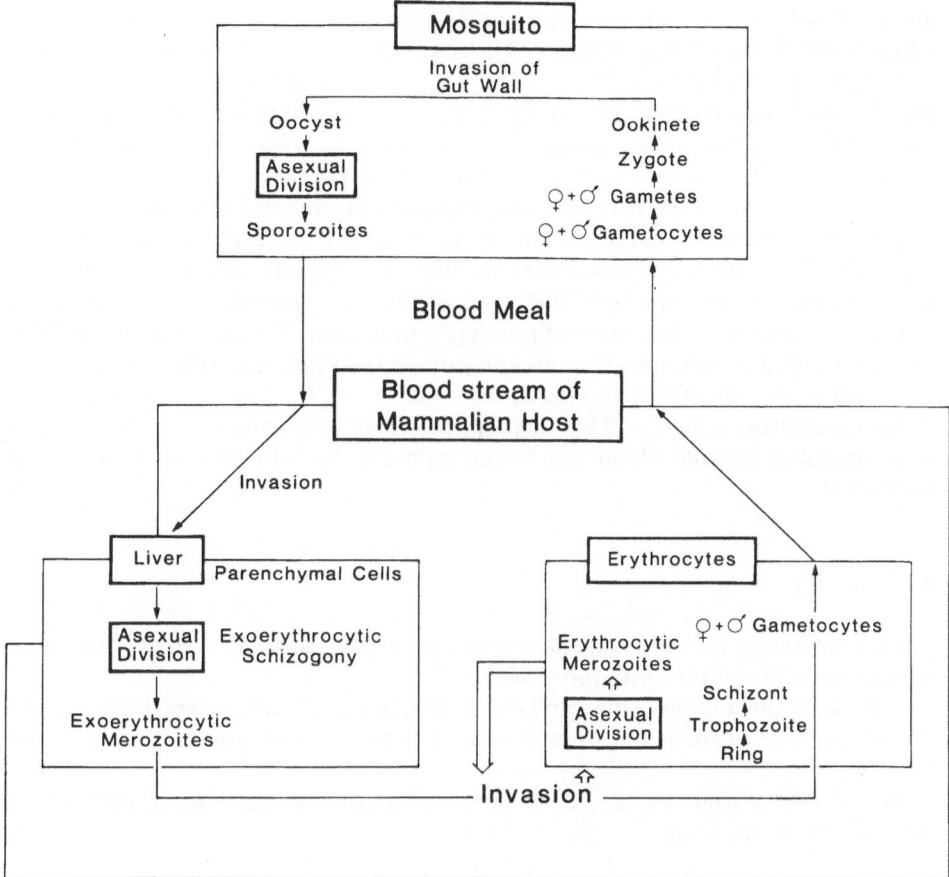

Fig. 1. Life cycle of *Plasmodium*

quickly by the free merozoites and serve as host cells for a repetitive cycle of intracellular asexual division. Within the erythrocyte, the merozoite develops from the "ring" stage through the trophozoite finally to the mature schizont containing daughter merozoites which, after red cell rupture, will initiate the next intraerythrocytic cycle. The parasite cell cycle within the red cell is in general synchronous, with a periodicity which varies with parasite species. It is also tied to the diurnal cycle of the host. Red cell rupture and merozoite release occur at approximately the same time of day for the duration of infection when the host experiences the acute "ague", whose periodicity is the best-known characteristic of malaria.

The life cycle is completed by the differentiation of a small proportion of intra-erythrocyte merozoites derived from the continuing asexual cycle into male and female gametocytes. These are ingested by a feeding mosquito into the insect midgut. Here the exflagellation of male gametocytes releases male gametes which can then fertilise released female gametes. Maturation of the zygote to an ookinete is followed by invasion of the gut wall to form an oocyst and further rounds of asexual division resulting in the release of sporozoites into the haemcoel and their final maturation in the insect salivary glands. Available evidence suggests that the organism is haploid for most of the life cycle, with meiosis occurring during one of the rounds of cell division in the mosquito (WALLIKER 1983). An interesting facet of the life cycle is the "repetitive nature" of the invasion/division process, with the three invasive phases (see Fig. 1) showing many ultrastructural features in common (CANNING and SINDEN 1973; AIKAWA 1971). Nevertheless, morphologically and functionally distinct stages of the parasite which are both intracellular and extracellular occur in the vertebrate host. This adds an additional dimension to the study of parasite antigens since it has been known for some time that each of these parasite stages is antigenically distinct in terms of the induction of protective immunity. For instance, potentiation of an antisporozoite immune response may under the right conditions confer protection against the bite of infected mosquitos, but leaves the host fully susceptible to blood-transmitted infection. This stage specificity of immunity means that antigens of sporozoites, asexual blood stages and gametocytes will have to be discussed separately.

1.2.2 Species

There are a large number of *Plasmodium* species producing malaria in a diverse range of vertebrate hosts (see GARNHAM 1966).

Only four species infect man and only one of these, *P. falciparum*, is responsible for the vast majority of deaths from malaria. The three others however cause serious morbidity.

In Table 1, the four human species together with the most commonly used laboratory infections are listed.

1.2.3 Patterns of Malaria Infection in Endemic Areas

In areas of hyperendemic malaria, there is a well-defined age-related acquisition of immunity. Once the passive immunity acquired from maternal antibody ceases at weaning, infants may suffer acute and repeated attacks of the disease. During this period they will either succumb to infection and die of malaria or another endemic pathogen; or slowly, over a period of years, build up an increasing degree of immunity. This immunity is such that it enables children (who are still subject to repeated reinfection) first to survive high parasitaemia without overt illness and then to keep parasite numbers down to subclinical levels. In hyperendemic areas where parasites are transmitted throughout the year this stable state will be achieved by the age of 5 or 6 years. If the transmission pattern is seasonal (epidemic) then depending on the length of the annual transmission period this state of stable immunity may take much longer to achieve or never occur.

Removal of an inhabitant from a hyperendemic area for a significant period or transfer of inhabitants from one epidemic area to another may render the individual susceptible to acute infection again.

The general characteristics of immunity to malaria among exposed populations

Table 1. Species of *Plasmodium*

	Comments
Human malarias	
P. falciparum	Only *P. falciparum* extensively investigated in the laboratory in vitro culture
P. malariae	or in certain experimental primates
P. vivax	
P. ovale	
Primate malarias	
P. knowlesi	Extensive laboratory investigations usually in unnatural laboratory hosts
P. cynomolgi	
P. inui	
P. brasilianum	
Rodent malarias	
P. berghei	Natural hosts are *Thamnomys* sp. Extensive laboratory studies carried out
P. chabaudi	in inbred and outbred rodents
P. vinkei	
P. yoelii	
Avian malarias	
P. lophorae	Important differences in life cycle from mammalian species since exoerythro-
P. gallinaceum	cyte schizogony can be initiated from erythrocytic merozoites

are thus that it is parasite species and at least partially parasite strain specific. It is slow to develop, usually associated with chronic, subclinical parasitaemia and wanes relatively quickly after the total elimination of infection (TALIAFERRO 1949; JEFFREY 1966; SADUN et al. 1966; GARNHAM 1966; POWELL et al. 1972; WILSON and PHILLIPS 1976; COHEN 1979; MOLINEAUX and GRAMICCIA 1980; BROWN 1982).

This chronicity of infection both in the short and long term is a feature it shares with many other parasitic diseases. Such a pattern of infection is the result of successful parasitism. If a large reservoir of disease is to be maintained then parasites must be capable of survival in individuals with a high degree of immunity. This must be achieved without the death of more than a small proportion of the host population.

Such an equilibrium has evolved over a long period. Descriptions of diseases which are almost certainly malaria go back to the most ancient writings (BRUCE-CHWATT and DE ZULUETA 1980). In fact malaria has been a problem to human populations long enough to have influenced their genotype. In Africa, the high incidence of individuals negative for the Duffy blood group in malarious areas coincides with the fact that such erythrocytes are not invaded by *Plasmodium vivax* (MILLER and CARTER 1976). Moreover, apparently disadvantageous genes such as that for haemoglobin S or the G6PD-deficient genotype are present at unusually high frequencies in endemic areas. Since such characteristics would normally be reduced to very low levels by natural selection, their abundance in malarious areas is presumably due to the resistance they confer to malaria infection (LUZZATTO 1974).

A length of time necessary to have had such an influence on the genetics of the host population has also ensured that the host/parasite equilibrium is finely tuned. The characteristics of immunity to malaria outlined in brief above have presumably arisen as an evolutionary compromise between survival of both the parasite and the host.

1.2.4   Variation and Diversity in the Parasite Population

The process of genetic selection as a result of infection has not only been one way. Since the parasite has an obligate sexual stage in its life cycle, the same capacity for the generation of diversity and subsequent selection according to environment is also present within the genotype of the parasite. The most striking manifestation of adaption by the parasite must be the alarmingly rapid spread of drug resistance among strains of *P. falciparum*. Parasites resistant to all antimalarials in common use are known and the rate at which the resistance is spreading suggests that if present trends continue, in the foreseeable future our present battery of drugs may become virtually redundant.

The example of drug resistance is merely the obvious tip of a very large iceberg. Variation has been found in almost all characteristics that have been examined (WALLIKER 1983). Enzymic diversity (CARTER 1970, 1973, 1978; CARTER and McGREGOR 1973; CARTER and VOLLER 1973; CARTER and WALLIKER 1975, 1976; SANDERSON et al. 1981) has been used as a basis of strain typing. Antigenic diversity extends from the serology of heat-stable antigens (WILSON 1980) to strain typing by monoclonal antibodies and to variation in antigenicity and structure in polypeptides implicated in the induction of protective immunity (McBRIDE et al. 1982; SCHOFIELD et al. 1982; NEWBOLD et al. 1984). It has also been found in erythrocyte surface antigens thought to be a target for antibody infections (HOMMEL et al. 1982; UDEINYA et al. 1983). Strain variation in polypeptides separated by two-dimensional electrophoresis has also been reported (TAIT 1981).

In view of our knowledge of the parasite life cycle, the very large numbers of organisms produced during infection and the species and strain specificity of protective immunity, this diversity is not in itself surprising. It is nevertheless relevant to the characterisation of *Plasmodium* antigens since this diversity will almost certainly extend to many antigens involved in the induction of protective immunity.

In addition to the mechanisms outlined above for the generation of diversity by mutation, sexual recombination and other mechanisms of genetic exchange, there is a strong body of evidence that demonstrates that malaria parasites undergo phenotypic antigenic variation (see Sect. 2.2.8).

1.2.5   The Parasite in the Laboratory

Field investigations of malaria infections are complicated by the uncertain history of the individuals under study. Not only is it difficult to assess the contribution of chronic infection or reinfection to the presence of parasites in the blood, but the number of other viral, bacterial, protozoan and worm pathogens endemic in malarious areas complicate the analysis of immunological and epidemiological data.

In order to overcome these problems and to obtain reproducible experimental systems open to the full range of biochemical, immunological and molecular biological techniques, a number of simian, rodent and avian malarias have been established in the laboratory. *P. falciparum* can now also be maintained in long-term in vitro cultures. The relative merits of each of these laboratory systems are an important consideration in experimental design. If the experimental question is one relating solely to the parasite then the use of cultured human material will give results most directly relevant to man. Unfortunately, although *P. falciparum* will infect certain primates in the laboratory, such infections are abnormal and the animals are outbred and often of uncertain history. It is also not yet clear what effect long-term in vitro cultivation may have on the parasite (WILSON and PHILLIPS 1976; LANGRETH et al. 1979; JENSEN et al. 1981; DUBOIS et al. 1983; LE BRAS et al. 1983).

Choices remaining among the other laboratory models are best governed by the type of experiment in question. For example, if these are immunological questions related to the host then the preferred system would be one where the pattern of infection was similar to that in the wild. The use of inbred rodents is necessary for detailed immunological studies. If large numbers of parasites are required for in vitro manipulation then an unrestricted, lethal infection may be most useful. In all cases however it is important to remember that the laboratory host is virtually never the natural one for the parasite and for this reason abnormal antigen recognition, or lack of recognition, may be occurring.

An equally important consideration for laboratory studies is the effect of the selection for and long-term maintenance of parasites under abnormal conditions. Under conditions of natural transmission, frequent exchange of genetic information is occurring (see WALLIKER 1983) and the true capacity of the parasite for adaptation, variation and diversity is realised. Laboratory isolates have been obtained at some point in time and maintained by blood or mosquito passage in experimental animals or in vitro. Some blood-passaged isolates may no longer even be infectious to mosquitos. Two conflicting interests are inherent in the use of such isolates. Long-term blood passage has removed the possibility of the exchange of genetic information between diverse parasite populations such that, after many years, these organisms may no longer be comparable to their wild counterparts. The use of unnatural or highly selected hosts (e.g. inbred mice) may also select certain restricted parasite populations or affect significantly their phenotype. However, this lack of heterogeneity (especially if cloned populations are employed) enhances the reproducibility of experimental data. Probably the best compromise is the use of cloned parasites from recent isolates. Under these conditions reproducibility of infection is optimised while at the same time comparison can be made between clones from different wild populations. It may then be possible to ascertain whether or not the conclusions from laboratory experiments may be complicated by the greater phenotypic or genetic diversity existing under conditions of natural transmission.

1.3 General Techniques Used for the Identification of Important Malarial Antigens

1.3.1 Criteria for Immunologically Important Antigens

Two main criteria have been the basis for most experiments aimed at the identification of parasite antigens involved in the induction of protective immunity.

Firstly there is the criterion of location. It is a reasonable assumption that only antigens accessible to the protective immune response located at the parasite surface or the surface of infected host cells will be vulnerable to immune effector mech-anisms. Thus if such antigens can be identified then they may be naturally immunogenic and therefore targets of an antiparasite immune response. If for some reason they are not immunogenic in vivo then presentation of purified antigen in the right way may be capable of eliciting a protective immune response that is normally absent.

The second criterion is that of differential analysis of antibodies present in sera of hosts with known history. The basis of this approach is that hosts which are functionally immune to infection should possess higher levels of antibodies directed against important parasite antigenic determinants than non-immune controls. One problem with this approach is that it relies purely on the presence of antibody. A number of immune effector mechanisms are known to operate in parasite killing during infection and while antibody is undoubtedly important in this regard, the presence or absence of specific antibody is not in itself a good indicator of functional immunity. As yet there is no reliable in vitro correlate for protective immunity to any stage of the parasite (see Sect. 2.1.3, 2.2.10).

A more detailed description of the methodologies involved will be given in the sections describing individual stages of the parasite life cycle. They can nevertheless be grouped arbitrarily by the basic approach.

1.3.2 Surface Exposure

1.3.2.1 Surface Radiolabelling

There are a number of well-documented techniques for the specific radiolabelling of cell surface components that have been applied to malaria parasites. These are basically the following (a) iodination catalysed by lactoperoxidase or the sparingly soluble reagent Iodogen (MARKWELL and FOX 1978; HUBBARD and COHN 1972), (b) reduction with tritiated sodium borohydride of cell surface components previous-ly oxidised with periodate or galactose oxidase (GAHMBERG 1976; LIAO et al. 1973) and (c) reduction with tritiated sodium borohydride of cell curface Schiff's bases formed between exposed lysine residues and the impermeable reagent pyridoxal phosphate (RIFKIN et al. 1972).

1.3.2.2 Surface Antibody Binding

An alternative approach has been to use antibody as a surface-specific reagent. Because of their size, immunoglobulins are not permeable to viable, intact cells. Incubation of metabolically labelled parasites or parasitised cells with immune serum results in the binding of antibody to surface determinants. Subsequent washing and immunoprecipitation can then identify the parasite molecules which were initially exposed to antibody (NEWBOLD et al. 1982b).

1.3.2.3 Monoclonal Antibodies

Monoclonal antibodies produced against parasites can identify surface antigenic determinants if screened by radioimmunoassay (RIA) or indirect fluorescent anti-body test (IFAT) on intact fixed or unfixed cells (e.g. YOSHIDA et al. 1980).

1.3.2.4 Use of Defined Sera

Serum taken from individuals from endemic areas can be used to immunoprecipitate metabolically labelled parasite antigens from short-term in vitro cultures. When compared with control or primary infection serum, the relative frequency of anti-bodies specific to certain polypeptides is correlated to the immune status of the serum donor (e.g. Brown GV et al. 1981, 1982a; Perrin et al. 1981a). Similar approaches can be used in controlled laboratory infections.

Because of the antigenic diversity of wild isolates of malaria, it is important in this type of experiment that the relationship between the sera employed and the geographical source of the parasite should be taken into account.

1.3.3 Molecular Biological Techniques

Since it is unlikely that parasites could ever be grown in sufficient quantity in vitro to provide a source of antigen for mass immunisation, the approach adopted by many laboratories is one of artificial antigen production. This can either be achieved by chemical synthesis of a suitably immunogenic peptide or by the expression of cloned parasite genes in a heterologous organism. In either case the identification of the relevant antigen relies on the techniques 1 and 2 described above. The contribution of molecular biology is either to provide the sequence of a relevant peptide (from cloned DNA sequences) or to allow the expression of large quantities of parasite antigen from cloned cDNA sequences in bacteria, eucaryotic organisms or viruses such as vaccinia.

The application of molecular biological techniques to the study of malaria has many other implications. By analogy with the African trypanosomes, the investigation of gene organisation and expression, especially of genes coding for immunologically important antigens, may be of fundamental scientific interest in addition to yielding information relevant to immunoprophylaxis. It may also enable an accurate estimate to be made of the degree of intraspecific and intrastrain variation in relevant polypeptides (see Sect. 2.2.8).

2 Parasite Antigens

Each stage of the parasite life cycle will be discussed separately. The more recent work involving the identification and characterisation of relevant antigens will be considered in terms of the relationship of the antigens to the induction of or evasion of immunity and their possible use in epidemiological studies.

2.1 Sporozoites

2.1.1 Sporozoite Antigens

The invasive form of malaria, the sporozoite, is the first point of contact between host and parasite. This contact is extremely brief (probably less than 20 min) and organisms which have not invaded liver parenchymal cells by this time do not go on to initiate an infection. The fate of sporozoites which fail to invade is uncertain but they are presumably cleared by the recticuloendothelial system.

Initial attempts to identify sporozoite surface components utilised the *P. berghei/* mouse laboratory model. By surface labelling purified intact sporozoites with ^{125}I/lactoperoxidase and subsequent solubilisation and immunoprecipitation with appropriate antisera, GWADZ et al. (1979) identified a single major surface antigen of 41 kd. YOSHIDA et al. (1980) prepared monoclonal antibodies against *P. berghei* sporozoites and were able to refine these data. They demonstrated that the surface of *P. berghei* sporozoites was covered with a 44-kd polypeptide which they term Pb44. The surface location was implicated by immune electron microscopy, by immuno-precipitation of surface-labelled sporozoites and by the ability of 3D11 (the anti Pb44 monoclonal antibody) to induce a circumsporozoite precipitation (CSP) reaction. This latter test was developed by VANDERBERG et al. (1969) and involves the in-cubation of intact sporozoites with serum. If antibodies of appropriate specificity are present then they bind to the entire surface of the sporozoite. A thread-like precipitate subsequently forms which is gradually shed from the anterior end of the parasite. It is presumed to be due to the interaction of specific antibody with antigenic determinants at the surface of the sporozoite and the subsequent shedding of antibody/antigen complexes. The same monoclonal antibody was capable of neutralising sporozoite infectivity in vitro and was specific both for the parasite species and stage. Subsequent publications by the same group demonstrated the following features of McAb 3D11 and Pb44.

Mice bearing ascitic tumours of hybridoma 3D11 were protected against homologous sporozoite challenge.

The monoclonal antibody (or Fab fragments of it) could be used passively to protect recipient mice against homologous sporozoite challenge but in a dose-dependent fashion, suggesting that the antibody was being titrated against the challenge infection.

3D11 produced a CSP reaction with the homologous parasite but not with the other rodent malaria species *P. yoelii* and *P. chabaudi*.

That in metabolic labelling experiments Pb44 was synthesised as a precursor of 52 kd which in pulse chase experiments was converted to Pb44.

An additional major band of 54 kd was also synthesised by sporozoites and could be immunoprecipitated with antibody 3D11. This may be the primary transcription product of the antigen.

Pb44 could be removed from intact sporozoites by trypsin treatment (POTOCNJAK et al. 1980; YOSHIDA et al. 1981).

Purified 3D11 or its Fab fragments were also shown to be capable of inhibiting in vitro the invasion of human embryonic lung or hepatoma cells by *P. berghei* sporozoites. A similar dose dependence of invasion and development was evident, suggesting again that saturating levels of antibody were required for complete inhibition. Using this in vitro system it is possible to get complete development of exoerythrocytic schizonts of *P. berghei* (HOLLINGDALE et al. 1982, 1983).

More recently, the scope of this research has been broadened by the preparation of monoclonal antibodies against the surface of sporozoites of a primate malaria (*P. knowlesi*) and two human malarias (*P. falciparum* and *P. vivax*) (COCHRANE et al.

1982; NARDIN et al. 1982). In all three cases, the monoclonal antibodies had very similar characteristics which were also similar to those described above for 3D11 and *P. berghei*.

In all species, monoclonal antibodies were directed against a surface antigenic determinant at the surface of the parasite as judged by their ability to induce a CSP reaction on intact sporozoites, by their reaction in an IFAT with the surface of intact glutaraldehyde-fixed sporozoites and by the ability of a number of them to abolish or reduce infectivity to experimental animals in an in vivo neutralisation assay (see later). Moreover, with *P. knowlesi* in metabolic labelling studies, the surface-specific monoclonal antibodies precipitated three polypeptides from the homologous parasite which could be shown by pulse chase experiments to have a similar precursor-product relationship to that described for *P. berghei*. The two higher molecular weight antigens were apparently precursors of the smallest, which was located on the surface since it was sensitive to trypsin. An as yet un-explained feature of some of these monoclonal antibodies is their cross-reaction with different *Plasmodium* species (COCHRANE et al. 1982). Three anti-*P. knowlesi* sporozoite (5H8, 8B8, 2G3) antibodies cross-reacted with *P. cynomolgi* and one (2G3) cross-reacted with *P. falciparum*. This cross-reaction involving a surface antigenic determinant occurs although immunity to sporozoite infection is species specific (see later). Another strange feature of these *P. knowlesi* antibodies was their behaviour in an in vitro sporozoite neutralisation assay. In two cases (6B8, 2G3) Fab fragments were active in abolishing infectivity of sporozoites in vitro whereas the intact purified antibodies were not (COCHRANE et al. 1982). Serum from hybridoma-bearing mice was effective in in vitro neutralisation, with five out of seven monoclonal antibodies directed against the sporozoite surface antigen.

A monoclonal antibody against *P. falciparum* sporozoites (3D6) recognised parasite polypeptides of 67 kd, 65 kd and 58 kd. Antibody 2F2, prepared against *P. vivax* sporozoites, precipitated labelled polypeptides of 51 kd and 45 kd from this parasite (NARDIN et al. 1982; SANTORO et al. 1983). These antibodies were only partially effective in an in vitro neutralisation assay.

A more detailed comparison of these circumsporozoite proteins (CS proteins) from *P. berghei, P. knowlesi, P. cynomolgi* and *P. falciparum*, by tryptic peptide mapping/IIPLC and two-dimensional electrophoresis, was carried out by SANTORO et al. (1983). These workers suggested that the transition from the two higher molecular weight compounds presumed to be intracellular precursors to the lowest molecular weight surface antigen occurred by the cleavage of basic amino acid residues since the pI values of the surface antigens were always lower than that of their precursors. Moreover high-pressure liquid chromatography (HPLC) separation of tryptic peptides revealed an apparent similarity of structure in the antigens from the species examined since peptides with identical retention times could be derived from any of the purified surface antigens. Thus not only were the sporozoite surface antigens similar in location and synthesis in the different species of parasite, but they also exhibited at least a degree of conserved structure.

The most recent experiments involving the circumsporozoite antigen have been carried out with *P. knowlesi*. GODSON and co-workers (ELLIS et al. 1983; LUPSKI et al. 1983; GODSON et al. 1983) have synthesised cDNA from mRNA extracted from infected mosquitos. Using an expression system for cloning this cDNA into

Escherichia coli and a two-site radioimmunoassay for the identification of recombinant bacterial clones synthesising antigen, they have isolated DNA sequences coding for the portion of the *P. knowlesi* antigen recognised by monoclonal antibodies 2G3, 8A8, 5H8 and 8B8. It was found that the same monoclonal antibody (2G3) could also be used both to bind antigen to plates and to assay for its presence, suggesting that the epitope was repeated. Further work involved transposon mapping of the DNA sequences contained in the plasmid which coded for the region of the molecule recognised by antibody and sequencing of one plasmid insert encoding this epitope. It proved to be a most unusual structure, consisting of twelve repeats of a 36-codon unit encoding 12 amino acid residues. Thus the intact CS protein contained a 12 amino acid peptide repeated 12 times (144 amino acids, approximately 13 kd), making up approximately one-quarter of the largest of the in vivo polypeptides (52 kd) recognised by monoclonal antibody. This dodecapeptide and the first repeat (a tetraeicosopeptide) were synthesised chemically and shown to inhibit the binding of radiolabelled monoclonal antibody to CS protein from *P. knowlesi*. Moreover, the tetraeicosopeptide was inferred to contain two epitopes since the same monoclonal antibody (2G3) could be used to bind it to antigen plates and assay for antigen (GODSON et al. 1983).

Repeated epitopes have also been inferred to be present in the CS proteins of *P. knowlesi*, *P. vivax* and *P. falciparum* by the same two-site/one-antibody radioimmunoassay. Monoclonal antibodies directed against sporozoites of one species were mutually inhibitory in the binding assay. They also inhibited the binding of a high proportion of antibodies from immune serum directed against the homologous parasite. This suggested that for each species there was an immunodominant, repeated epitope at the sporozoite surface that was recognised by monoclonal antibody (ZAVALA et al. 1983). Since a number of these monoclonal antibodies were species specific, it was possible with the use of a sensitive radioimmunoassay to use them in the field to determine which species of malaria was present in infected mosquitos and to what level the mosquitos were infected (ZAVALA et al. 1982). This may have extremely useful epidemiological applications in terms of correlating transmission rates with magnitude and species of vector infection.

2.1.2 Natural Immunity to Sporozoite Infection

Epidemiological studies suggest that adults who have been continuously resident in hyperendemic malarious areas are still prone to reinfection by mosquito-transmitted parasites (e.g. MOLINEAUX and GRAMICCIA 1980). However, since the general level of parasitaemia in the adult population is extremely low and subclinical, it is difficult to assess how this functional immunity is distributed between sporozoites and asexual stages. More than 90% of adults from endemic areas have detectable levels of antisporozoite antibody either by CSP reaction or IFAT with intact fixed parasites (NARDIN et al. 1979b). It is not clear however how these antibody levels correlate with functional immunity (SPITTALNY and NUSSENZWEIG 1973; CLYDE et al. 1973a, b, 1975; MCCARTHY and CLYDE 1977).

2.1.3 Induction of Antisporozoite Immunity in the Laboratory

In the laboratory it is possible to obtain high levels of immunity to sporozoite challenge using the *P. berghei*/mouse model. This can be induced by the intravenous injection of multiple doses of X-irradiated sporozoites. Under the right conditions 100% of animals can be protected against sporozoite challenge. The level of induced immunity has been found to depend on the number of immunising organisms, the level of irradiation and the number and route of the immunisations (NUSSENZWEIG et al. 1967, 1969a, b; SPITTALNY and NUSSENZWEIG 1973; BEAUDOIN et al. 1976; ORIJH et al. 1981). Animals vaccinated in this manner show a time-dependent acquisition of immunity to sporozoite challenge but remain fully susceptible to blood-transmitted infection. If left unchallenged, the level of immunity falls off with time. In this rodent model it has also been possible to demonstrate some cross-immunity to challenge with other rodent species.

Attempts to correlate the immune status of experimental animals with simple in vitro tests have relied either on an indirect fluorescent antibody test (IFAT) or the CSP reaction.

Circumsporozoite precipitation antibodies can always be detected eventually in appropriately immunised hosts, but the kinetics of their appearance do not always correlate with functional immunity (SPITTALNY and NUSSENZWEIG 1973). Serum containing CSP antibody shows strict parasite stage specificity if intact parasites are used but cross-reacts with asexual stages in IFAT if internal antigens are rendered accessible (NUSSENZWEIG et al. 1969b, 1973; NUSSENZWEIG and CHEN 1974; NARDIN et al. 1979a). In the rodent system, CSP antibodies do not react with sporozoites of avian, primate or human malarias but do react with sporozoites of other rodent species at lower antibody titre (VANDERBERG et al. 1969; NARDIN and NUSSENZWEIG 1978).

The general conclusions about immunity to sporozoite challenge in the rodent system are summarised below:

1. High levels of immunity to challenge can be induced experimentally.

2. Parameters important in the immunisation suggested that a successful immune response required some development of the attenuated parasites within the host. Dead parasites in adjuvant were poor immunogens in vaccination experiments.

3. Functionally immune animals had circulating antibody directed against the sporozoite surface but the kinetics of appearance and level of these antibodies did not necessarily correlate with functional immunity.

4. Immunity was strictly stage specific and partially species specific.

2.1.4 Relevance of CS Proteins to Induction of Immunity

The application of modern techniques to a difficult experimental problem has resulted in the identification of a family of sporozoite surface antigens (the CS proteins) in a variety of malaria species including those which infect man. DNA sequence data have provided a complete amino acid sequence for a primate (*P. knowlesi*) CS antigen (OZAKI et al. 1983). The presence of a tandemly repeated sequence (representing an immunodominant epitope) in the middle of the protein suggests that such antigens may be ideal candidates for the production of a synthetic vaccine. Indeed if such a tandemly repeated structure is a feature of all CS proteins (which is the implication of recent research) then the chemical synthesis of the repeats for a variety of malaria species should soon be a realisable goal. What is at present uncertain is whether or not such peptides could successfully be used to vaccinate and prevent malaria infection in man. Results from the *P. berghei*/mouse model are extremely encouraging. Not only is it possible to induce 100% immunity by vaccination with irradiated parasites, but mice can be passively protected against challenge by a monoclonal antibody which recognises CS protein in this species. The same monoclonal antibody also blocks the entry of sporozoites into cultured cells (HOLLINGDALE et al. 1982). Attempts to obtain similar results with either primate or human malarias have met with far more limited success (CLYDE et al. 1973a, b, 1975; REICKMANN et al. 1974; McCARTHY and CLYDE 1977). In some cases no protection against challenge infection was induced (WARD and HAYES 1972; COLLINS and CONTACOS 1972). In other cases up to half of the subjects were protected against homologous species challenge for up to 4 months. Although cross-strain immunity was observed, cross-species protection was never demonstrated. One experiment with serum from a monkey successfully protected against *P. knowlesi* challenge showed marked strain specificity in exposed sporozoite antigenic determinants (GWADZ et al. 1979). Monoclonal antibodies raised against *P. cynomolgi* also showed strain specificity. Other data (NUSSENZWEIG et al. 1973; NUSSENZWEIG and CHEN 1974; CHEN et al. 1976) suggest that sporozoite surface antigenic determinants in primate and human malarias are strictly species specific but not strain specific.

There are a number of possible reasons why immunisation of higher mammals has been less effective to date than immunisation of rodents.

In the first place, the difficulties in obtaining sufficient parasites and hosts (or human volunteers!) suggests that experimental parameters may not have been optimised in the primate and human trials. On the other hand, the *P. berghei*/AJ mouse system is not a natural host/parasite combination. Sporozoite infectivity in AJ mice is only 2% (0.06%–6%) compared with 45% in the natural host (VANDERBERG et al. 1968). Whilst laboratory passage may have increased this figure, both immunisation and challenge experiments at such low infectivity may give results which are difficult to reproduce when infectivity is 1–2 orders of magnitude higher.

As a more general overview, there are a number of questions concerning the biology of the sporozoite which may have to be answered before a successful antisporozoite vaccine is forthcoming.

Why does the CSP reaction occur? Such a dramatic shedding of bound antibody is unlikely to be without biological significance. Pulse chase and structrual studies with a number of malaria species indicate that CS protein is initially synthesised as a high molecular weight intracellular precursor. This is cleaved through an intermediate to surface CS protein. Since it is very difficult to chase label out of the primary gene produce (YOSHIDA et al. 1981; COCHRANE et al. 1982) it is likely that there is a large intracellular pool. Thus the CSP reaction may be a method od shedding potentially harmful antibody whilst continual replenishment of surface antigen is occurring. If the pool is large enough then this replenishment may continue for sufficient time to allow parenchymal cell invasion to occur.

Some of the monoclonal antibodies produced against primate CS protein show cross-species reactivity (COCHRANE et al. 1982). One of these (2G3) is known to recognise the tandemly repeated sequence in *P. knowlesi* CS protein.

With *P. falciparum* and *P. vivax*, cross-reactions have also been observed with antibodies from protected human volunteers. While 2G3 (anti-*P. knowlesi*) reacted with the intermediate-sized CS precursor (P67) of *P. falciparum*, serum from a human protected against *P. vivax* reacted with the largest *P. falciparum* CS precursor (NARDIN et al. 1982). These findings are somewhat surprising in view of the strict species specificity of immunity in man and primates.

Many of these questions may be answered when vaccination studies with chemically synthesised peptides have been carried out. If these are successful in the *P. knowlesi* system then it should not be long before the results can be translated into human trials with synthetic vaccines against the human malarias. However, to date the evidence implicating the repeated epitopes as a target of protective immunity relies on in vitro neutralisation data. It is not certain at present whether this assay reflects in vivo mechanisms in primate and human malarias. The cross-reactivities with some monoclonal antibodies and the ability of Fab fragments sometimes to be more effective in in vitro parasite killing than intact antibody remain to be explained.

Also remaining to be fully explained is the presence, in a molecule of apparent functional importance to the parasite, of a repetitive, highly immunogenic amino acid sequence. If this sequence is indeed the target of neutralising antibody or other protective immune effector mechanisms, then why do the populations of endemic areas not become immune to sporozoite challenge even after 50 years? Perhaps the extremely encouraging results with vaccination studies in rodents have led to a premature false optimism about the possibility of vaccination with chemically synthesised repeated sequences. It is possible that the tandem repeat has evolved as a good target for the immune response, diverting the attention of the host from other portions of the molecule which are functionally indispensable to the parasite. Since the evidence indicates that CS protein is shed after antibody binding but may be quickly replaced from a large intracellular pool, then immune recognition of an irrelevant area of an important molecule may not prevent parasite invasion of host liver cells. The other possibility is one of strain variation. Cross-protection between strains and species apparently occurs in the rodent system, where sporozoite infectivity is extremely low. Monoclonal antibodies to sporozoites of primate malaria cross-react between species although cross-protection is never observed. What may therefore be important is the level of cross-reactivity or level of immune recognition of commoness. In a high-infectivity system this may not be sufficient to prevent infection whereas it is totally effective in a rodent model. Despite these unanswered questions it is at present certain that research with sporozoites has taken us much closer to the concept of a realisable vaccine than work with any other stage of the life cycle. CS proteins are undoubtedly the target of protective immune responses. Whether the repeated sequence is an immunological decoy, is the site of attachment of neutralising antibody, or whether other parts of the molecule are important or show strain variation remains to be firmly established.

It is also important to remember that a sporozoite vaccine would need to be 100% effective to prevent infection. Because of the stage specificity of immunity, a single sporozoite would be capable of initiating a blood infection.

2.2 Asexual Intraerythrocytic Cycle

In recent years, the literature concerning antigens produced by asexual intraerythro-cytic parasites has expanded at a tremendous rate. For this reason, this section will be limited to the discussion of antigens which have been characterised at least partially at the molecular level.

Since the growth of malaria parasites within erythrocytes is in general synchro-nous, it is possible to examine various points of the cell cycle in terms of their antige-nicity and biochemical/biosynthetic properties. Protein synthesis within a single cyc-le of intraerythrocytic growth has been shown to vary both in terms of the quantity and type of polypeptides produced (NEWBOLD et al. 1982a; BROWN GV et al. 1982b; HOLDER and FREEMAN 1982; DEANS et al. 1983a, b; KILEJIAN 1980; BOYLE et al. 1983b). The rate of RNA synthesis and the time of DNA synthesis have also been shown to be tied to the developmental stage of the parasite (Newbold et al. 1982a). These data are consistent with a number of other studies which demonstrate that the spectrum of pa-rasite antigens detectable by immunoprecipitation of metabolically labelled parasi-tes with immune serum varies dramatically with the degree of parasite maturation within the red cell (e.g. DEANS et al. 1978; BROWN et al. 1982; BOYLE et al. 1982; PERRIN and DAYAL 1982). Expression of antigens by intraerythrocytic *Plasmodium* is thus stage specific in terms of the parasite cell cycle.

Since most of the available evidence suggests that specific antibody-mediated mechanisms of parasite destruction in vivo or in vitro are directed mainly at antigens exposed on schizont-infected erythrocytes or merozoites (COHEN et al. 1969; DIGGS and OSLER 1969; BROWN et al. 1970a, b; FREEMAN at al. 1980; BOYLE et al. 1982; SCHMIDT-ULLRICH et al. 1983) the majority of this section will deal with the consider-able number of "late-stage" parasite antigens that have been identified in a number of parasite species. Since some antigens expressed at this time also exhibit intraspecies antigenic diversity or phenotypic antigenic variation (BROWN and BROWN 1965; BROWN 1971; WILSON 1980; MCBRIDE et al.1982; NEWBOLD et al. 1984), the relationship of this to the induction and expression of the protective immunity will also be discus-sed. Finally, mechanisms exist in vivo which are capable of eliciting intracellular pa-rasite death at much earlier parts of the growth cycle (e.g. JENSEN et al. 1983; HAIDARIS et al. 1983). Since the mechanism of induction of such mechanisms is poorly under-stood and the effector mechanisms involved are largely non-specific they will not be considered here.

As pointed out in the introduction, the main focus of experimental effect has been on parasite antigens which are somehow exposed to the host's immune response at a surface location. The methods that have been used to identify such antigens have already been summarised. Overall, the results of such investigations in a variety of host/parasite combinations or using in vitro cultured *P. falciparum* have produced a bewildering array of antigen molecular weights. In order to attempt to rationalise the large amount of data in the literature the various antigens will be considered firstly in terms of cases where good correlations have been obtained between different labora-tories and between different parasite species and secondly with regard to antigens which as yet have no correlate in other systems.

2.2.1 High Molecular Weight Schizont-Specific Polypeptides

A number of laboratories have identified high molecular weight parasite-coded poly-peptides which are synthesised late in the intraerythrocytic cycle. These share most of the following features in common.

They are all between 195 kd and 250 kd (EPSTEIN et al. 1981; HOLDER and FREEMAN 1981, 1982; NEWBOLD et al. 1982b; BOYLE et al. 1983a, b; HALL et al. 1983).

All are synthesised late in the parasite cycle and are lost between final schizont maturation and the next round of merozoite invasion. This has been shown by pulse chase experiments (BOYLE et al. 1983a, b or by lack of monoclonal antibody reactivity with ring stages (FREEMAN et al. 1980; EPSTEIN et al. 1981; MCBRIDE et al. 1982; HOLDER and FREEMAN 1982; BOYLE et al. 1982). Monoclonal antibodies directed against these antigens show a characteristic pattern of indirect immunoflourescence on me-thanol- or acetone-fixed blood smears. Fluorescence is observed around the mem-brane enclosing developing merozoites and on free merozoites (FREEMAN et al. 1980; EPSTEIN et al. 1981; HOLDER and FREEMAN 1982; BOYLE et al. 1982; MCBRIDE et al. 1982). They are cleaved post-translationally in a specific fashion to a well-defined se-ries of smaller polypeptides as evidenced by pulse chase experiments and/or mono-clonal antibody immunoprecipitations (HOLDER and FREEMAN 1981, 1982; BOYLE et al. 1982). One of these processing products ends up on the surface of the merozoite (83 kd in *P. falciparum*) and this explains the monoclonal antibody reactivity with merozoites (FREEMAN and HOLDER 1983). They are identified specifically in a surface antibody binding and immunoprecipitation assay (see Sect. 1 and NEWBOLD et al. 1982b, HOWARD et al. 1982a).

In at least three cases they have been implicated in the induction of protective immunity by inhibition of parasite growth in vitro with monoclonal antibody (EP-STEIN et al. 1981), by passive transfer of monoclonal antibody (BOYLE et al. 1982) or by immunisation experiments with purified antigen (HOLDER and FREEMAN 1981; SCHMIDT-ULLRICH et al. 1983; NEWBOLD, SCHRYER, JARRA and BROWN, unpublished observations).

Comparison of antigens between different parasite species indicates that while most determinants are species specific, cross-species-reacting regions of the molecu-le exist (HOLDER et al. 1983). They have also been shown to exhibit considerable anti-genic and structural diversity within a single parasite species as well as maintaining some conserved structure (MCBRIDE et al. 1982; HOLDER et al. 1983; NEWBOLD et al. 1984). The majority of them appear to be glycoproteins (EPSTEIN et al. 1981; NEWBOLD et al. 1982a). The equivalent antigen has been demonstrated in *P. yoelii, P. chabaudi, P. berghei, P. knowlesi* and *P. falciparum.* This family of antigens will henceforth be re-ferred to as large, schizont, processed antigens or LSPA. Another feature of LSPA in the rodent malaria *P. chabaudi* is a very high frequency of intragenic recombination of epitopes recognised by monoclonal antibodies. When two cloned rodent lines are crossed in a mosquito and the progeny cloned, nearly all of the recombinants for iso-zyme markers are apparently recombinants for LSPA as well (MCLEAN, WALLIKER and BOYLE, unpublished observations). This apparent frequency of intragenic recom-bination is several orders of magnitude higher than would be expected on a random recombination event model. It may therefore represent a parasite mechanism for the rapid generation of diversity in an immunologically important polypeptide. Recently,

Schmidt-Ullrich et al. have vaccinated rhesus monkeys with a purified 74-kd parasite-coded polypeptide and obtained protection against death from an otherwise lethal *P. knowlesi* challenge (SCHMIDT-ULLRICH et al. 1983). Serum from the vaccinated monkeys immunoprecipitated parasite-coded polypeptides ranging from 74 to 230 kd. These polypeptides all showed extensive homology in tryptic peptide maps, suggesting that the 74-kd antigen was in fact a cleaved product of a higher molecular weight precursor. It therefore seems likely that this polypeptide is a natural cleavage product of the *P. knowlesi* LSPA although further data will be needed to clarify this point.

Data involving surface-specific immunoprecipitation reactions, or attempts at purifying host erythrocyte membrane from infected erythrocytes, suggested that LSPAs or their processed products were inserted into the host cell membrane (NEW-BOLD et al. 1982b; HOWARD et al. 1982a; SCHMIDT-ULLRICH et al. 1983). Neither of these approaches are reliable however. In the first place it is extremely difficult to obtain a population of schizont-infected erythrocytes where a degree of merozoite release is not occurring. Thus the definition of host red cell surface or merozoite surface becomes blurred in this type of investigation. Secondly, since membrane purification involves homogenisation of infected cells and because no reliable enzymatic markers exist for determining parasite contamination of purified membranes, it is impossible to be sure of the in vivo location of particular proteins by these techniques at present.

Immune electron microscopy with monospecific sera or monoclonal antibodies is probably the most reliable approach for antigen localisation. To date, this technique reinforces the immunofluorescence data in that antigenic determinants recognised by LSPA-specific antisera reside on the surface of merozoites or on the parasite plasma membrane enclosing developing merozoites (OKA et al. 1983; FREEMAN and HOLDER 1983).

2.2.2 Merozoite Surface Antigens

Two other merozoite/schizont surface-associated antigens have been identified in *P. knowlesi* either by monoclonal antibodies or by surface-specific radiolabelling techniques. DEANS et al. (1982) have produced two monoclonal antibodies against a very late stage merozoite surface-associated antigen of 66 kd. These antibodies inhibit merozoite invasion in an in vitro reinvasion inhibition assay. Miller's group (HUDSON et al. 1983) have identified a 140-kd parasite polypeptide at the surface of *P. knowlesi* merozoites by a combination of surface-labelling techniques (MILLER et al. 1980; JOHNSON et al. 1981) and the production of monospecific antisera (HUDSON et al. 1983). LSPAs and *P. knowlesi* merozoite surface components of 66 and 140 kd at present seem to be unrelated immunologically or by any other criteria.

2.2.3 Merozoite Organelle-Associated Antigens

In *P. yoelii* (FREEMAN et al. 1980); HOLDER and FREEMAN 1981) and *P. lophurae* (KILEJIAN 1974, 1976, 1978) merozoite organelle-associated antigens have been implicated in the induction of protective immunity to malaria. Using *P. yoelii*, partial protection against challenge infection was afforded by vaccination of mice with a monoclonal antibody affinity purified parasite antigen of 235 kd. Interestingly, the challenge infection was confined to reticulocytes. Immunoflourescence with the monoclonal antibody localised at two bright-staining dots within merozoites which were assumed to be the paired organelles. Using *P. lophurae*, KILEJIAN (1978) chemically purified an unusual parasite-coded polypeptide which contained 70% histidine. High specific activity labelling of parasitised erythrocytes with radiolabelled histidine, followed by autoradiography at the electron microscopic level, localised incorporated label in cytoplasmic granules and in rhopteries and micronemes of the merozoite (KILEJIAN 1976). The purified protein when used with adjuvant to immunise ducklings protected against death from an otherwise lethal *P. lophurae* challenge infection (KILEJIAN 1978). Some dispute exists concerning the interpretation of these immunisation experiments (KILEJIAN 1981; MCDONALD et al. 1981; SHERMAN 1981). A histidine-rich protein has also been reported to be associated with "knobs" in *P. falciparum* (KILEJIAN 1979; HADLEY et al. 1983) and proteins enriched in histidine have been detected in *P. chabaudi* (NEWBOLD et al. 1982a).

2.2.4 Other Antigens Identified Using Monoclonal Antibodies

Perrin's group have produced a number of monoclonal antibodies which were shown to inhibit the growth of *P. falciparum* in vitro (PERRIN et al. 1981b). The antibodies recognised parasite-coded polypeptides of 41 kd, 41 and 82 kd or 140 kd (PERRIN and DAYAL 1982) which were present in schizont-infected erythrocytes. The precise localisation of these antigens is at present unclear although the 41-kd component is inferred to be on the surface of the schizont-infected erythrocyte from surface-labelling studies. Growth inhibition in vitro was obtained with ascitic fluid or a purified IgG fraction. Strangely, one of the inhibitory monoclonal antibodies specific for the 41-kd component was prepared by fusion of spleen cells of mice infected with *P. berghei*. This antibody reacted with both parasite species in an immunofluorescence assay. As yet it is not clear how these antigens are related to other late-stage antigens described by other laboratories.

SCHOFIELD et al. (1982) have also produced a number of monoclonal against Papua New Guinea lines of *P. falciparum*. These antibodies bound to schizonts of all Papua New Guinea (PNG) isolates tested but failed to react with *P. falciparum* from Thailand, Nigeria, Ghana and the Netherlands. Some of these antibodies were also growth inhibitory for in vitro cultures of PNG lines, but not for the other lines tested. It seemed that the antigens recognised by these antibodies had a molecular weight in excess of 200 kd.

2.2.5 Late-Stage Antigens Identified by Differential Immunoprecipitation with Defined Sera

A different approach to the identification of immunologically important antigens has been taken by some laboratories. They have reasoned that serum from functionally immune hosts or serum that demonstrably inhibits parasite multiplication in vitro should have significant levels of antibodies directed against important parasite antigens. Thus by comparative immunoprecipitation with serum of known in vitro growth inhibitory activity or sera from patients of known immune status, antibody specificities can be detected which may correlate with the activity of the serum. Using this approach, parasite antigens of molecular weight 95–96 kd have been found to be selectively precipitated from *P. falciparum* by growth inhibitory sera or sera from functionally immune donors (BROWN et al. 1981; BROWN GV et al. 1982a). In a different study, PERRIN et al. (1981a) and PERRIN and DAYAL (1982) found *P. falciparum* antigens of 200, 140, 115, 105 and 76 kd which were precipitated more consistently by African serum than European primary infection serum. GYSIN et al. (1982) also with *P. falciparum* found that antibodies to a 75-kd parasite antigen correlated with the immune status of experimental monkeys.

2.2.6 Parasite Antigens Involved in Red Cell Recognition

A detailed discussion of the complex and remarkably specific process by which a free merozoite recognises and invades an appropriate host erythrocyte is outside the scope of this chapter. Nevertheless, some progress has recently been made in identifying parasite molecules which bind specifically to host erythrocyte components. This is of relevance since such antigens with well-defined function are potential targets for a successful host immune response.

Experiments from a number of laboratories have suggested that glycophorin in human erythrocytes acts at least in part as a ligand for *P. falicparum* merozoite binding prior to erythrocyte invasion (DEAS and LEE 1981; WEISS et al. 1981; PERKINS 1981; HOWARD and MILLER 1981; HOWARD et al. 1982c; PASVOL et al. 1982; BREUER et al. 1983; JUNGERY et al. 1983b). By using a sialoglycoprotein extract of human erythrocytes as a solid-phase absorbent, Jungery et al. (1983a) were able to identify metabolically labelled schizont components of 210 kd, 140 kd and 35 kd which bound specifically. If instead an absorbent containing terminal *N*-acetyl D-glucosamine was used, polypeptides of 140 kd, 70 kd and 35 kd were retained. This latter sugar has been shown to inhibit merozoite invasion in vitro either in solution (WEISS et al. 1981; JUNGERY et al. 1983b) or when immobilised on albumin (JUNGERY et al. 1983b). It has also been shown to be toxic to the parasite when used in solution in the reinvasion inhibition assay (HOWARD et al. 1982c). Thus while evidence implicating both glycophorin and possible carbohydrate determinants on it as ligands for merozoite binding is strong, the fine specificity of the interaction remains to be elucidated. Nevertheless, the possibility of the purification of the parasite erythrocyte receptor by affinity absorption is a promising avenue of research which may lead to the solution of this problem.

2.2.7 Results from Surface Labelling Experiments

By labelling either schizont-infected erythrocytes or purified merozoites with surface-specific radiolabelling techniques, numerous laboratories have attempted to identify exposed antigenic determinants which might be involved in the induction of protective immunity (WALLACH and CONLEY, 1977; SCHMIDT-ULLRICH and WALLACH 1978; SCHMIDT-ULLRICH et al. 1979, 1981, 1982; MILLER et al. 1980; JOHNSON et al. 1981; PERKINS 1982; HOWARD et al 1982a, b; NEWBOLD et al. 1982b). This subject has recently been reviewed (HOWARD 1982). The main problems in interpreting results from such experiments have been alluded to before. In the case of schizont-infected erythrocytes the fragility of the cells and the possibility of schizont rupture occurring during the labelling period make it extremely difficult to be certain that the label is genuinely surface specific. Purified merozoites on the other hand have a very short life in vitro. Molecules identified by labelling merozoites may therefore be surface located or unnaturally exposed by the rapid degeneration of the organisms occurring during purification or the labelling procedure itself. To date such approaches have not yielded consistent results which are repeatable in many laboratories. Indeed the only consistency in such experiments has been the labelling of LSPAs in schizont-infected cells (NEWBOLD et al. 1982b; HOWARD et al. 1982a, b; PERRIN and DAYAL 1982). Since there is no definitive evidence which indicates that they are exposed at the surface of the infected erythrocyte (see Sect. 2.2.1) the caution needed in interpretation of surface labelling is obvious.

2.2.8 Antigenic Variation and Diversity

A number of pathogenic agents such as the African trypanosomes (VICKERMAN 1974) and Borrelia (STOENNER et al. 1982) and non-pathogenic agents such as *Paramecium* (BEALE 1954) have been found to exhibit the phenomenon of antigenic variation. This term is used here to describe the ability of certain cloned organisms to change their antigenic phenotype. Since such antigenic changes occur in a population of organisms derived from a clone, the genetic information encoding the multiple antigenic phenotypes is present in the genome of a single organism. The variation usually involves cell surface determinants and has presumably developed to enable free-living as well as parasitic organisms to survive as populations in adverse environmental conditions. In the case of pathogenic agents, these conditions are usually an ongoing immune response whereas with free-living organisms a variety of other changes, e.g. in temperature, pH or salinity have been implicated. Variation is distinguished from antigenic diversity by the competence of a single organism to change its phenotype. Diversity in this context therefore describes presence of multiple antigenic phenotypes within a population of organisms where each individual organism is only capable of expressing one antigenic type.

Since diversity exists in all populations of organisms as a result of mutation, selection or sexual recombination, it is only relevant to this discussion of malaria parasites where the diversity extends to immunologically, or epidemiologically, important antigens.

2.2.8.1 Antigenic Variation

The evidence for true phenotypic antigenic variation in *Plasmodium* is now overwhelming. It was first detected reproducibly using a simple serological assay by BROWN and co-workers in *P. knowlesi* infections of rhesus monkeys (BROWN and

BROWN 1965, 1968). In this experimental model, a normally lethal infection is transformed into a chronic infection by subcurative drug therapy. Parasites isolated at different times during a chronic infection can be distinguished from each other by a simple haemagglutination assay. Serum taken from the animal will agglutinate schizont-infected erythrocytes (SICA-schizont-infected cell agglutination) from all parasites preceding the serum simple but none of the succeeding parasite populations. Thus each wave of parasitaemia expresses distinct serologically defined SICA antigens at the red cell surface. This work has recently been reparated with cloned *P. knowlesi* parasites, where it was demonstrated that infections derived initially from a single organism were also capable of undergoing antigenic variation (BARNWELL et al. 1983). Moreover, parasite-coded polypeptides that appear to be implicated as the SICA antigens have now been identified as a family of proteins of approximately 200 kd (HOWARD et al. 1983; NEWBOLD, SCHRYER and BROWN, unpublished). The expression of these antigens is modulated by the presence or absence of the host spleen. After several passages in splenectomised monkeys, cloned parasites cease to express SICA antigens at the erythrocyte surface. This change is reversible since subsequent inoculation of SICA negative parasites into intact animals results in the reappearance of SICA reactivity (BARNWELL et al. 1982, 1983). Antigenic variation in determinants exposed at the surface of schizont-infected erythrocytes has also recently been demonstrated in *P. falciparum* infections of squirrel monkeys. HOMMEL et al. (1982, 1983) using a sensitive surface immunofluorescence assay have shown that in chronic infections successive waves of parasitaemia can be differentiated by their surface fluorescence with defined sera, in an analogous fashion to the variation found in *P. knowlesi*. Such antigens are also strain specific. The expression of these antigens is again modulated by the presence of the spleen, but in this instance passage in splenectomised animals induces a switch to a serotype which is different from any occurring in intact animals. The switch can also occur with cloned parasites.

Evidence for antigenic variation in other host parasite systems relies either on the sensitivity to defined sera of defined parasite populations in passive transfer experiments or in immunisation and challenge experiments with suspected parasite variants. BRIGGS et al. (1968) and WELLDE and DIGGS (1978) used immune serum passively to protect mice infected with *P. berghei*. The parasites surviving the serum treatment were shown to be resistant to the same serum in subsequent experiments. WERY et al. (1979) using cloned *P. berghei* parasites and VOLLER and ROSSAN (1969) using *P. cynomolgi bastianelli* isolated several populations of parasites from chronic infections. These populations were then used to infect naive animals. After cure, these animals were found to be more resistant to challenge with the immunising population than to challenge with other populations of parasites taken from the same animal at different times. CORWIN et al. (1970) using uncloned *P. lophurae* were able to isolate parasite populations that were capable of growth in ducklings immunised against the parent population. MCLEAN et al. (1982) showed that parasites from acute-phase infection in a cloned line of a naturally chronic *P. chabaudi* were more sensitive to serum taken immediately after crisis than were parasites taken from later recrudescences.

Thus by direct evidence from a primate and a human malaria and by inference from results with avian and rodent malarias antigenic variation is a common feature of *Plasmodium* infections.

2.2.8.2 Antigenic Diversity

In the introduction it was pointed out that genetic diversity is an inevitable consequence of mutation, selection and sexual recombination in random mating populations of parasites. Diversity in a number of plasmodial characteristics has been demonstrated and limited genetic analysis of some of these characteristics including strain-specific immunity has been carried out (WALLIKER et al. 1971, 1975; OXBROW 1973; ROSARIO 1976; PADUA 1981). This section will concentrate on diversity that has been demonstrated in antigens involved in the induction of protective immunity or to be of likely epidemiological importance.

S-Antigens. These are a family of heat-stable parasite antigens that are found in the plasma of humans infected with *P. falciparum* (WILSON et al. 1969). Different S-antigens are often found in separate malarial episodes in individuals exposed to reinfection (WILSON et al. 1975). The expression of a given S-antigen by a line of parasites is stable to blood passage or long-term in vitro culture and therefore can be used as one marker for serotyping isolates of *P. falciparum* (WILSON 1980). Recently, biosynthetic labelling studies and molecular biological analysis has identified the molecules responsible for S-antigen reactivity (WINCHELL et al. 1984; KEMP et al. 1983; ANDERS et al. 1983). They have a wide range of molecular weights (approximately 150–220 kd) in different isolates and are synthesised late in the erythrocytic cycle. Part of the gene coding for one S-antigen has been cloned and sequenced. This particular clone coded for an 11 amino acid peptide tandemly repeated 23 times (COPPEL et al. 1983). By analogy with the sporozoite CS protein, the apparent molecular weight calculated from sodium dodecyl sulphate (SDS) gels is likely to be seriously in error and it is therefore probable that variations in the number and composition of the repeats account for the observed diversity in molecular weight. Interestingly, the clone coding for this particular repeat within an S-antigen did not hybridise in Southern blots to DNA from other strains of *P. falciparum*, suggesting that the repeat sequence was at least one source of interstrain variability. As yet, no function has been assigned to these proteins, nor is there any evidence for their involvement in the induction of protective immunity. This question will be discussed later.

LSPAs. This family of antigens has been discussed previously (Sect. 2.2.1). Evidence for considerable diversity in LSPAs exists both in *P. falciparum* from monoclonal antibody (MCBRIDE et al. 1982) and peptide mapping studies (NEWBOLD et al. 1984) and in *P. chabaudi* by similar techniques (NEWBOLD et al. 1984). Since much of this work was carried out with cloned populations of parasites, it is probable that diversity in LSPAs in the field is far more widespread. In the case of *P. chabaudi* infections of CBA mice, the antigenic differences between cloned parasite populations have been tentatively linked to the strain-specific component of protective immunity. Immunisation with purified 250-kd antigen from one clone protects more effectively against challenge from the clone from which the antigen was derived than against heterologous challenge (NEWBOLD, SCHRYER and BROWN, in preparation).

For LSPAs then, it is likely that the antigenic diversity demonstrated in two parasite species will extent to many others and may be in some way linked to the induction of protective immunity and so to the evasion of the immune response by different infecting parasite populations.

2.2.9 Erythrocyte Surface Antigens on Parasitised Cells

Under this heading it is necessary to reconsider the "variant antigens" described for *P. knowlesi* and *P. falciparum*. Since these antigens undergo phenotypic variation they will also show strain specificity. It is unlikely that two different parasite populations will express the same antigen at the same time. Different strains of *P. knowlesi* also express different SICA antigens (BROWN et al. 1970a).

However, in the case of *P. falciparum*, an alternative assay has identified schizont-infected erythrocyte surface determinants that show strain specificity. Schizonts of *P. falciparum* are not generally seen in the peripheral circulation since they are seques-tered to endothelial cells lining small blood capillaries. An amelanotic melanoma cell line has been shown to bind schizont-infected erythrocytes of *P. falciparum* but not ring or early trophozoite containing red cells (SCHMIDT et al. 1982). Cultured human endothelial cells behaved similarly (UDEINYA et al. 1981). This is thought to model vascular sequestration in vivo. Such binding can be inhibited by immune serum in a strain-specific fashion (UDEINYA et al. 1983), suggesting that the molecules involved in sequestration are both antigenic and stage specific. Moreover, the binding in vitro to the melanoma cell line is abolished if parasites are first passaged through splen-ectomised animals (DAVID et al. 1983). This observation links the *P. falciparum* variant antigen to the antigen involved in sequestration since serum directed against parasites which had been passaged in splenectomised animals was ineffective in blocking binding of parasites from intact animals to the cell line in vitro. Explana-tions for these observations centre at present around the hypothesis that vascular sequestration protects parasites against the filtering action of the spleen. Specific antibody can reverse this binding both in vitro and in vivo (DAVID et al. 1983) and thus render the parasite susceptible to splenic damage. Antigenic variation in such a mole-cule would therefore evade the immune response by enabling parasites of new variant types continually to sequester. This is probably not the whole story since there are a number of primate and rodent malarias where vascular sequestration does not occur (GARNHAM 1966).

Electron microscopic evidence suggested that in *P. falciparum* the erythrocyte surface deformations known as "knobs" that occur in trophozoite/schizont-infected cells were the point of attachment to vasclar endothelium (TRAGER et al. 1966; LUSE and MILLER 1971; GRUENBERG et al. 1983) and also the point where antibody binding was localised (LANGRETH and REESE 1979). Whilst this is no doubt the case, it seems that a parasite-derived erythrocyte surface antigen can be expressed on "knobby" parasites which are not sequestered (DAVID et al. 1983). Moreover, knobless variants seem to be better at inducing protection against challenge infection than wild-type "knobby" parasites (TRAGER et al. 1983).

It is obvious therefore that the relationship between erythrocyte surface strain-specific/variant antigens and the presence of knobs on the parasitised erythrocyte is not yet fully understood. It is also not clear whether or not strain specificity in this context is simply a manifestation of phenotypic antigenic variation. Many species of malaria do not have knobs or do not undergo vascular sequestration (GARNHAM 1966) and whether or not these express parasite-derived erythrocyte surface antigens of this type remains to be established.

2.2.10 Conclusions

Immunity to the asexual intraerythrocytic cycle of *Plasmodium* infections is an extremely complex phenomenon. It is in general specific for this stage of the parasite, is largely specific for parasite species and has components which are both strain and variant specific. There are also non-specific components of the immune response which can confer limited protection not only against other malaria species but also against infection with *Babesia* species (Cox 1978; Cox and MILAR 1968; COHEN 1979; JEYWARDENA 1981; BROWN 1982). Since nearly all known immune effector mechanisms have been implicated in parasite destruction, this section will be mainly concerned with the induction phase of protective immunity. The many facets of immunity to blood-stage malaria are unlikely to be the result of an immune response to a single parasite antigen. It is therefore equally unlikely that immunisation with a single antigen will reproduce the results of immunisation by chronic infection. Functional immunity is in general associated with chronic, low-grade parasitaemia (see "Introduction") and so in analysing the results from experiments with purified antigen it must be remembered that sterile immunity is unlikely to be achieved.

It has been known for 40 years that crude parasite preparations (usually schizont-infected erythrocytes or partially purified merozoites) are effective immunising agents when given with Freund's complete adjuvant (FREUND et al. 1948; BROWN et al. 1970a; MITCHELL et al. 1977; SIDDIQUI 1977). In the case of *P. knowlesi*, sterile immunity can be induced by vaccination with a merozoite preparation, providing that animals are challenged with the homologous variant. After challenge with homologous or heterologous variants, such animals are rendered immune to heterologous strains (BROWN et al. 1970a, b; MITCHELL et al. 1975). More recently, similar antigen preparations have been shown to protect against death from potentially lethal challenge when other adjuvants such as muramyl dipeptide (MDP) or stearoyl MDP are used (REESE et al. 1978; SIDDIQUI et al. 1978). Irradiated parasites were also shown to be good immunogens without adjuvant (WELLDE et al. 1979). In most cases however the challenged animals suffered substantial parasitaemias. While this in itself is not discouraging, the presence of significant amounts of host erythrocyte material in such antigen preparations makes it impossible to contemplate their use in humans. It is also unclear what contribution antiparasite or antihost/modified host responses make to the protective effect (BROWN and HILLS 1979; BROWN et al. 1980; BROWN 1982; BROWN KN et al. 1982). The use of purified parasite components overcomes this problem and to date has implicated at least two classes of antigens in the induction of immunity: the LSPAs and merozoite organelle-associated polypeptides. All these immunisation experiments have nevertheless employed Freund's complete adjuvant and have only protected against death from an otherwise lethal challenge.

Other antigens implicated in the induction of immunity have been identified by in vitro growth inhibition with serum or monoclonal antibodies. This assay is known not to be a reliable correlate of immune status of the serum donor (WILSON and PHILLIPS 1976; CHULAY et al. 1981; DUBOIS et al. 1983; BROWN et al. 1983; PATARPOTIKUL et al. 1983). It therefore remains for many of these antigens to be tested in vivo before firm conclusions can be drawn about their relevance. Purification of sufficient material from parasitised erythrocytes is also a formidable problem. Thus the rapid advances in the application of molecular biological techniques to *Plasmodium* is a more promising avenue of research which may lead to the production of large quantities of antigenic material. Gene sequences can be amplified and expressed in bacteria or used to provide polypeptide sequence data for the manufacture of synthetic peptide immunogens.

With the explosion of information about asexual plasmodial antigens it is easy to be optimistic about the production of a usable malaria vaccine. It is important to bear in mind however that experiments to date have been carried out in unnatural host/parasite combinations with laboratory isolates which may bear little resemblance to the parasite in its natural environment (see "Introduction").

The characteristics of infections in animals immunised with purified antigen or dead parasites are usually a reduction in the peak parasitaemia during acute infection and a longer prepatent period. In general, immunisation by infection and drug cure, or by irradiated parasites, is a more effective method of protection against homologous parasite challenge. This is similar to the results with sporozoite immunisation, suggesting again that induction of an appropriate immune response is more effective if some development of the parasite within the host occurs. These observations may be of critical importance in understanding precisely how induction of protective immunity to malaria occurs and suggests that antigen presentation is likely to be at least as vital a parameter as antigen identification.

These conclusions have so far neglected the contribution of antigenic diversity and phenotypic variation in parasite evasion of the immune response. Many authors have suggested that the acquisition of immunity to malaria in endemic areas involves the exposure of susceptible hosts to the diverse range of parasite strains or antigenic types in a particular location. Evidence is emerging that at least one class of antigens involved in the induction of immunity to malaria (LSPAs) shows a high degree of diversity between isolates. The problem of identifying the true extent of the diversity in this single antigen has not yet been properly addressed. It may well be that LSPAs (in combination with or in addition to S-antigen) will provide extremely useful markers for epidemiological studies aimed at typing parasite isolates and determining their geographical distribution.

Antigenic variation has been demonstrated in a variety of parasite species and associated with immune evasion by immunisation and challenge studies, by passive serum transfer or by association of the presence of variant-specific opsonising antibody with immune status (Brown et al. 1970a, b; Brown 1971). Despite the fact that considerable evidence links the chronicity of malaria infections with the appearance of new antigenic variants, the implications of phenotypic variation for vaccine development have in general been considered secondary to the identification of "protective" antigens. One problem has been that the search for immunologically important antigens has proceeded quite often in isolation from studies associated with parasite function. The possibility that a parasite antigen with defined function (such as that involved in vascular sequestration of *P. falciparum*) may be equivalent to an antigen known to undergo phenotypic variation may help to resolve this point. It is also difficult to envisage why natural selection has resulted in the presence of variable antigens if they are not important in parasite survival. The fact that sterile immunity is rarely achieved by immunisation with a single variant may be associated with the rate and mechanism of variant appearance rather than lack of immunological importance.

BROWN (1971) has suggested that the acquisition of immunity may be associated with T-cell recognition of common structure between different variants.

It is clear that a number of questions concerning the type, mode of presentation and possible variability in potentially protective antigens remain unanswered. Nevertheless, the search will continue for "protective antigens" which can be used in acceptable immunisation regimes and which will transcend strain and variant differences. Whether such a goal is realistic remains to be established.

2.3 Gametocytes

Gametocytes are produced by a separate pathway of differentiation in a small proportion of asexual intraerythrocytic parasites. They develop into male (micro) or female (macro) gametocytes within erythrocytes of the mammalian host and are only liberated as functional gametes in the midgut of the mosquito vector. Sexual stages of plasmodial parasites have been the subject of recent review (SINDEN 1983). Whilst a considerable amount is known about the ultrastructure and biology of this stage in the life cycle, there is very little information to date at the molecular level.

It has been demonstrated in a number of *Plasmodium* species that it is possible to induce transmission-blocking immunity by immunisation of experimental animals with partially purified gametocytes (CARTER and CHEN 1976; GWADZ 1976; GWADZ and GREEN 1978; CARTER et al. 1979a; MENDIS and TARGETT 1979). In the case of the avian malaria *P. gallinaceum* such immunisation, in the absence of adjuvant, has little effect on the development of asexual parasitaemia or gametocytes in the immunised hosts but renders the animals non-infective to mosquitos. It seems that antibodies to free gametes present in the serum of these animals are ingested by a feeding mosquito and prevent gamete fusion in the mosquito midgut. Such antibodies appear to block fertilisation but have no effect on subsequent zygote maturation (CARTER et al. 1979b). In the case of simian (GWADZ and GREEN 1978) or rodent (MENDIS and TARGETT 1979) malarias, immunisation with partially purified gametes was effective in blocking transmission, but also suppressed significantly the asexual parasitaemia. Since gametocytogenesis was relatively normal in these animals, the most likely explanation for this phenomenon was the presence of contaminating asexual parasites in the gametocyte preparation.

Monoclonal antibodies have been produced to gametes of *P. gallinaceum* (RENER et al. 1980; KAUSHAL et al. 1983b) and *P. falciparum* (RENER et al. 1983). In both cases, two different antibodies have been produced which react with the surface of male and female gametes and act synergistically in blocking their infectivity to mosquitos. Each antibody against *P. falciparum* precipitated the same three proteins from female gametes of 255, 59 and 53 kd. Evidence was also presented that suggested that the individual monoclonal antibodies were active in the presence of complement. The antibodies against *P. gallinaceum* recognised components of 240, 56 and 54 k in male and female gametes. In this case a mixture of the two antibodies induced microgamete agglutination in vitro. Results from previous immunisation experiments (CARTER et al. 1979b) suggested that transmission-blocking immunity was not associated with agglutinating antibody but with an antibody-mediating fixation of gametes to glass (SF antibody). This may represent a difference between the in vivo and in vitro assays used in the two experiments.

KAUSHAL et al. (1983a) have purified zygotes of *P. gallinacem* and surface labelled them with ^{125}I. Antigenic parasite surface components of 40–240 kd were identified. In addition, it appeared that a chicken serum protein of 180 kd was specifically absorbed to the parasite surface. The antiserum that precipitated these components was also claimed to be effective in blocking infectivity to mosquitos.

2.3.1 Conclusions

A gamete-directed vaccine against malaria has a number of attractions and some drawbacks. Successfully vaccinated individuals would not be capable of transmitting the disease but would probably suffer normal clinical symptoms. Since gametes are not normally liberated in the mammalian host it is possible that gamete antigens are not normally exposed. Thus there is no selection pressure on the parasite to diversify or vary these molecules. Moreover gametes must be capable of fusion to other gametes of the same species, suggesting common antigenic determinants which would not be subject to strain variation. So far this seems to be the case (GWADZ 1976; RENER et al. 1983). Lack of normal exposure to these antigens however also suggests that immunised hosts would not have their antibody responses boosted by infection. Vaccination may therefore have to be extremely effective or frequent. In any event gamete determinants would form a useful part of any proposed mixed stage vaccine preparation directed against malaria.

3   Overall Conclusions

Huge advances have been made in recent years in the identification and characterisation of immunologically important antigens from all stages of the parasite life cycle. With modern techniques, the large-scale production of these antigens (either by chemical synthesis of appropriate peptides or by expression in bacteria of genes coding for relevant polypeptides) is a realisable goal. However, this progress has not been associated with similar rapid advances in understanding structure/function relationships of these antigens or answers to more basic biological and immunological questions relating to the chronicity and pathology of malaria infection and the difficulty in induction of and instability in a state of immunity to the disease.

It must be hoped that when answers to some of these more fundamental questions are forthcoming then the important information derived from molecular studies can be applied rapidly to produce a vaccine preparation containing components from all parasite stages. Providing that problems associated with antigen presentation and/or the finding of suitable adjuvants can be overcome, then such a preparation might be capable of preventing death and alleviating the morbidity of malaria infection. This would be a remarkable achievement. The present state of knowledge however suggests that it is unlikely in the forseeable future that a state of sterile immunity to malaria infection will be induced by immunoprophylaxis.

Acknowledgements. The author wishes to thank W. Jarra, R. J. M. Wilson and particularly K. N. Brown for invaluable discussion and criticisms and comments on this manuscript. C. I. N. was supported by the UNDP/World Bank/WHO Special Programme for Research and Training in Tropical Diseases under the Scientific Working Group on the Immunology of Malaria during the preparation of this manuscript.

References

Aikawa M (1971) *Plasmodium*: the fine structure of malarial parasites. Exp Parasitol 30:284–320

Anders RF, Brown GV, Edwards A (1983) Characterisation of an S-antigen synthesised by several isolates of *P. falciparum*. Proc Natl Acad Sci USA 80:6652–6656

Barnwell JW, Howard RJ, Miller LH (1982) Altered expression of *Plasmodium knowlesi* variant antigen on the erythrocyte membrane in splenectomised rhesus monkeys. J Immunol 128:224–226

Barnwell JW, Howard RJ, Coon HG, Miller LH (1983) Splenic requirement for antigenic variation and expression of the variant antigen on the erythrocyte membrane in cloned *P. knowlesi* malaria. Infect Immun 40:985–994

Beale GH (1954) The genetics of *Paramecium aurelia*. University Press, Cambridge

Beaudoin RL, Strome CPA, Tubergen TA, Mitchell F (1976) *Plasmodium berghei*: irradiated sporozoites of the ANKA strain as immunising antigens in mice. Exp Parasitol 39:438–443

Boyle DB, Newbold CI, Smith CC, Brown KN (1982) Monoclonal antibodies that protect *in vivo* against *Plasmodium chabaudi* recognised at 250,000-Dalton parasite polypeptide. Infect Immun 38:94–102

Boyle DB, March JC, Newbold CI, Brown KN (1983a) Parasite polypeptides lost during schizogony and erythrocyte invasion by the malaria parasites *Plasmodium chabaudi* and *P. knowlesi*. Mol Biochem Parasitol 7:9–18

Boyle DB, Newbold CI, Wilson RJM, Brown KN (1983b) Intraerythrocytic development and antigenicity of *Plasmodium falciparum* and comparison with simian and rodent malaria parasites. Mol Biochem Parasitol 9:227–240

Breuer WV, Kahane I, Baruch D, Ginsburg H, Cabantchik ZT (1983) Role of internal domains of glycophorin in *Plasmodium falciparum* invasion of human erythrocytes. Infect Immun 42:133–140

Briggs NT, Wellde BT, Sadun EH (1968) Variants of *Plasmodium berghei* resistant to passive transfer of immune serum. Exp Parasitol 22:338–345

Brown GV, Anders RF, Stace JD, Alpers MP, Mitchell GF (1981) Immunoprecipitation of biosynthetically labelled proteins from different Papua New Guinea *Plasmodium falciparum* isolates by sera from individuals in the endemic area. Parasite Immunol 3:283–298

Brown GV, Anders RF, Mitchell GF, Heywood PF (1982a) Target antigens of purified human immunoglobulins which inhibit growth of *Plasmodium falciparum* in vitro. Nature 297:591–593

Brown GV, Coppel R, Vrbova H, Grumont RJ, Anders RF (1982b) *Plasmodium falciparum*: comparative analysis of erythrocyte stage dependent protein antigens. Exp Parasitol 53:279–284

Brown GV, Anders RF, Knowles G (1983) Differential effect of immunoglobulin on the *In vitro* growth of several isolates of *Plasmodium falciparum*. Infect Immun 39:1228–1235

Brown IN, Brown KN (1968) Immunity to malaria: the antibody response to antigenic variation by *Plasmodium knowlesi*. Immunology 14:127–138

Brown KN (1971) Protective immunity to malaria provides a model for the survival of cells in an immunologically hostile environment. Nature 230:163–167

Brown KN (1982) Host resistance to malaria. In: Chandra RK (ed) Critical reviews in tropical medicine. Plenum, New York

Brown KN, Brown IN (1965) Immunity to malaria: antigenic variation in chronic infections of *Plasmodium knowlesi*. Nature 208:1286–1288

Brown KN, Hills LA (1979) The possible role of isoantigens in protective immunity to malaria. Bull WHO [Suppl 1] 57:135–138

Brown KN, Brown IN, Hills LA (1970a) Immunity to malaria 1. Protection against *Plasmodium knowlesi* shown by monkeys sensitised with drug suppressed infections or by dead parasites in Freund's adjuvant. Exp Parasitol 28:304–317

Brown KN, Brown IN, Trigg PI, Phillips RS, Hills LA (1970b) Immunity to malaria II. Serological response of monkeys sensitised by drug suppressed infection or by dead parasitised cells in Freund's complete adjuvant. Exp Parasitol 28:318–338

Brown KN, Grundy MS, Hills LA, Jarra W (1980) Cold isohaemagglutinins in *Plasmodium berghei* infected rats reacting with parasitised reticulocytes. Bull WHO 58:449–457

Brown KN, McLaren DJ, Hills LA, Jarra W (1982) The binding of antibodies from *Plasmodium berghei* infected rats to isoantigenic and parasite specific antigen sites on the surface of infected reticulocytes. Parasite Immunol 4:21–31

Bruce-Chwatt LJ, de Zulueta J (1980) The rise and fall of malaria in Europe. A historico-epidemiological study. Oxford University Press, Oxford, pp 1–27

Canning EV, Sinden RE (1973) The organisation of the ookinete and observations on nuclear division in oocysts of *Plasmodium berghei*. Parasitology 67:29–40

Carter R (1970) Enzyme variation in *Plasmodium berghei*. Trans R Soc Trop Med Hyg 64:401–406

Carter R (1973) Enzyme variation in *Plasmodium berghei* and *Plasmodium vinkei*. Parasitology 66:297–307

Carter R (1978) Studies on enzyme variation in murine malaria parasites *Plasmodium berghei, P. yoelii, P. vinkei* and *P. chabaudi* by starch gel electrophoresis. Parasitology 76:241–267

Carter R, Chen DH (1976) Malaria transmission blocked by immunisation with gametes of the malaria parasite. Nature 263:57–60

Carter R, McGregor IA (1973) Enzyme variation in *Plasmodium falciparum* in the Gambia. Trans R Soc Trop Med Hyg 67:830–837

Carter R, Voller A (1973) Enzyme variants in primate malaria parasites. Trans R Soc Trop Med Hyg 67:14–15

Carter R, Walliker D (1975) New observations on the malaria parasites of rodents of the Central African Republic *Plasmodium vinkei petterii* subsp. nov. and *Plasmodium chabaudi* London 1965. Ann Trop Med Parasitol 69:187–196

Carter R, Walliker D (1976) The malaria parasite of rodents of the Congo Brazzaville *Plasmodium chabaudi adami* subsp. nov. and *Plasmodium vinkei lentum*. Ann Parasitol Hum Comp 51:637–646

Carter R, Gwadz RW, McAuliffe FM (1979a) *Plasmodium galliniceum*: transmission blocking immunity in chickens 1. Comparative immunogenicity of gametocyte and gamete containing preparations. Exp Parasitol 47:185–193

Carter R, Gwadz RW, Green I (1979b) *Plasmodium gallinaceum*: transmission blocking immunity in chickens II. The effect of antigamete antibodies *in vitro* and *in vivo* and their elaboration during infection. Exp Parasitol 47:194–208

Chen DH, Nussenzweig RS, Collins WE (1976) Specificity of the circumsporozoite precipitation antigen(s) of human and simian malarias. J Parasitol 62:636–637

Chulay JD, Haynes D, Diggs CL (1981) Inhibition of *in vitro* growth of *Plasmodium falciparum* by immune serum from monkeys. J Infect Dis 144:270–278

Clyde DF, McCarthy VC, Miller RM, Hornick RB (1973a) Specificity of protection of man immunized against sporozoite induced falciparum malaria. Am J Med Sci 266:398–403

Clyde DF, Most H, McCarthy VC, Vanderberg JP (1973b) Immunisation of man against sporozoite induced falciparum malaria. Am J Med Sci 266:169–177

Clyde DF, McCarthy VC, Miller RM, Woodward WE (1975) Immunisation of man against falciparum and vivax malaria by use of attenuated sporozoites. Am J Trop Med Hyg 24:397–401

Cochrane AH, Santoro F, Nussenzweig V, Gwadz RW, Nussenzweig RS (1982) Monoclonal antibodies identify the protective antigens of sporozoites of *Plasmodium knowlesi*. Proc Natl Acad Sci USA 79:5651–5655

Cohen S (1979) Immunity to malaria. Proc R Soc Lond (Biol) 203:323–336

Cohen S, Butcher GA, Crandall RB (1969) Action of malarial antibody *in vitro*. Nature 223:368–371

Collins WE, Contacos PF (1972) Immunisation of monkeys against *Plasmodium cynomolgi* by x-irridated sporozoites. Nature (New Biol) 236:176–177

Coppel RL, Cowman AF, Lingelbach KR, Brown GV, Saint RB, Kemp DJ, Anders RF (1983) Isolate-specific S-antigen of *Plasmodium falciparum* contains a repeated sequence of eleven amino acids. Nature 306:751–756

Corwin RM, Cox HW, Ludford CG, McNett SL (1970) Relapse mechanisms in malaria: emergence of variant strains of *P. lopurae* from serum antigen immunised ducks. J Parasitol 56:431–438

Cox FEG (1978) Concomitant infections in rodent malaria. In: Killick-Kendrick R, Peters W (eds) Rodent Malaria. Academic, London, pp 309–325

Cox HW, Milar R (1968) Cross protective immunisation by *Plasmodium* and *Babesia* infections of rats and mice. Am J Trop Med Hyg 17:173–179

David P, Hommel M, Miller LH, Udeinya IJ, Olgino LD (1983) Parasite sequestration in *Plasmodium falciparum* malaria: spleen and antibody modulation of cytoadherence of infected erythrocytes. Proc Natl Acad Sci USA 80:5075–5079

Deans JA, Dennis ED, Cohen S (1978) Antigenic analysis of sequential erythrocytic stages of *Plasmodium knowlesi*. Parasitology 77:333–344

Deans JA, Alderson T, Thomas AW, Mitchell GH, Lennox ES, Cohen S (1982) Rat monoclonal antibodies which inhibit the *in vitro* multiplication of *Plasmodium knowlesi*. Clin Exp Immunol 49:297–309

Deans JA, Thomas AW, Cohen S (1983a) Stage specific protein synthesis by asexual blood stages of *Plasmodium knowlesi*. Mol Biochem Parasitol 8:31-44

Deans JA, Thomas AW, Inge PM, Cohen S (1983b) Stage specific protein synthesis by asexual blood stage parasites of *Plasmodium falciparum*. Mol Biochem Parasitol 8:45-51

Deas JE, Lee LT (1981) Competitive inhibition by soluble erythrocyte glycoproteins of penetration by *Plasmodium falciparum*. Am J Trop Med Hyg 30:1164-1167

Diggs CL, Osler AG (1969) Humoral immunity in rodent malaria: II. Inhibition of parasitaemia by serum antibody. J Immunol 102:298-305

Dubois P, Paulliac S, Fandeur T, Dedet J-P, Pereira da Silva L (1983) Characterisation of protective antigens from erythrocytic stages of *Plasmodium falciparum*. J Cell Biochem [Suppl] 7A:12

Ellis J, Ozaki LS, Gwadz RW, Cochrane AH, Nussenzweig V, Nussenzweig RS, Godson GN (1983) Cloning and expression in *E. coli* of the malaria sporozoite surface antigen gene from *Plasmodium knowlesi*. Nature 302:536-538

Epstein N, Miller LH, Kaushel DC, Udeinya IJ, Renes J, Howard RJ, Asofsky R, Aikawa M, Hess RL (1981) Monoclonal antibodies against a specific surface determinant on malaria (*Plasmodium knowlesi*) merozoites block erythrocyte invasion. J Immunol 127:212-217

Freeman RR, Holder AA (1983) Surface antigens of malaria. J Exp Med 158:1647-1653

Freeman RR, Trejdosiewicz AJ, Cross GAM (1980) Protective monoclonal antibodies recognising stage specific merozoite antigens of a rodent malaria parasite. Nature 284:366-368

Freund J, Jefferson-Thomson K, Sommer HE, Walter AW, Pisani TM (1948) Immunisation of monkeys against malaria by means of killed parasites with adjuvants. Am J Trop Med 28:1-22

Gahmberg CG (1976) External labelling of human erythrocyte glycoproteins. Studies with galactose oxidase and fluorography. J Biol Chem 251:510-515

Garnham PCC (1966) Malaria parasites and other hemosporidia. Blackwell Scientific, Oxford

Godson GN, Ellis J, Svec P, Schlesinger DH, Nussenzweig V (1983) Identification and chemical synthesis of a tandemly repeated immunogenic region of a *Plasmodium knowlesi* circumsporozoite protein. Nature 305:29-33

Gruenberg J, Allred DR, Sherman IW (1983) Scanning electron microscope analysis of the protusions (knobs) present on the surface of *Plasmodium falciparum* infected erythrocytes. J Cell Biol 97:795-802

Gwadz RW (1976) Malaria: successful immunisation against the sexual stages of *Plasmodium gallinaceum*. Science 193:1150-1151

Gwadz RW, Green I (1978) Malaria immunisation in rhesus monkeys. J Exp Med 148:1311-1323

Gwadz RW, Cochrane AH, Nussenzweig V, Nussenzweig RS (1979) Preliminary studies on vaccination of rhesus monkeys with irradiated sporozoites of *Plasmodium knowlesi* and characterisation of the surface antigens of these parasites. Bull WHO [Suppl] 1:165-173

Gysin J, Dubois P, Pereira da Silva L (1982) Protective antibodies against erythrocytic stages of *Plasmodium falciparum* in experimental infection of the squirrel monkey *Saimiri sciureus*. Parasite Immunol 4:421-430

Hadley TJ, Leech JH, Green TJ, Daniel WA, Wahlgren M, Miller LH, Howard RJ (1983) A comparison of knobby (K+) and knobless (K-) parasites from two strains of *Plasmodium falciparum*. Mol Biochem Parasitol 9:271-278

Haidaris CG, Haynes JD, Meltzer ME, Allison AC (1983) Serum containing tumor necrosis factor is cytotoxic for the human malaria parasite *Plasmodium falciparum*. Infect Immun 42:385-393

Hall R, McBride J, Morgan G, Tait A, Zolg JW, Walliker D, Scaife J (1983) Antigens of the erythrocytic stages of the human malaria *Plasmodium falciparum* detected by monoclonal antibodies. Mol Biochem Parasitol 7:247-265

Holder AA, Freeman RR (1981) Immunisation against blood stage rodent malaria using purified parasite antigens. Nature 294:361-364

Holder AA, Freeman RR (1982) Biosynthesis of a *Plasmodium falciparum* schizont antigen recognised by immune serum and a monoclonal antibody. J Exp Med 156:1528-1538

Holder AA, Freeman RR, Newbold CI (1983) Serological cross reaction between high molecular weight proteins synthesised in blood schizonts of *Plasmodium yoelii*, *Plasmodium chabaudi* and *Plasmodium falciparum*. Mol Biochem Parasitol 9:191-196

Hollingdale MR, Zavala F, Nussenzweig RS, Nussenzweig V (1982) Antibodies to the protective antigen of *Plasmodium berghei* sporozoites prevent entry into cultured cells. J Immunol 128:1929-1930

Hollingdale MR, Leland P, Leef JL, Schwartz AL (1983) Entry of *Plasmodium berghei* sporozoites into cultured cells and their transformation into trophozoites. Am J Trop Med Hyg 32:685-690

Hommel M, David PH, Olgino LD, David JR (1982) Expression of strain specific surface antigens on *Plasmodium falciparum* infected erythrocytes. Parasite Immunol 4:409–419

Hommel M, David PH, Olgino LD (1983) Surface alterations of erythrocytes in *Plasmodium falciparum* malaria. J Exp Med 157:1137–1148

Howard RJ (1982) Alterations in the surface membrane of red blood cells during malaria. Immunol Rev 61:67–107

Howard RJ, Miller LH (1981) Invasion of erythrocytes by malaria merozoites: evidence for specific receptors involved in attachment and entry. In: Adhesion and microorganism pathogenicity. Ciba Foundation Symposium 80. Pitman Medical, Tunbridge Wells, pp 202–219

Howard RJ, Barnwell JW, Kao V, Daniel WA, Aley SB (1982a) Radioiodination of new protein antigens on the surface of *Plasmodium knowlesi* schizont-infected erythrocytes. Mol Biochem Parasitol 6:343–368

Howard RJ, Barnwell JW, Kao V (1982b) Tritiation of protein antigens of *Plasmodium knowlesi* schizont-infected erythrocytes using pyridoxal phosphate-sodium boro (^{3}H)hydride. Mol Biochem Parasitol 6:369–388

Howard RJ, Haynes JD, McGinnis MH, Miller LH (1982c) Studies on the role of red blood cells glycoproteins as receptors for invasion by *Plasmodium falciparum* merozoites. Mol Biochem Parasitol 6:303

Howard RJ, Barnwell JW, Kao V (1983) Antigenic variation in *Plasmodium knowlesi* malaria: identification of the variant antigen on infected erythrocytes. Proc Natl Acad Sci USA 80:4129–4133

Hubbard AL, Cohn ZA (1972) The enzymatic iodination of the red cell membrane. J Cell Biol 55:390–405

Hudson DE, Miller LH, Richards RL, David PH, Alving CR, Gitler C (1983) The malaria merozoite surface: a 140,000 m.w. protein antigenically unrelated to other surface components on *Plasmodium knowlesi* merozoites. J Immunol 130:2886–2890

Jaywardena AN (1981) Immune responses in malaria. In: Mansfield JM (ed) Parasitic diseases, vol 1. The immunology. Dekker, New York, pp 86–122

Jeffrey GM (1966) Epidemiological significance of repeated infections with homologous and heterologous strains and species of *Plasmodium*. Bull WHO 35:873–882

Jensen JB, Capps TC, Carlin JM (1981) Clinical drug resistant falciparum malaria acquired from cultured parasites. Am J Trop Med 30:523–525

Jensen JB, Boland MT, Allan JS, Carlin JM, Vande Waa JA, Divo AA, Akood MAS (1983) Association between human serum-induced crisis forms in cultured *Plasmodium falciparum* and clinical immunity in malaria in Sudan. Infect Immun 41:1302–1311

Johnson JG, Epstein N, Schiroishi T, Miller LH (1981) Identification of surface proteins on viable *Plasmodium knowlesi* merozoites. J Protozool 78:160–164

Jungery M, Boyle D, Patel T, Pasvol G, Weatherall DJ (1983a) Lectin-like polypeptides of *P. falciparum* bind to red cell sialoglycoproteins. Nature 301:704–705

Jungery M, Pasvol G, Newbold CI, Weatherall DJ (1983b) A lectin-like receptor is involved in the invasion of red cells by *P. falciparum*. Proc Natl Acad Sci USA 80:1018–1022

Kaushal DC, Carter R, Howard RJ, McAuliffe FM (1983a) Characterisation of antigens on mosquito midgut stages of *Plasmodium gallinaceum*. 1. Zygote surface antigens. Mol Biochem Parasitol 8:53–69

Kaushal DC, Carter R, Rener J, Grotendorst CA, Miller LH, Howard RJ (1983b) Monoclonal antibodies against surface determinants on gametes of *Plasmodium gallinaceum* block transmission of malaria parasites to mosquitos. J Immunol 131:2557–2562

Kemp DJ, Coppel RL, Cowman AF, Saint RB, Brown GV, Anders RF (1983) Expression of *Plasmodium falciparum* blood stage antigens in *Escherichia coli*: detection with antibodies from immune humans. Immunology 80:3783–3791

Kilejian A (1974) A unique histidine-rich polypeptide from the malaria parasite *Plasmodium lophorae*. J Biol Chem 249:4650–4655

Kilejian A (1978) Does a histidine rich protein from *Plasmodium lophurae* have a function in merozoite penetration. J Protozool 23:272–277

Kilejian A (1978) Histidine rich protein as a model malaria vaccine. Science 201:922–924

Kilejian A (1979) Characterization of a protein correlated with the production of knob like protusion on membranes of erythrocytes infected with *Plasmodium falciparum*. Proc Natl Acad Sci USA 76:4650–4653

Kilejian A (1980) Stage specific proteins and glycoproteins of *Plasmodium falciparum* identification of antigens unique to schizonts and merozoites. Proc Natl Acad Sci USA 77:3695-3699

Kilejian A (1981) *Plasmodium lophurae*: immunogenicity of a histidine rich protein. Exp Parasitol 52:291

Langreth SG, Reese RT (1979) Antigenicity of the infected erythrocyte and merozoite surfaces in falciparum malaria. J Exp Med 150:1241-1254

Langreth SG, Reese RT, Motyl MR, Trager W (1979) Loss of knobs on the infected erythrocyte surface after long term culture. Exp Parasitol 48:213-219

Laveran A (1880) Note sur un nouveau parasite trouvé dans le sang de plusiers malades attients de fievre palustre. Bull Med Soc Med Hosp Paris (2nd set) 17:158-164

Le Bras J, Ricour A, Andrieu B, Savel J, Couland JP (1983) *Plasmodium falciparum*: drug sensitivity *in vitro* of isolates before and after adaption to continuous culture. Exp Parasitol 56:9-14

Liao TH, Gallop PM, Blumenfeld OO (1973) Modification of the sialyl residues of sialoglycoproteins of the human erythrocyte surface. J Biol Chem 248:8247-8253

Lupski JR, Ozaki LS, Ellis J, Godson GN (1983) Localisation of a *Plasmodium* surface antigen epitope by Tn5 mutagenesis mapping of a recombinant cDNA clone. Science 220:1285-1287

Luse SA, Miller LH (1971) *Plasmodium falciparum* malaria: ultrastructure of parasitised erythrocytes in cardiac vessels. Am J Trop Med Hyg 20:655-660

Luzzatto L (1974) Genetic factors in malaria. Bull WHO 50:195-202

Markwell MAK, Fox CF (1978) Surface-specific iodination of membrane proteins of viruses and eucaryotic cells using 1, 3, 4, 6 tetrachloro-3a-6a diphenylglycoluril. Biochemistry 17:4807-4817

McBride JS, Walliker D, Morgan G (1982) Antigenic diversity in the human malaria parasite *Plasmodium falciparum*. Science 217:254-257

McCarthy VC, Clyde DF (1977) *Plasmodium vivax*: correlation of circumsporozoite precipitation (CSP) reaction with sporozoite-induced protective immunity in man. Exp Parasitol 41:167-171

McLean SA, Pearson CD, Phillips RS (1982) *Plasmodium chabaudi*: evidence of antigenic variation by the parasite during recrudescent parasitaemias in mice. Exp Parasitol 54:296-302

McDonald V, Hannon M, Tanigoshi L, Sherman IW (1981) *Plasmodium lophurae*: immunisation of Peking ducklings with different antigen preparations. Exp Parasitol 51:195-203

Mendis KN, Targett GAT (1979) Immunisation against gametes and asexual erythrocytic stages of a rodent malaria parasite. Nature 277:389-391

Miller LH, Carter R (1976) Innate resistance in malaria. Exp Parasitol 40:132-146

Miller LH, Johnson JG, Schmidt-Ullrich R, Haynes J, Wallach DFH, Carter R (1980) Determinants on surface proteins of *Plasmodium knowlesi* merozoites common to *Plasmodium falciparum* schizonts. J Exp Med 151:790-798

Mitchell GH, Butcher GA, Cohen S (1975) Merozoite vaccination against *Plasmodium knowlesi* malaria. Immunology 29:397-407

Mitchell GH, Butcher GA, Langhorne J, Cohen S (1977) A freeze-dried merozoite vaccine effective against *Plasmodium knowlesi* malaria. Clin Exp Immunol 28:276-279

Molineaux L, Gramiccia G (1980) The Garki project. WHO, Geneva

Nardin EH, Nussenzweig RS (1978) Stage specific antigens on the surface membrane of sporozoites of malaria parasites. Nature 274:55-57

Nardin E, Gwadz RW, Nussenzweig RS (1979a) Characterisation of sporozoite surface antigens by indirect immunofluorescence: detection of stage and species specific antimalarial antibodies. Bull WHO [Suppl] 1:211-217

Nardin EH, Nussenzweig RS, McGregor IA, Bryan JH (1979b) Antibodies to sporozoites: their frequent occurrence in individuals living in an area of hyperendemic malaria. Science 206:597-599

Nardin EH, Nussenzweig V, Nussenzweig RS, Collins WE, Harmasita KT, Tapchairsm P, Chomcham Y (1982) Circumsporozoite proteins of human malaria parasites. *Plasmodium falciparum* and *Plasmodium vivax*. J Exp Med 156:20-30

Newbold CI, Boyle DB, Smith CC, Brown KN (1982a) Stage specific protein and nucleic acid synthesis during the asexual cycle of the rodent malaria *Plasmodium chabaudi*. Mol Biochem Parasitol 5:33-44

Newbold CI, Boyle DB, Smith CC, Brown KN (1982b) Identification of a schizont and species specific surface glycoprotein on erythrocytes infected with rodent malarias. Mol Biochem Parasitol 5:45-54

Newbold CI, Schryer M, Boyle DB, McBride JS, McLean A, Wilson RJM, Brown KN (1984) A possible molecular basis for strain specific immunity to malaria. Mol Biochem Parasitol 11:337-347

Nussenzweig RS, Chen D (1974) The antibody response to sporozoites of simian and human malaria parasites, its stage and species specificity and strain cross reactivity. Bull WHO 50:293-294

Nussenzweig RS, Vanderberg J, Most H, Orton C (1967) Protective immunity produced by the injection of x-irradiated sporozoites of *Plasmodium berghei*. Nature 216:160–162

Nussenzweig RS, Vanderberg J, Most H (1969a) Protective immunity produced by the injection of x-irradiated sporozoites of *Plasmodium berghei*. IV. Dose response, specificity and humoral immunity. Milit Med [Suppl] 134:1176–1182

Nussenzweig RS, Vanderberg JP, Most H, Orton C (1969b) Specificity of protective immunity produced by x-irradiated *Plasmodium berghei* sporozoites. Nature 222:488–489

Nussenzweig RS, Montuori W, Spitalny GL, Chen D (1973) Antibodies against sporozoites of human and simian malaria produced in rats. J Immunol 110:600–601

Oka M, Aikawa M, Holder AA, Freeman RR (1983) Ultrastructural localisation of protective antigens of *Plasmodium yoelii* by the use of monoclonal antibodies and ultrathin cryomicrotomy. Paper presented at the 2nd international conference on malaria and babesiosis, Annecy, France 19–22 September

Orjih AV, Cochrane AH, Nussenzweig RS (1981) Active immunisation and passive transfer of resistance against sporozoite induced malaria in infant mice. Nature 291:331–333

Oxbrow AI (1973) Strain specific immunity to *Plasmodium berghei*: a new genetic marker. Parasitology 67:17–27

Ozaki LS, Svec P, Nussenzweig RS, Nussenzweig V, Godson GN (1983) Structure of the *Plasmodium knowlesi* gene coding for the circumsporozoite protein. Cell 34:815–822

Padua RA (1981) *Plasmodium chabaudi*. Genetics of resistance to chloroquine. Exp Parasitol 52:419–426

Pasvol G, Jungery M, Weatherall DJ (1982) Glycophorin as a possible receptor for *Plasmodium falciparum*. Lancet 2:947–950

Patarapotikul J, Tharavarij S, Poothang C (1983) Multiple strains of *Plasmodium falciparum* are necessary for the growth inhibition assay. Southeast Asian J Trop Med Public Health 14:149–153

Perkins M (1981) Inhibitory effects of erythrocyte membrane proteins on the *in vitro* invasion of the human malarial parasite (*Plasmodium falciparum*) in its host cell. J Cell Biol 90:563–567

Perkins M (1982) Surface proteins of schizont infected erythrocytes and merozoites of *Plasmodium falciparum*. Mol Biochem Parasitol 5:55–64

Perrin LH, Dayal R (1982) Immunity to asexual erythrocytic stages of *Plasmodium falciparum*: role of defined antigens in the humoral response. Immunol Rev 61:245–270

Perrin LH, Dayal R, Rieder H (1981a) Characterisation of antigens from erythrocytic stages of *Plasmodium falciparum* reacting with human immune sera. Trans R Soc Trop Med Hyg 75:163–165

Perrin LH, Ramirez E, Lambert PH, Miescher PA (1981b) Inhibition of *P. falciparum* growth in human erythrocytes by monoclonal antibodies. Nature 289:301–303

Potocnjak P, Yoshida N, Nussenzweig RS, Nussenzweig V (1980) Monovalent (Fab) fragments of monoclonal antibodies to a sporozoite surface antigen (Pb44) protect mice against malaria infection. J Exp Med 151:1504–1513

Powell RD, McNamara J, Rieckman KH (1972) Clinical aspects of acquisition of immunity to falciparum malaria. Proc Helminthol Soc Wash 39:51–66

Reese RT, Trager W, Jensen JB, Miller DA, Tantravaki R (1978) Immunisation against malaria with antigen from *Plasmodium falciparum* cultivated *in vitro*. Proc Natl Acad Sci USA 75:5665–5669

Rener J, Carter R, Rosenberg Y, Miller LH (1980) Anti gamete monoclonal antibodies synergistically block transmission of malaria by preventing fertilisation in the mosquito. Proc Natl Acad Sci USA 77:6797–6799

Rener J, Graves PM, Carter R, Williams JL, Burkot TR (1983) Target antigens of transmission blocking immunity on gametes of *Plasmodium falciparum*. J Exp Med 158:976–981

Rieckmann KH, Carson PE, Beaudoin RL. Cassells JS, Sell KW (1974) Sporozoite induced immunity in man against an Ethiopian strain of *Plasmodium falciparum*. Trans R Soc Trop Med Hyg 68:258–259

Rifkin DB, Compans RW, Reich E (1972) A specific labelling procedure for proteins on the outer surface of membranes. J Biol Chem 247:6432–6437

Rosario VE (1976) Genetics of chloroquine resistance in malaria parasites. Nature 261:585–586

Sadun EH, Hickman RL, Wellde BT, Moon AP, Udeozo OK (1966) Active and passive immunisation of chimpanzees infected with West African and Southeast Asian strains of *Plasmodium falciparum*. Milit Med [Suppl] 131:1250–1262

Sanderson A, Walliker D, Molez J-F (1981) Enzyme typing of *Plasmodium falciparum* from some African and other old world countries. Trans R Soc Trop Med Hyg 75:263–267

Santoro F, Cochrane AH, Nussenzweig V, Nardin EH, Nussenzweig PS, Gwadz RW, Ferreira A (1983)

Structural similarities among the protective antigens of sporozoites from different species of malaria parasites. J Biol Chem 257:3341-3345

Schmidt JA, Udeinya IJ, Leech JH, Hay RJ, Aikawa M, Barnwell J, Green I, Miller LH (1982) *Plasmodium falciparum* malaria. An amelanotic melanoma cell line bears receptors for the knob ligand on infected erythrocytes. J Clin Invest 7:379-386

Schmidt-Ullrich R, Wallach DFH (1978) *Plasmodium knowlesi*-induced antigens in membranes of parasitised rhesus monkey erythrocytes. Proc Natl Acad Sci USA 10:4949-4953

Schmidt-Ullrich R, Wallach DFH, Lightholder J (1979) Two *Plasmodium knowlesi* specific antigens on the surface of schizont infected rhesus monkey erythrocytes induce antibody production in immune hosts. J Exp Med 150:86-99

Schmidt-Ullrich R, Miller LH, Wallach DFH, Lightholder J, Powers KG, Gwadz RW (1981) Rhesus monkeys protected against *Plasmodium knowlesi* malaria produce antibodies against a 65,000 Mr *P. knowlesi* glycoprotein at the surface of infected erythrocyte. Infect Immun 34:519-525

Schmidt-Ullrich R, Miller LH, Wallach DFH, Lightholder J (1982) Immunogenic antigens common to *Plasmodium knowlesi* and *Plasmodium falciparum* are expressed on the surface of infected erythrocytes. J Parasitol 68:185-193

Schmidt-Ullrich R, Lightholder J, Monroe TM (1983) Protective *Plasmodium knowlesi* Mr 74,000 antigen in membranes of schizont infected rhesus monkey erythrocytes. J Exp Med 158:146-158

Schofield L, Saul A, Myler P, Kidson C (1982) Antigenic differences among isolates of *Plasmodium falciparum* demonstrated by monoclonal antibodies. Infect Immun 38:893-897

Sherman IW (1981) *Plasmodium lophurae*: protective immunogenicity of the histidine-rich protein? Exp Parasitol 52:292-295

Siddiqui WA (1977) An effective immunisation of experimental monkeys against a human malaria parasite *Plasmodium falciparum*. Science 197:388-389

Siddiqui WA, Taylor DW, Kan S-C, Kramer K, Richmond-Grum SM, Kotani S, Shiba T, Kusumota S (1978) Vaccination of experimental monkeys against *Plasmodium falciparum*: a possible safe adjuvant. Science 201:1237-1239

Sinden RE (1983) The cell biology of sexual development in *Plasmodium*. Parasitology 86:7-28

Spittalny GL, Nussenzweig RS (1973) *Plasmodium berghei*. Relationship between protective immunity and anti-sporozoite (CSP) antibody in mice. Exp Parasitol 33:168-178

Stoenner HG, Dodd T, Larsen C (1982) Antigenic variation of *Borrelia hermsii*. J Exp Med 156:1297-1311

Tait A (1981) Analysis of protein variation in *Plasmodium falciparum* by two dimensional gel electrophoresis. Mol Biochem Parasitol 2:205-218

Taliaferro WH (1949) Immunity to the malaria infect. In: Boyd MF (ed) Malariology, vol II. Saunders, Philadelphia, pp 835-965

Trager W, Rudzinska MA, Bradbury PC (1966) The fine structure of *Plasmodium falciparum* and its host erythrocytes in natural malarial infections in man. Bull WHO 35:883-885

Trager W, Lanners HN, Stanley HA, Langreth SG (1983) Immunisation of owl monkeys to *Plasmodium falciparum* with merozoites from cultures of a knobless clone. Parasite Immunol 5:225-236

Udeinya IJ, Schmidt JA, Aikawa M, Miller LH, Green I (1981) Falciparum malaria-infected erythrocytes specifically bind to cultured human endothelial cells. Science 213:555-557

Udeinya IJ, Miller LH, McGregor IA, Jenson JB (1983) *Plasmodium falciparum* strain-specific antibody blocks binding of infected erythrocytes to amelanotic melanoma cells. Nature 303:429-431

Vanderberg JP, Nussenzweig RS, Most H (1968) Further studies on the *Plasmodium berghei - Anopholes stephensi* rodent system of mammalian malaria. J Parasitol 54:1009-1016

Vanderberg J, Nussenzweig RS, Most H (1969) Protective immunity produced by the injection of x-irradiated sporozoites of *Plasmodium berghei* V. *In vitro* effects of mice serum on sporozoites. Milit Med [Suppl] 134:1183-1190

Vickerman K (1974) Antigenic variation in African trypanosomes. Ciba Foundation Symposium 25. Elsevier Holland, Amsterdam, pp 53-80

Voller A, Rossan RN (1969) Immunological studies with simian malarias: 1. Antigenic variation of *Plasmodium cynomolgi bastianelli*. Trans R Soc Trop Med Hyg 63:46-56

Wallach DFH, Conley M (1977) Altered membrane proteins of monkey erythrocytes infected with simian malaria. J Mol Med 2:119-136

Walliker D (1983) The contribution of genetics to the study of parasitic protozoa. Research Studies Press, Letchworth, Herts, England

Walliker D, Carter R, Morgan S (1971) Genetic recombination in malaria parasites. Nature 232:561-562

Walliker D, Carter R, Sanderson A (1975) Genetic studies on *Plasmodium chabaudi*: recombination between enzyme markers. Parasitology 70:19–24

Ward RA, Hayes DE (1972) Attempted immunisation of rhesus monkeys against cynomolgi malaria with irradiated sporozoites. Proc Helminthol Soc Wash (special issue) 311:525–529

Weiss MW, Oppenheim JD, Vanderberg JP (1981) *Plasmodium falciparum*: assay *in vitro* for inhibitors of merozoite penetration of erythrocytes. Exp Parasitol 51:400–407

Wellde BT, Diggs CL (1978) *Plasmodium berghei*: biological variation in immune serum treated mice. Exp Parasitol 44:197–201

Wellde BT, Diggs CL, Anderson S (1979) Immunization of *Aotus trivirgatos* against *Plasmodium falciparum* with irradiated blood forms. Bull WHO [Suppl 1] 57:153–157

Wery M, Weyn J, Timperman G, Hendrix L (1979) Observations on the virulence and the antigenic characters of cloned and uncloned lines of the ANKA isolate of *Plasmodium berghei*. Ann Soc Belge Med Trop 59:347–360

Wilson RJM (1980) Serotyping *Plasmodium falciparum* malaria with S-antigens. Nature 284:451–452

Wilson RJM, Phillips RS (1976) Method to test inhibitory action of human sera to wild populations of *Plasmodium falciparum*. Nature 263:132–134

Wilson RJM, McGregor JA, Hall P, Williams K, Bartholomew R (1969) Antigens associated with *Plasmodium falciparum* infections in man. Lancet II:201–205

Wilson RJM, McGregor IA, Hall PJ (1975) Persistence and recurrence of S-antigens in *Plasmodium falciparum* infections in man. Trans R Soc Trop Med Hyg 64:460–467

Winchell E, Ling IT, Wilson RJM (1984) Metabolic labelling and characterisation of S-antigens, heat stable, strain specific antigens of *Plasmodium falciparum*. Mol Biochem Parasitol 10:287–296

Yoshida N, Nussenzweig RS, Potocnjak P, Nussenzweig V, Aikawa M (1980) Hybridoma produces protective antibodies directed against the sporozoite stage of malaria parasite. Science 207:71–73

Yoshida N, Potocnjak P, Nussenzweig RS, Nussenzweig V (1981) Biosynthesis of Pb44, the protective antigen of sporozoite of *Plasmodium berghei*. J Exp Med 154:1225–1235

Zavala F, Gwadz RW, Collins FH, Nussenzweig RS, Nussenzweig V (1982) Monoclonal antibodies to circumsporozoite proteins identify the species of malaria parasite in infected mosquitos. Nature 299:737–738

Zavala F, Cochrane AH, Nardin EH, Nussenzweig RS, Nussenzweig V (1983) Circumsporozoite proteins of malaria parasites contain a single immunodominant region with two or more identical epitopes. J Exp Med 157:1947–1957

Toxoplasmosis: The Need for Improved Diagnostic Techniques and Accurate Risk Assessment

H.P.A. Hughes

1 Introduction

Toxoplasma gondii is a coccidian parasite of cats (HUTCHISON et al. 1971a, b) which has an extremely wide range of secondary hosts. With the advent of efficient screening methods, there is an increasing awareness of the organism as an important pathogen in human and veterinary medicine. The parasite has a wide geographical distribution, with almost 500 million people worldwide showing serological evidence of infection.

Department of Immunology, St. George's Hospital Medical School, Cranmer Terrace, London SW17 0RE, United Kingdom

Present address: United States Department of Agriculture, Agricultural Research Service, Animal Parasitology Institute, Building 1040, BARC-East, Beltsville, MD 20705, USA

Current Topics in Microbiology and Immunology, Vol. 120
© Springer-Verlag Berlin · Heidelberg 1985

T. gondii exists in two distinct cycles: an enteroepithelial sexual cycle in cats and an extraintestinal cycle in most warm-blooded animals.

There are three forms of *Toxoplasma:* the tachyzoite (or proliferative form), the tissue cyst and the oocyst. The tachyzoite is present during the acute stages of infection and invades all types of mammalian cell except perhaps the anucleate red blood cell. Following invasion (usually in the gastrointestinal tract), parasites multiply locally and then are carried to other sites via the reticuloendothelial system, either as free parasites or in infected cells.

Tissue cysts are formed after repeated endodyogeny of the invading organism within the host cell cytoplasm. The cyst wall is then formed, with the disruption of the host cell. The cyst may contain up to several thousand parasites. Cysts are demonstrable as early as 8 days after infection, and persist for the life of the host. They may be present in almost every organ, though the brain and heart or skeletal muscles appear to be favoured sites.

Following invasion of the cat intestinal mucosa by tachyzoites, a typical coccidian sexual cycle occurs, with schizogony, micro- and macrogamogony and ultimately the formation of oocysts. During the acute stages of infection, thousands of oocysts may be shed in faeces each day. Sporogony then occurs with the formation of two sporocysts, each containing four sporozoites.

Oocysts and tissue cysts are the principal modes of transmission to man and animals. Congential toxoplasmosis occurs through the passage of extra- or intracellular parasites across the placenta, during the acute stage of maternal infection (Fig. 1).

The symptoms of toxoplasmosis often mimic those of other more common diseases, although the manifestations in adults are variable and include symptomatic and asymptomatic lymphadenopathy, myocarditis, pericarditis, hepatitis, encephalitis, myalgia and maculopapular rash (REMINGTON and DESMONTS 1976). The lymphadenopathic form of toxoplasmosis is recognised as the most common clinical manifestation of the disease in adults (REMINGTON 1974; PALMER 1974). Since some cases may exactly mimic the haematological picture of infectious mononucleosis, accurate diagnosis of toxoplasmosis is essential for differential diagnosis.

Although an acquired infection may persist for the life of the host, recurrences of acute toxoplasmosis are rare. However, in patients who are immunocompromised either due to an underlying disease [such as Hodgkin's disease, non-Hodgkin's lymphoma, systemic lupus erythromatosus (SLE) and AIDS] or as a result of antitumour therapy or chemical immunosuppression, fatal dissemination of encysted organisms may occur (LUNA and LICHTIGER 1971; REMINGTON 1982; KRAHENBUHL and REMINGTON 1982; GRANSDEN and BROWN 1983). Accurate and rapid diagnosis of an active *Toxoplasma* infection in these cases is essential for patient management and prognosis.

Since the first description of congenital toxoplasmosis, *Toxoplasma* has become recognised as an important pathogen during pregnancy and the perinatal period. Recent seroepidemiological surveys have suggested that the incidence of congenital toxoplasmosis is approximately one in every thousand births (STRAY-PEDERSEN and LORENTZEN-STYR 1979; STRAY-PEDERSEN 1980), and the annual cost for the aftercare of children and adults with the sequelae of congenital toxoplasmosis is in the hundreds of millions of dollars (KRAHENBUHL and REMINGTON 1982). Congenital toxoplasmosis probably also leads to severe economic loss in the agricultural

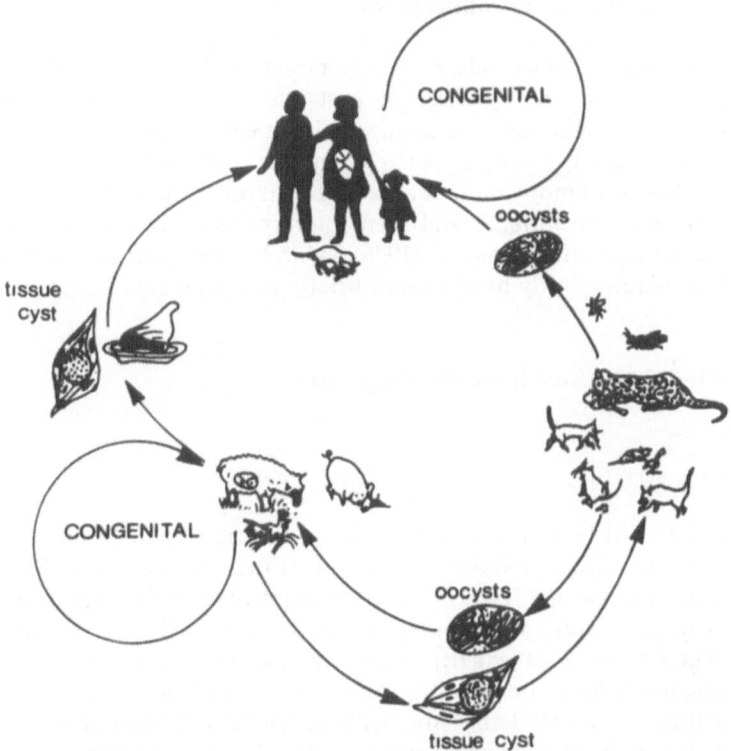

Fig. 1. Transmission of toxoplasmosis. The primary modes of transmission to man are through the ingestion of oocysts or tissue cysts. The transmission to domestic animals (particularly herbivores) is probably through the ingestion of oocysts on contaminated pasture. This cycle may be perpetuated by feeding cats infected meat. Tachyzoite transmission may occur through the drinking of infected milk, but is of prime importance in transplacental infections which can lead to severe congenital toxoplasmosis in man and animals. There is almost certainly a large reservoir of infected wild felids and secondary hosts

industry, particularly in sheep and pig farming. The cost of screening large numbers of farm animals by present methods is prohibitive, so an accurate assessment of the economic consequences of toxoplasmosis is not available.

The majority of research on toxoplasmosis has been directed towards improving diagnostic methods and elucidating the biology and immunology of intracellular parasitism. Although a large number of serodiagnostic tests have been described for toxoplasmosis, laboratories throughout the world still ultimately rely on the Toxoplasma dye exclusion test of SABIN and FELDMAN (1948). Accurate epidemiological information and policy decisions on vaccination programmes in both man and animals will only be possible when serological methods become rapid, reliable and specific.

2 Diagnosis of Toxoplasmosis

The diagnosis of toxoplasmosis can be established by three different criteria: (a) the demonstration of parasites or cysts in tissues by conventional histological or immunocytochemical techniques following biopsy, (b) by subinoculation of infected tissues (usually biopsy or postmortem material) into experimental animals or (c) by serological demonstration of specific antibody or antigens in serum and body fluids. Serological techniques and their interpretation are comprehensively discussed in REMINGTON and DESMONTS (1976) and in KRAHENBUHL and REMINGTON (1982), and will therefore only be discussed briefly in the present paper.

2.1 Existing Serodiagnostic Procedures

2.1.1 Dye Test

Since its description in 1948, the dye test has been standardised (BEVERLEY and BEATTIE 1958), adapted to a microscale (FELDMAN and LAMB 1966; WALDELAND 1976; BALFOUR et al. 1982) and remains the definitive test for toxoplasmosis. The test relies on live parasites and an abundant supply of normal human serum, and requires a high degree of technical expertise both for its safe execution and accurate interpretation. The test is based on the observation that when living organisms from the peritoneal exudate of infected mice are incubated in the presence of immune serum and complement (SCHREIBER and FELDMAN 1980), immune lysis occurs and the parasites are unable to incorporate the vital stain methylene blue. Parasites exposed to non-immune serum under identical conditions appear deeply stained when observed microscopically. The titre reported is the dilution at which 50% of the parasites remain stained and, by comparison with a WHO international reference serum (LYNG and SIIM 1982), may be expressed in international units per millilitre.

2.1.2 Indirect Fluorescent Antibody Test

The conventional indirect fluorescent antibody test (IFA) uses whole *Toxoplasma* organisms, either fixed or dried onto slides, which are incubated with serial dilutions of test serum. Antibody is detected using a fluorescent-labelled antispecies IgG or whole immunoglobulin and viewed by fluorescent microscopy. Numerous workers have found quantitative agreement between the conventional IFA and the dye test and the test is usually specific (WALTON et al. 1966; SULZER and HALL 1967; KRAMER et al. 1970).

The IgM IFA has been used successfully in the diagnosis of congenital toxoplasmosis (REMINGTON et al. 1968a, b; KARIM and LUDLAM 1975a). The methodology of the test is identical to that of the conventional IFA, except that a fluorescent-labelled antispecies IgM is used for the detection of *Toxoplasma*-specific IgM. The test was developed for the early detection of congenital and acute acquired toxoplasmosis, in which the presence of IgM signifies an active neonatal response or recently acquired infection (KARIM and LUDLAM 1975a, b; CAMARGO et al. 1978b; KRAHENBUHL and REMINGTON 1982). Nonspecific (false-positive) reactions may occur in patients who have high levels of antinuclear antibody (ANA) (for example, in SLE), antiglobulin antibodies or high levels of anti-*Toxoplasma* IgG. These reactions may be avoided by separation of IgM from the serum sample by using a small gel filtration column (ORDONEZ et al. 1982) or by absorption of sera with heat-aggregated IgG (HYDE et al. 1975).

2.1.3 Agglutination Tests

The indirect haemagglutination test (IHA) was first described by JACOBS and LUNDE in 1957, using tanned red blood cells sensitised with *Toxoplasma* antigen prepared by water lysis. Titres in the IHA may lag by several weeks or months behind those determined by the dye test or IFA, but tend to remain at high levels even longer than the dye test. For this reason, the IHA is not satisfactory as a screening test during pregnancy (CAMARGO and LESER 1976; CAMARGO et al. 1976; KARIM and LUDLAM 1975a). Positive titres in the IHA have also been detected in patients with a negative dye test (BALFOUR et al. 1980).

2.1.4 Enzyme-Linked Immunosorbent Assay

Unlike the tests described above, the enzyme-linked immunosorbent assay (ELISA) is rapid, requires relatively little technical expertise, can be assessed objectively and lends itself to automation. ELISA has been used for detection of specific IgG (VAN LOON and VAN DER VEEN 1980; CARLIER et al. 1980; HUGHES et al. 1982), IgM (CAMARGO et al. 1978a; MINEO et al. 1980; NAOT et al. 1981) and total specific antibody (LIN et al. 1980; MILATOVIC and BRAVENY 1980). However, despite the plethora of studies which have appeared in recent years, the qualitative and quantitative agreement between the ELISA and more established diagnostic tests varies between laboratories. Good agreement between the ELISA and conventional tests has been reported in some instances (WALLS et al. 1977; LIN et al. 1980; VAN LOON and VAN DER VEEN 1980) but not in others (BALSARI et al. 1980; MILATOVIC and BRAVENY 1980).

NAOT and REMINGTON (1980) recently developed a double-sandwich IgM ELISA for the detection of *Toxoplasma* antibodies. The methodology of this test relies on the isolation of IgM in situ on the test plate, so the need for prior separation of IgM is avoided. This variation of the ELISA is also more specific and sensitive than a conventional ELISA or IFA (WIELAARD et al. 1983). VAN LOON et al. (1983) have also reported a specific and sensitive ELISA which used enzyme-labelled antigen and a similar capture technique to that described for the double-sandwich IgM ELISA.

2.1.5 Serological Spectrum of Human Toxoplasmosis

Antibody as assessed by the dye test and conventional (IgG and IgM or IgG alone) IFA antibodies generally appear within 1–2 weeks after infection and can rise rapidly over the ensuing weeks to titres ≥ 1:1000. Antibody titre has no bearing on the severity of the illness and asymptomatic cases may have titres ≥ 1:32 000 or as low as 1:64. High titres can persist for years and then decline. The dye test in man is unusually specific; false-positive results due to antigenic cross-reactions are unknown.

Complement-fixing antibodies have been studied in acquired toxoplasmosis and generally appear later than dye test-reactive antibodies, in common with the IHA. Therefore these tests may only have a useful role in assessing the stage of infection when dye test and IgG-IFA titres have already stabilised (CAMARGO et al. 1976; WELCH et al. 1980; KARIM and LUDLAM 1975b; KRAHENBUHL and REMINGTON 1982). The Sabin-Feldman dye test, indirect fluorescent antibody test (IFA), indirect haemagglutination test (IHA) and direct agglutination test are the methods used most widely in *Toxoplasma* serodiagnosis. However, the latex agglutination test and ELISA have recently become available and are now used in parallel with conventional serological tests in both routine and research laboratories.

Serological tests using soluble antigen rather than intact parasites have been reported in a variety of systems by many workers. In the majority of cases, the antigens used were prepared by simple extraction methods such as water lysis, freezing and thawing or sonic disruption of *Toxoplasma* tachyzoites (VOLLER et al. 1976; WALLS et al. 1977; VON LOON and VAN DER VEEN 1980). In these instances, membrane antigens can only be present in the test system if (a) they are easily soluble in water or buffer [e.g. certain polysaccharides (PANDE et al. 1961) or proteins (JOHNSON et al. 1983a)] or (b) if the final centrifugation step in the procedure leaves small membrane fragments suspended in the solution (BJERRUM 1977). Variation in the amount of fragmented membranes in such preparations may explain why some workers obtain good agreement with the dye test or IFA (WALLS et al. 1977; CAMARGO et al. 1978a; LIN et al. 1980), while others do not (LUNDE and JACOBS 1967; BALFOUR et al. 1980).

The ELISA must become the method of choice for all future diagnosis of toxoplasmosis. The replacement of the dye test with a reproducible and accurate ELISA is regarded by many workers as the principal step towards improving diagnostic capabilities. However, tests which detect IHA or complement fixation test (CFT) reactive antibody are also useful in indicating the infection status in patients when no follow-up sera are available.

3 Intracellular Antigens of *Toxoplasma* and Their Putative Role in Diagnosis

Early studies on the kinetics of antibody formation using the dye test and the IHA led to the hypothesis that these tests were detecting different antigens (KARIM and LUDLAM 1975a). It is generally accepted that the dye test measures antibody against the membrane, leading to immune lysis and parasite killing. It is thought that the release of intracellular components then stimulates the production of antibodies that are reactive in tests which use simple aqueous extracts. Evidence for this hypothesis has been gathered by using immunoabsorption, serological and immunochemical techniques.

In 1961, FLECK observed that dye test, IHA and CFT activities could be removed from human sera by absorption with live organisms of the virulent RH strain of *T. gondii* grown in mice.

Absorption of the same serum with red cells coated with IHA antigen prepared from the soluble fraction from water-lysed organisms (JACOBS and LUNDE 1957) removed the IHA activity, but the dye test and CFT activity remained intact. Later studies (FLECK and PAYNE 1963) suggested that there were "heavy" and "light" antigens which were specific to the dye test and IHA respectively. Studies on water lysates of the RH strain showed that *Toxoplasma* membrane ghosts elicited a dye test response in rabbits but no IHA response. Antigen recovered in the supernatants of a low-speed centrifugation elicited both dye test and IHA antibodies, whereas the high-speed supernate stimulated the formation of predominantly IHA antibodies (LUNDE and JACOBS 1967).

Analysis of water-lysed *Toxoplasma* by density gradient centrifugation and gel filtration estimated the IHA activity to be present in both 135 000 and 800 000 mw fractions respectively (FUJITA et al. 1969). More recently, HUGHES and BALFOUR (1981) studied *Toxoplasma* antigens prepared from water-lysed, sonicated or frozen-thawed parasites by gel filtration, isoelectric focussing and two-dimensional immunoelectrophoresis (2-DIEP). All three preparations from *Toxoplasma* yielded between two and three antigens (antigens 4, 5 and 6; Fig. 2), in agreement with pre-

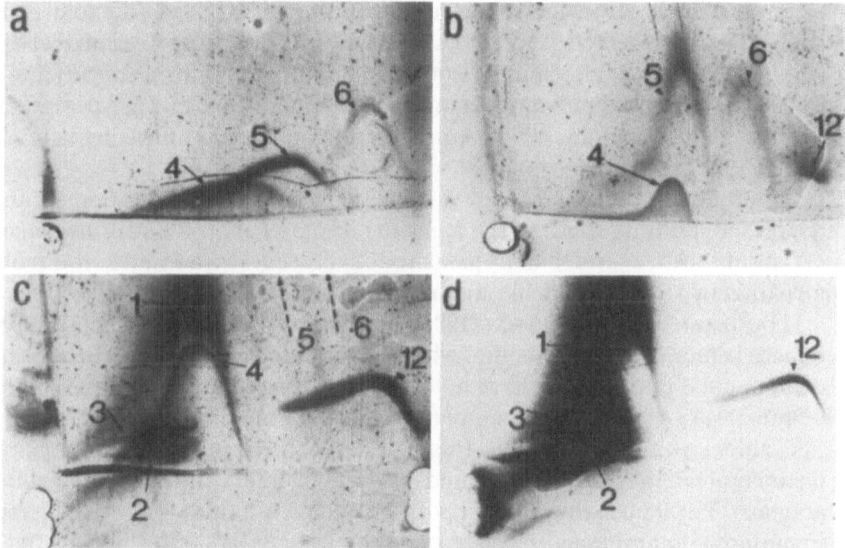

Fig. 2a–d. Immunoelectrophoresis of **a** a freeze-thaw preparation. Three antigens are present after Coomassie-blue staining (*4, 5, 6*). **b** Tween-ether preparation. Four antigens (*4, 5, 6, 12*) are present. **c** Electrophoretic pattern of a ^{125}I-membrane-labelled preparation solubilised using the detergents saponin and octyl glucoside. Antigens *1, 2, 3, 4, 5, 6* and *12* are present. The autoradiograph **d** shows that antigens *1, 2, 3* and *12* are of membrane origin. Antigens *4* and *5* are present as components of circulating antigen during acute toxoplasmosis. In this figure, the first-dimension anode is to the right and the second-dimension anode is to the top. (From HUGHES et al. 1982)

vious gel-diffusion studies (CHORDI et al. 1964). Antigen-sensitised cells can be prepared by heating fixed red cells and antigen together at 56 °C. Only antigen 6 was found to be heat stable at 56 °C in these preparations, which suggested that it could be a major component in the IHA. This antigen had a mw of 10 000 estimated by gel filtration and identified as a protein using a multiple-staining technique (HUGHES and BALFOUR 1981; HUGHES et al. 1980). Later studies showed that antigens 4, 5 and 6 were not labelled by tachyzoite surface ^{125}I-iodination (HUGHES et al. 1982) nor were they retarded in a hydrophobic gel (HUGHES and VAN KNAPEN 1982). These results implied that these antigens were of intracellular origin, with little or no membrane association.

Analysis of *Toxoplasma* polypeptides from sonicated organisms using sodium dodecyl sulphate polyacrylamide gel electrophoresis (SDS-PAGE) and ^{125}I-concanavalin A (Con-A) binding studies has detected nine intracellular polypeptides of parasite origin. The molecular weight range of these polypeptides was estimated as between 20 000 and 98 000, and none were glycosylated (JOHNSON et al. 1981b). Iodination of a similar ultrasonic preparation and subsequent precipitation by immune mouse serum showed that ten polypeptides were antigenic.

With the advent of the lymphoid cell hybridisation technique descibed by KOHLER and MILSTEIN (1975), monoclonal antibodies have been developed as powerful tools for probing the antigenic structure of *Toxoplasma*. JOHNSON et al. (1981a) described two monoclonal antibodies which reacted in the IHA. One of these, FMC 19, reacted strongly in the IFA and was thought to be detecting an antigen that is a water-extractable *Toxoplasma* membrane component. This was confirmed by immuno-electron microscopy, which showed that this antibody reacted exclusively with the parasite membrane (JOHNSON et al. 1983a). FMC 20 reacted exclusively in the IHA and appeared to be detecting a cytoplasmic antigen. This antibody precipitated a 98 000 mw component from a cytoplasmic preparation, but did not precipitate any peptides or glycopeptides from a membrane preparation of the parasite (JOHNSON et al. 1983c).

Immunoelectron microscopy showed that this antibody also detects antigens that are present on the surface of the parasite (JOHNSON et al. 1981a; JOHNSON et al. 1983a, c). The authors suggest that the parasite actively secretes a small amount of the 98 000 mw antigen onto the parasite membrane.

HANDMAN and REMINGTON (1980a) and SETHI and BRANDIS (1981) have also described six monoclonal antibodies which are directed against *Toxoplasma* cytoplasmic components. The antigens have not as yet been adequately characterised, so it would be premature to speculate on any role that they may have in diagnosis.

Table 1 summarises those intracellular components which appear to have a distinct role in the IHA or associated tests which detect antibody against intracellular antigens. The 98 000 mw antigen polypeptide precipitated by FMC 20 appears to be important in membrane-associated diagnostic tests, due to its binding to the parasite surface. The authors suggest that this antigen is present in such small amounts on the *Toxoplasma* membrane that it is not detected in the IFA (JOHNSON et al. 1983a). The FMC 20 clone secretes the IgG1 isotype (JOHNSON et al. 1981a), which has been implicated as a blocking antibody (FELDMAN 1972) with a possible parasite-protective effect (MITCHELL et al. 1977). It is possible that this 98 000 mw *Toxoplasma* polypeptide could play a role in the evasion of the host's immune response.

Table 1. Intracellular antigens of *Toxoplasma* involved in the IHA

MW	Estimation	Nature (method)	Orientation (method)	Reference
800 000/ 135 000	Sucrose density gradient centrifugation, G-200 gel filtration and IHA activity	Protein (trypsin sensitive)	Probably intracellular (differential centrifugation)	Fujita et al. (1969)
98 000	Binding of IHA + ve monoclonal antibody	Polypeptide (did not bind to ^{125}I-Con-A	Intracellular and membrane (IHA, IFA test and immuno-histology)	Johnson et al. (1981a, b), Johnson et al. (1983a, c)
10 000	S-300 gel filtration and binding to immune human serum	Protein (multiple staining)	Intracellular (iodination and two-dimensional hydro-phobic interaction immuno-electrophoresis)	Hughes et al. (1980), Hughes and Balfour (1982), Hughes and van Knapen (1982)

See text for full details

The IHA in its present form serves a useful purpose in the accurate diagnosis of the infection status of the host (Karim and Ludlam 1975b). Although the replacement of the this test is not a priority requirement, it is obviously important to know the nature and disposition of the antigens that are involved for a full understanding of the basis of this assay. This information will enable the rational development of more specific diagnosis of subacute and chronic infection to take place, as well as helping to eliminate false results in tests detecting membrane-specific antibody or circulating antigen.

4 Membrane Components of *Toxoplasma gondii*

The complexity of the *Toxoplasma* membrane and the manner in which its various antigenic components relate to diagnosis, immunity or escape from the host's immune response is only beginning to be understood. Preliminary studies suggested that there were only four or five membrane antigens on the parasite surface, and that very few if any of these were glycoproteins (Handman et al. 1980b; Hughes et al. 1982). More recent studies using the sensitive western blot technique and radio-labelled lectins have shown that the parasite surface has a number of carbohydrate moieties and that immune sera applied to western blots can recognise at least four antigens during acute infection and up to 20 antigens during the convalescent phase (Sharma et al. 1983). While it may be premature to delineate specific roles for defined antigens using polyspecific or monoclonal antibodies and immunoprecipitation or western blotting, studies on the *Toxoplasma* membrane over the past 5 years have indicated that only relatively few antigens may have important roles in diagnosis and immunity.

4.1 Immunochemical Analysis of *Toxoplasma* Membrane Antigens

A variety of non-ionic detergents have been studied for their ability to liberate antigens from *Toxoplasma* detected by serum from an acute case of toxoplasmosis using a 2-DIEP system (HUGHES and BALFOUR 1981). When used singly, the detergents yielded a pattern of soluble proteins that was essentially similar to that obtained from sonicated, frozen-thawed or water-lysed parasites. However, a combination of saponin and octyl glucoside was found to liberate 11 antigens, one of which was identified as a glycoprotein and another as a lipopolysaccharide (HUGHES et al. 1980; HUGHES and BALFOUR 1981). Further studies showed that at least four of these antigens were of membrane origin using ^{125}I surface iodination and hydrophobic gels (HUGHES et al. 1982; HUGHES and VAN KNAPEN 1982). The majority of *Toxoplasma* antigens appeared to have a mw of between 100 000 and 150 000, and all were iso-electric at acid pH (HUGHES and BALFOUR 1981).

POKORNY et al. (1972, 1979) described a Tween-ether extract of *Toxoplasma* which underwent serological trials as the basis of a complement fixation test (CFT). This extract was compared with a *Toxoplasma* freeze-thaw preparation (WARREN and SABIN 1942) and an alkali extract (PETTERSEN 1968). Serological studies showed that the Tween-ether preparation gave results that were both more specific and sensitive in the CFT when compared with the dye test. When a similar preparation was analysed and compared with a freeze-thaw antigen by 2-DIEP, there was a single antigen (antigen 12) that was common to the Tween-ether preparation and the freeze-thaw preparation (HUGHES et al. 1982 and Fig. 2). This antigen could be extracted in greater quantity by using high concentrations (5% w/v) of the saponin and octyl glucoside combination. It was shown to be an externally disposed protein using ^{125}I-membrane iodination (Fig. 2) and hydrophobic interaction chromatography (HUGHES et al. 1982; HUGHES, unpublished).

The reactivity of different sera against antigen 12 was measured by microdensitometry of autoradiographs following 2-DIEP of an ^{125}I-labelled preparation. The results showed that with rising dye test titre the density of the antigen 12 arc on the autoradiograph increased (HUGHES 1982). The implication that antigen 12 was a major dye test component was reinforced by testing a saponin/octyl glucoside solubilised preparation by ELISA against a freeze-thaw antigen, using the IgG-IFA as a standard. This preparation showed an enhanced sensitivity and specificity compared with a freeze-thaw preparation which was assessed using polynomial regression analysis (HUGHES et al. 1982).

When *Toxoplasma* organisms were biosynthetically labelled using [^{35}S] methionine, up to 1000 proteins of parasite origin could be identified after two-dimensional PAGE. Approximately 70 of these proteins were precipitated by antisera from infected mice, and it was suggested that the majority of these proteins were of intracellular origin (HANDMAN et al. 1980b). Owing to the complexity of these patterns obtained after electrophoresis, ^{125}I-iodination of viable tachyzoites was used and produced a more realistic profile of the major membrane proteins. Solubilisation of ^{125}I-labelled tachyzoites followed by immunoprecipitation and SDS-PAGE yielded only five major membrane proteins (p), designated according to their molecular weight: p 43 000, p 40 000, p 35 000, p 27 000 and p 14 000. p 40 000 occurred sporadically in some precipitates and it was suggested that it was a proteolytic breakdown product of p43 000. The other four major proteins were consistently precipitated by sera from chronically infected mice and acutely infected humans (Fig. 3). Lectin-binding studies failed to detect any carbohydrate moieties either on whole *Toxoplasma* tachyzoites or in the solubilised membrane fraction.

4.2 Monoclonal Antibodies Directed Against *Toxoplasma* Membrane Antigens

A total of 32 monoclonal antibodies have been described which react against *Toxoplasma* membrane antigens (HANDMAN et al. 1980b; HANDMAN and REMINGTON 1980a; JOHNSON et al. 1981a; SETHI et al. 1980; SETHI and BRANDIS 1981; KASPER et al. 1982, 1983). All have been serologically defined; four showed reactivity against both membrane and cytoplasmic components and 28 were positive only in tests recognising membrane antigens.

Most of these monoclonal antibodies have been tested by immunoprecipitation and SDS-PAGE to define the antigens against which they react (Table 2). The majority of them appeared to detect components that had a mw of 35 000, 27 000 or 14 000. This would suggest that either these antigens comprise the bulk of the membrane components of *Toxoplasma* or that they are highly immunogenic. All three of these membrane antigens share properties that are ascribed to three of the four major membrane proteins discussed in Sect. 4.1 above (HANDMAN et al. 1980b).

The four clones described by HANDMAN et al. (1980b) have been defined serologically using conventional tests as well as ELISA systems incorporating subcellular fractions of *Toxoplasma* (NAOT and REMINGTON 1981). Antibody from the 3E6 and 2G11 clones both precipitated 35 000 and 14 000 mw antigens. There was variability in the intensity of the reaction, and it was proposed that the 14 000 mw antigen is a monomeric form of the 35 000 mw component. These two polypeptides are not linked by disulphide bonds, since there was no difference in electrophoretic patterns produced in both non-reducing and reducing conditions, though they may be linked by non-covalent forces.

JOHNSON et al. (1981a) described four monoclonal antibodies (FMC 18, FMC 21, FMC 22, FMC 23) which showed distinct membrane specificity and one (FMC 19) which showed specificty for an antigen which they propose as being a water-extractable membrane component.

Immunoelectronmicroscopy (JOHNSON et al. 1983a) confirmed that FMC 18, FMC 19, FMC 22, FMC 23 as well as the 2G11 and 3E6 clones of HANDMAN et al. (1980b) were specific for tachyzoite membranes. Later studies showed that FMC 19 was similar to 2G11, in that it strongly precipitated a 35 000 mw antigen with a limited reactivity against a 14 000 mw component (Table 2). It was found that FMC 19 could protect mice against challenge (discussed more fully in Sect. 8) while 2G11 could not. This is probably because the monoclonal antibodies are detecting different epitopes on the same antigen (JOHNSON et al. 1983b).

KASPER et al. (1982, 1983) have recently described two precipitable antigens of mw 22 000 (P22) and 30 000 (P30). P22 was found to be present on wild-type RH strain tachyzoites but not in chemically mutagenised parasites. The significance of these findings will be discussed in a later section. The P30 component has been established as an intrinsic membrane protein by charge-shift electrophoresis; it does not exist as a

Table 2. Monoclonal antibodies against defined *Toxoplasma* antigens

Clone	Isotype	MW[a]	Titre		
			DT[b]	IFA	m-RIA[c]
IE3[d]	IgG2	43 000	ND[e]	ND	ND
TGA-62-4[f]	IgG2	43 000	0	64	+
2G11[d]	IgG2	35 000	2048	4096	+
		14 000			
3E6[d]	IgG2	35 000	2048	256	+
		14 000			
FMC19[g]	IgG3	35 000	ND	8192	ND
		14 000			
B[h]	IgG	30 000	+	ND	ND
LE11[d]	IgG3	27 000	2048	64	+
TGB-69-2[f]	IgM	27 000	0	320	+
TGB-70-3[f]	IgM	27 000	0	64	+
TG-15-B7[f]	IgG2a	27 000	64	2560	ND
TG-15-C10[f]	IgG2a	27 000	128	2560	ND
G[h]	IgG	22 000	+	ND	ND
FMC18[g]	IgG1	14 000	ND	512	ND
FMC22[g]	IgG1	14 000	ND	512	ND
FMC23[g]	IgG1	14 000	ND	8192	ND
2B7[d]	IgG2	Unknown	0	64	+
2F8[d]	IgG2	Unknown	0	256	+
5B6[d]	IgG2	Unknown	64	256	+
TGA-71-1[f]	IgG2	Unknown	32	128	+
TG-21-D2[f]	IgM	Unknown	16	640	ND
TG-32-D6[f]	IgM	Unknown	< 2	320	ND
TG-24-A7[f]	IgG1	Unknown	< 2	1280	ND

[a] Molecular weight of precipitated antigen
[b] Dye test
[c] RIA-measuring activity against membrane components
[d] HANDMAN et al. (1980b), HANDMAN and REMINGTON (1980a)
[e] Not done
[f] SETHI et al. (1980), SETHI and BRANDIS (1981)
[g] JOHNSON et al. (1981a, b, c)
[h] KASPER et al. (1982, 1983)

disulphide-linked homo- or heterodimer and is thought to comprise 3%–5% of the total protein of *Toxoplasma* RH strain tachyzoites.

SETHI et al. (1980), SETHI and BRANDIS (1981) and HANDMAN et al. (1980b) have described monoclonal antibodies that were serologically defined as reacting against the tachyzoite membrane, yet did not precipitate ^{125}I-labelled membrane antigen. It is possible that these monoclonal antibodies react with antigens that either have few if any exposed tyrosine residues or that they are either polysaccharide or glycolipid antigens. Alternatively, it is possible that these antigens may be sufficiently masked by other components that labelling could not take place due to allosteric inhibition.

Table 2 summarises those monoclonal antibodies that have been tested for their activity against membrane components by both diagnostic tests and immunoprecipitation. There is a remarkable homogeneity in both the charge and mw of *Toxoplasma* antigens, though the mw estimated by SDS-PAGE is considerably reduced when compared with the studies of FUJITA et al. (1969) or HUGHES and BALFOUR (1981). This disparity is almost certainly due to differences in the migration of solubilised intrinsic membrane proteins through SDS-PAGE gels when compared with gel filtration columns.

The passive immunisation studies carried out by JOHNSON et al. (1983b) have shown that monoclonal antibodies which react against the same membrane antigen may recognise different epitopes. Similar studies could be carried out on other monoclonal antibodies that recognise a shared antigen, and could be expanded to include studies to define the molecular structure of the epitopes concerned.

4.3 Membrane Antigens of Possible Diagnostic Significance

Almost all of the studies using polyspecific antisera from mice, acute and convalescent sera from humans and the monoclonal antibodies described above have relied on the RH strain of *Toxoplasma* for their source of antigen. This strain was first isolated by SABIN (1941), has undergone innumerable serial passages in mice, is unable to complete the sexual reproduction cycle in cats and thus cannot be regarded as being typical of *Toxoplasma* strains present in the wild. Two relevant questions need to be asked: (a) are those antigens detected on the RH strain present on all *Toxoplasma* strains and (b) are they expressed at the same stage of the infection? The studies of POKORNY et al. (1972, 1979), HUGHES (1982) and HUGHES et al. (1982) would suggest that at least one antigen (Ag12) may be of serological significance, being detected by sera from patients with both acute and chronic infection induced by a variety of strains.

Studies on the *Toxoplasma* membrane have indicated that all strains are homogeneous with regard to the composition of membrane proteins, but differ with regard to the kinetics of their expression. Evidence for this has been gathered by direct comparison of major membrane proteins and by studying the precipitation of antigens from strains of differing virulence. Many workers have chosen to compare *Toxoplasma* strains that differ widely with regard to their virulence in mice, as they would represent both extremes of a *Toxoplasma* infection. It is now generally accepted that virulence does not correlate with the antigenic profile of the parasite but with generation time (HANDMAN et al. 1980a).

Initial studies showed that different strains of *Toxoplasma* had no significant differences in the properties of their major proteins detectable by gel electrophoresis (BLOOMFIELD and REMINGTON 1970). The ability of immune sera to precipitate membrane proteins on two strains of differing virulence was first described by HANDMAN and REMINGTON (1980b). They found that IgG from mice infected with an avirulent (C37) strain precipitated the same proteins as IgG from mice infected with a virulent (C56) strain. The precipitated proteins were also found to be similar to the highly virulent RH strain. However, precipitation of these antigens by IgG from sera from mice infected with the (virulent) C56 strain appeared 2 days before precipitating antibodies induced by the (avirulent) C37 strain (Fig. 3). These observations have been confirmed by JOHNSON et al. (1983c), who found that the antigenic profiles of two other unrelated strains were identical to those described by HANDMAN and REMINGTON (1980b). When sera from acutely infected humans were used in precipitation studies,

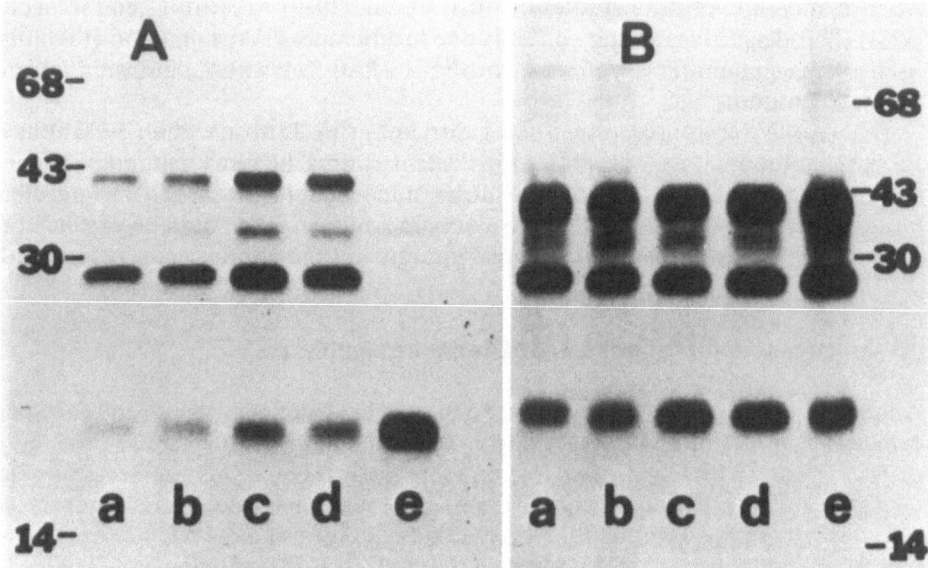

Fig. 3A, B. One-dimensional SDS-PAGE pattern of ^{125}I-labelled *Toxoplasma* membrane antigens bound by antibodies from humans acutely infected and mice chronically infected with *Toxoplasma*. **A** Proteins precipitated by sera from five humans acutely infected with *Toxoplasma*. **B** Proteins precipitated by sera from mice chronically infected with one of the following *Toxoplasma* strains: *a*, CML; *b*, RJ; *c*, Sterling; *d*, DH; *e*, RH. *Numbers* represent molecular weight standards. (HANDMAN et al. 1980b)

the same antigens were detected, though there was variability in the manner and intensity of the membrane antigens detected.

The 30 000 mw component (P30) recognised by the clone B has been shown by double diffusion to be present in a variety of *Toxoplasma* strains and the monoclonal antibody which recognises it is lethal for a strain freshly isolated from cat faeces. Preliminary studies on human sera have shown that convalescent sera have high titres of antibody against P30 (KASPER et al. 1983). However, there is no information as to whether sera recognise this component during acute infection.

The variable reactivity of human sera against individual membrane antigens and the description of membrane glycoconjugates has initiated further studies to delineate those antigens that are recognised by sera from patients with acute and chronic toxoplasmosis. At first the biochemical nature of antigens detected in IgG and IgM ELISA systems were defined (NAOT et al. 1983). IgM and IgG classes were found to react with membrane components during the acute phase of infection. Pretreatment of membrane antigen preparations with lipase, DNase or RNase had no effect on antibody titre in both IgG and IgM double sandwich and conventional ELISA systems. Sodium periodate, heat or pronase treatment all yielded decreased titres. These confirmed that acute-phase IgM and IgG antibody react against both protein and polysaccharide antigens, and that there is no apparent activity against lipid moieties. The reactions of both IgM and IgG were reduced against protein and carbohydrate isolates when compared with an unfractionated preparation. This may imply that acute-phase IgG and/or IgM react with separate polysaccharide and protein epitopes either present as distinct molecules or on a single glycoprotein complex. These observations differ from those reported by MINEO et al. (1980), who concluded that IgM is directed to both polysaccaride and protein moieties of glycoprotein complex and that the IgG response to polysaccharide determinant is low. Despite the relative paucity of glycoprotein or polysaccharide antigens (MAURAS et al. 1980; JOHNSON et al. 1981b), it was concluded that the integrity of both carbohydrate and protein structures is necessary for an optimal reactivity of IgG and IgM antibodies (NAOT et al. 1983).

The membrane proteins and glycoproteins reactive in acute toxoplasmosis have been further characterised using the western blot technique (SHARMA et al. 1983). The results showed that both IgG and IgM present in the sera of patients infected with *Toxoplasma* recognise common determinants. IgM antibodies were found to recognise four major antigens corresponding to MWs of 45 000, 32 000, 22 000 and 6000. IgG was found to recognise four major polypeptides of MWs 43 000, 32 000, 27 000 and 21 000, though in excess of 20 antigens could be detected, including those recognised by IgM. The pattern observed after western blotting appears to differ from previous reports on IgG immunoprecipitation studies (HANDMAN et al. 1980b). Differences both in technique and sensitivity might account for this disparity. Binding of IgM to the 32 000 and 22 000 mw components was found to be reduced after pretreatment of the antigen with trypsin and pronase.

Antibodies from both acute and chronic stages of infection bind to a number of polysaccharide components as well as with the major protein antigens described by HANDMAN and REMINGTON (1980b). This has been demonstrated using ELISA systems in which polysaccharide antigens are present (MINEO et al. 1980; NAOT et al. 1983) as well as by the immunoblot and Con-A binding studies (SHARMA et al. 1983). It is unlikely that any of the polysaccharide components were present as glycolipids, because pretreatment of antigen with lipase prior to immunoblot analysis had no effect on the number of bands detected or their intensity.

The studies described above suggest that a limited number of membrane antigens are recognised by both IgG and IgM. These antigens appear to be present on all strains of *Toxoplasma* and are recognised by both acute and chronic infection sera. However, the kinetics of antigen expression by the parasite and/or antibody production by the host probably varies between strains. This would imply that it would be unreasonable to expect any single antigen isolated by a monoclonal antibody, or produced by gene cloning, to be a useful alternative to the dye test. A more pragmatic approach might be to use a "cocktail" of purified membrane antigens, each of which has a proven specific reactivity with antibodies from animals or humans at different stages of infection with different strains of *Toxoplasma*. This should be an improvement over using whole *Toxoplasma* extracts because only those membrane antigens which are proven to be of diagnostic significance would be present. This would confer an enhanced specificity over whole extracts of *Toxoplasma* as contamination by cytoplasmic antigens which may be active in the IHA would not be present. Also, a test using these antigens may be more sensitive, because more of the component(s) recognised by antibody from acute infections could be present in the test system than are available *in situ* on the parasite membrane.

The exact nature of many membrane antigens precipitated by anti-*Toxoplasma* antibody has yet to be defined, and until recently there was considerable controversy over the existence of polysaccharides on the *Toxoplasma* membrane. Although it has been established for some time that the tissue cyst wall is periodic-acid-schiff (PAS) positive and binds a variety of lectins (SETHI et al. 1977), it was thought that these polysaccharides are of host cell origin and may play a role in immune escape. The use of more sensitive detection methods has allowed a number of workers to detect between 3 (JOHNSON et al. 1981b) and 15 (SHARMA et al. 1983) different glycosylated polypeptides on the surface of RH strain *Toxoplasma* tachyzoites. Whether these antigens are present solely on the tachyzoite membrane or whether they are also responsible for the continued stimulation encountered in a chronic infection has yet to be established. As well as any putative role that such glycoconjugates or membrane polysaccharides may have to play in diagnosis, their possible role in immune escape merits further investigation.

5 Detection and Significance of Circulating Antigen

Infection with *Toxoplasma gondii* in the normal course of events leads to a rapid humoral response demonstrable by seroconversion in the IgM-IFA followed by a rising dye test titre. These conventional serological tests or their successors may still not give the clinician sufficient information, when the presence of an acute infection has to be quickly and accurately diagnosed from one serum sample. This is especially important in the case of an immunosuppressed patient when there is a danger of dissemination, or if there is an underlying hypogammaglobulinaemia. In these cases, as well as during the acute phase of infection, there is a need for the diagnostican to confirm the presence of an active infection by detecting antigenaemia. In congenital toxoplasmosis, the demonstration of antigen in body fluids such as the amniotic or cerebrospinal fluids can also provide an indication of the severity of *Toxoplasma* infection, thus allowing decisions to be made on therapy and prognosis.

Antigenaemia was first detected by RAIZMAN and NEVA (1975) in experimentally infected animals. VAN KNAPEN and PANGGABEAN (1977) have since developed an ELISA for circulating antigen and were able to detect antigenaemia in two groups of patients, viz: those with a recent (acute) infection and those with a reinfection or recrudescence of a latent infection. In experimental models, circulating antigen can be detected as early as the 1st day following infection (VAN KNAPEN and PANGGABEAN 1977; TURUNEN 1983). ARAUJO and REMINGTON (1980) demonstrated the presence of circulating antigen in 63.6% of sera from patients with recently aquired lymphadenopathic toxoplasmosis. In all but one of these cases, the dye test titre was ≧1:2048, showing that circulating antigen can be present with concurrent high membrane-specific antibody titres. *Toxoplasma* antigen could also be detected in the amniotic fluid and CSF of infected infants, indicating the presence of an active infection.

In an animal model of *Toxoplasma*-derived choroiditis, antigen could be detected in the vitreous humour at the height of clinical activity in the lesion (ROLLINS et al. 1983). Aqueous humour and serum both proved to be unreliable for the detection of *Toxoplasma* antigen presumably because of the intermittent seeding of intraocular antigen, rapid turnover of aqueous humour and the relatively high uveal blood flow, resulting in a rapid dissipation of antigen throughout the blood stream. One might infer from these studies that the absence of *Toxoplasma* antigenaemia need not necessarily be a true indicator of the state of infection, and that in certain cases a more accurate assessment of the infection status may be made by the detection of antibodies and antigens in situ (for example, in the CSF or eye).

There is some controversy regarding the composition of circulating antigen, and a number of theories have been proposed regarding its formation in vivo. Among these are that antigenaemia results from (a) immune lysis of parasites, (b) active secretion by tachyzoites and release into the blood stream, (c) the sloughing off of membrane components or (d) a combination of two or more of the above. The complexity and seemingly multifunctional roles of secretory components of *Toxoplasma* are only beginning to be elucidated. There is now evidence that circulating antigens are involved in pathogenesis, immunity and immune escape, as well as being an extremely useful indicator of acute or active infection.

5.1 Circulating Antigens of Intracellular Origin

The first description of soluble antigens produced by *Toxoplasma* was in 1962 by TON-JUM. Supernates of peritoneal washes from acutely infected animals were analysed by double diffusion against serum from patients with acute toxoplasmosis. Two antigens could be detected, both of which lost activity after heating at 56 °C for 30 min. Circulating antigen was detected by RAIZMAN and NEVA (1975) in the serum of mice with acute toxoplasmosis. Antigen could not be detected in infected rabbit sera, although it could be isolated by affinity chromatography. The affinity-purified circulating antigen was partially characterised as a heat-labile (56 °C/30 min) protein with an estimated mw >100 000. Double diffusion studies showed that circulating antigen shared components with water-lysed *Toxoplasma* extracts as used in the IHA. These observations were substantiated by HUGHES (1981), who studied the development of circulating antigen in cotton rats acutely infected with RH strain toxoplasmosis. The production of circulating antigen was studied in parallel with conventional serology, and it was found that antigenaemia could be detected by the 3rd day after infection. Circulating antigen was found to consist of two distinct antigens (antigens 4 and 5) which had been defined as being heat-labile proteins with a mw of approximately 150 000 and 324 000 respectively (HUGHES et al. 1980; HUGHES and BALFOUR 1981). However, circulating antigen was not identical to either a water-lysed preparation or supernates from infected peritoneal washes, both of which contained a third antigen (antigen 6, Fig. 2). The disparity between these findings and the earlier studies of TONJUM (1962) and RAIZMAN and NEVA (1975) is probably due to the enhanced sensitivity which the 2-DIEP system has compared with gel diffusion studies. All three studies described circulating antigen as being heat-labile proteins with a mw ≥100 000 with at least partial identity with IHA antigen (Table 3). Later studies by HUGHES and VAN KNAPEN (1982) confirmed that these proteins were of intracellular origin using both hydrophobic interaction immunoelectrophoresis and ^{125}I-labelling of viable tachyzoite membranes.

5.2 Penetration-Enhancing Factor

The mechanism by which tachyzoites invade host cells has been a controversial topic for a number of years. While there is evidence that *Toxoplasma* secretes substances in vivo and in vitro that will enhance their infectivity, other studies have suggested that host cell phagocytosis is the primary mechanism by which *Toxoplasma* tachyzoites gain entry into host cells. It is now generally accepted that a balance exists between these two modes of entry, though active penetration by the parasite appears to be an important factor in its survival.

LYCKE and NORRBY (1966) and NORRBY (1971) described a factor which could be isolated from *Toxoplasma* extracts which enhanced the infectivity of parasites both in vivo and in vitro. This factor was named penetration enhancing factor (PEF) and was described as a protein with enzymatic activity with two components with an estimated mw of 70 000-150 000 and 700 000-1 000 000. PEF was found to act upon the membrane of host cells, faciliating parasite entry, and immunofluorescent studies suggested that it is actively produced by the rhoptries (paired organelles) at the anterior end of the parasite.

RYNING and REMINGTON (1978), however, suggested that the host cell played a major role in parasite invasion. They found that cytochalasin D prevented penetration but not attachment of *Toxoplasma* to host cells. This inhibition of entry was totally reversible upon removal of cytochalasin D. It was proposed that cytochalasin D was having an effect on host cell microtubule function, which was essential for *Toxoplasma* entry.

WERK and BOMMER (1980) studied the entry of *Toxoplasma* in Friend leukaemia cells. They found that active energy-dependent cooperation between both host cells and parasites was required for optimal infectivity.

VAN KNAPEN (1982) has carried out further studies on PEF and has detected a polysaccharide (PAS-positive) component which shows at least partial immunological identity to secretory antigen. He suggested that tachyzoites secrete this component of PEF before they come into contact with host cells, and that excess PEF produced by the parasite binds non-specifically to serum proteins. This PAS-positive material appeared to have a similar parasite distribution to the lysosomal particles described by NORRBY et al. (1968) and could also be detected on the membranes of infected and ruptured cells at the initial site of entry.

Electron microscopic studies by NICHOLS et al. (1983) have supplied further evidence for an active role by the parasite in host cell entry. They have confirmed that the rhoptries of *Toxoplasma* are secretory organelles and are actively involved in all stages of the cell cycle, including penetration. They proposed that rhoptries secrete their contents *during* penetration and that the secretory products lyse the host cell plasmalemma, but not intracellular membranes. Lysosomal vesicles similar to those seen by NORRBY et al. (1968) were observed around the anterior end of tachyzoites as they entered cells, suggesting that the rhoptries are the source of lytic enzymes first described by NORRBY (1971) as PEF.

The role that PEF may have as a circulating antigen of diagnostic significance remains undefined. The localised effect of PEF implies that it may not be released in sufficient quantity to be a readily detectable component of circulating antigen and penetration-enhancing properties have not been ascribed to factors isolated from sera from acute infection. PEF, as originally described by NORRBY (1971), differs from the PAS-positive material described by VAN KNAPEN (1982) in that it is a protein which is not precipitated by immune sera. It appears that PEF consists of a complex mixture of enzymes and glycosylated proteins which aid host cell entry and intracellular survival.

5.3 Antigens Secreted by *Toxoplasma*

Concurrent with the increasing interest in *Toxoplasma* antigenaemia, a number of secretory antigens of parasite origin have been described. DESGEORGES et al. (1980) studied the kinetics of secretory antigen in vitro and showed that in a 12-day culture of RH strain parasites in either VERO or MRC-5 cells there was a peak of antigen production at about the 6th day. Later studies (AMBROISE-THOMAS et al. 1982) have shown that antigen can be isolated from culture supernates which is suitable for use in the skin test for toxoplasmosis, and is presently undergoing clinical trials.

HUGHES and VAN KNAPEN (1982) characterised antigens secreted by RH strain tachyzoites in vitro and showed that a single antigen reacted with serum from a case of acute toxoplasmosis. This antigen (antigen 5) was identical to one of the two components present during acute-stage antigenaemia in rats (HUGHES 1982). This secretory component was different from (a) PEF (NORRBY 1971) as it was precipitated by immune sera, (b) PEF (VAN KNAPEN 1982) as it was not PAS positive and (c) skin test antigen as is was denatured by heat treatment. Further studies have shown that the avirulent T1003 strain secretes a different antigen to the RH strain in vitro. The secreted antigen was immunologically identical to the second component (antigen 4) detected in acute toxoplasmic antigenaemia (HUGHES, unpublished data). These preliminary observations suggest that different *Toxoplasma* strains secrete different antigens in vitro, but express the same antigens in vivo. This is consistent with the hypothesis that different strains express the same antigens, but at different stages of infection (HANDMAN and REMINGTON 1980b).

5.4 Detection of Circulating Antigens Using Monoclonal Antibodies

Attempts to use monoclonal antibodies for the detection of circulating antigen have met with little success. ARAUJO et al. (1980) found that four monoclonal antibodies against defined membrane antigens (Table 3 and Sect. 4.2) were able to detect *Toxoplasma* antigenaemia in patients and infected mice, but with a low sensitivity. The $F(ab')_2$ fragment of a rabbit hyperimmune serum consistently detected high titres of circulating antigen in both human and mouse sera. These studies imply that membrane antigens, though present in the circulation during acute toxoplasmosis, are at such a low concentration that they have little or no serological significance in the detection of antigenaemia.

5.5 The Kinetics of Circulating Antigen Production

Table 3 summarises those antigens or factors which are known to be present during acute toxoplasmosis. They fall into three major catagories: membrane antigens, PEF and cytoplasmic/secretory antigens other than PEF.

Preliminary observations suggested that circulating antigen may be released in two ways; by active secretion by the parasite and by immune lysis. Circulating antigen does not show immunological identity with simple aqueous extracts or peritoneal washes, implying that antigenaemia is not merely a result of dying parasites or release of antigen from infected cells.

Table 3. Antigens and factors associated with acute *Toxoplasma* infection

MW	Orientation	Nature	Heat[a] lability	Detected in			Identity with IHA Ag
				Animal		Human serum	
				Peritoneum	Serum		
Unknown[b]	? S[c]	Protein	Yes	+[d]	ND[e]	ND	ND
Unknown[b]	? S	Protein	Yes	+	ND	ND	ND
> 100 000[f]	C	Protein	Yes	+	+	ND	+
150 000[g]	C/S	Protein	Yes	+	+	ND	Partial
324 000[g]	C/S	Protein	Yes	+	+	ND	Partial
> 70 000[h]	PEF (S)	Protein	ND	ND	ND	ND	ND
> 700 000[h]	PEF (S)	Protein	ND	ND	ND	ND	ND
Unknown[i]	PEF (S)	PAS + ve	ND	ND	ND	ND	ND
27 000[j]	M	Protein	ND	±	ND	±	±
35 000[j]	M	Protein	ND	±	ND	±	±
Unknown[j]	M/C	Unknown	ND	±	ND	±	±
Unknown[j]	M	Unknown	ND	±	ND	±	±

See text for full details
[a] Sensitivity to 56 °C
[b] TONJUM (1962)
[c] S, secretory; C, cytoplasmic; M, membrane; PEF, penetration-enhancing factor
[d] +, strong reaction in gels or ELISA; ±, weak reaction in gels or ELISA
[e] ND, not done
[f] RAIZMAN and NEVA (1975)
[g] HUGHES (1981)
[h] NORRBY (1971)
[i] VAN KNAPEN (1982)
[j] ARAUJO et al. (1980), HANDMAN and REMINGTON (1980a). Recognised by clones 1E11, 2G11, 5B6 and 3E6 respectively

HUGHES (1981) investigated the in vivo release of circulating antigen during acute RH strain *Toxoplasma* infections in rats. Antigenaemia was found to be present only in rats with a normal humoral response, as sublethal doses of cyclophosphamide inhibited the expression of circulating antigen. There was also a concurrent suppression of the antibody response, manifested by a failure to seroconvert in the dye test. Cyclophosphamide did not appear to have any direct effect on parasite proliferation or antigen secretion either in vivo (HUGHES 1981) or in vitro (HUGHES and VAN KNAPEN 1982). This initially led to the hypothesis that circulating antigen release was induced by lysis of the parasite mediated by membrane-specific antibody and complement.

However, there are two flaws in this argument, namely that it does not account for (a) the detection of antigenaemia as early as 1 or 2 days following infection (RAIZMAN and NEVA 1975; VAN KNAPEN and PANGGABEAN 1977; TURUNEN 1983) and (b) that the concentration of antigen present in serum and body fluids is far in excess of that expected from the lysis of available parasites (RAIZMAN and NEVA 1975; HUGHES 1981). In vitro culture of *Toxoplasma* tachyzoites with HEp2 cells resulted in the release of large amounts of antigen detectable by 2-DIEP, which showed immunological identity with one of the components of circulating antigen (antigen 5).

Although membrane antigens can be detected in small amounts in the serum and body fluids of patients with acute infection, the diagnostic relevance of these findings must be questioned. Firstly, antigenaemia can be detected in patients and experimental animals with high and rising dye test titres and, secondly, free membrane antigens can only be detected in small quantities when compared with the detection of cytoplasmic or secretory antigens. Furthermore, high dye test (i.e. membrane specific) with low-IHA (i.e. cytoplasmic-specific) titres are the normal serological profile during acute infection. This would favour the expression of cytoplasmic or secretory but not membrane antigenaemia. Circulating immune complexes (CIC) have recently been described by SIEGEL and REMINGTON (1983) using a ^{125}I-Clq binding assay. They found that CIC could be detected during acute toxoplasmosis and, in an experimental model, CIC appeared concurrently with membrane-specific antibody. Significantly, the presence of CIC in serum also appeared to parallel disease activity. The authors suggest that CIC may either have a causative role in more severe infections or be its result. The antigen present in CIC has not been defined, though as they occur at the same time as membrane-specific antibody, membrane antigens may be primary constituents of CIC.

6 The Roles of Antibody- and Cell-Mediated Immunity in Resistance to *Toxoplasma*

The relationship and significance of cell-mediated immunity (CMI) and humoral antibodies in the pathogenesis of and immunity against toxoplasmosis have been discussed in depth by KRAHENBUHL and REMINGTON (1982) and SHARMA and REMINGTON (1981). With these reviews as background, this section is intended as an introduction to the relative importance of antibody, macrophages and T-lymphocytes in *Toxoplasma* immunity.

High titres of serum antibody to *Toxoplasma* occur in humans and experimentally infected animals, but the role of antibody alone as a major effector mechanism in immunity to *Toxoplasma* remains questionable. HULDT (1966) was able to obtain high titres of antibody after immunising rabbits with heat-killed *Toxoplasma*, which were equivalent to titres obtained from chronically infected animals. On challenge with virulent organisms the immunised rabbits did not show any resistance, though those that were chronically infected were found to be immune. Passive transfer experiments using homologous or heterologous antisera also met with little or no success, though KRAHENBUHL et al. (1972) later found that repeat inoculations of immune mouse serum and graded doses of *Toxoplasma* resulted in a significant decrease in mortality. More recent studies showed that sublethal doses of cyclophosphamide prior to infection with an avirulent *Toxoplasma* strain produced high mortality in mice (HAFIZI and MODABBER 1978). When similar drug-treated and infected mice were passively immunised, all those that had an IFA titre $\geq$ 1:512 survived. The authors suggest that the high titres of specific antibody induced by vaccination with (for example) heat-killed organisms may not be protective, and that a delicate balance between the humoral response and CMI has to exist for the efficient control of a *Toxoplasma* infection.

Cell-mediated responses against *Toxoplasma* have been demonstrated both in vivo and in vitro, by delayed-type hypersensitivity (DTH) (FRENKEL 1948; HANDMAN et al. 1980a) and by *Toxoplasma* antigen-induced lymphocyte transformation (ANDERSON et al. 1979; WILSON et al. 1980a). FRENKEL (1967) demonstrated that cells from spleen and lymph nodes transferred specific immunity to *Toxoplasma* in syngeneic animals. Antiserum showed a slight protective effect by delaying the onset of death, and also enhanced the immunity transferred by cells.

Considerable evidence has accumulated which implicates macrophages as the major effector cell population in the cell-mediated response against *Toxoplasma* (REMINGTON et al. 1972; ANDERSON and REMINGTON 1974; SHARMA and REMINGTON 1981; HOF 1983) and other unrelated organisms including bacteria (RUSKIN and REMINGTON 1968), fungi (GENTRY and REMINGTON 1971) and helminths (WING et al. 1979). Infection with *Toxoplasma* results in the generation of activated macrophages, which are able to kill intracellular parasites (JONES et al. 1975). The mechanisms that lead to macrophage activation have not yet been defined, though the cell transfer experiments of FRENKEL (1967) demonstrated that T-lymphocytes have an essential role to play. Although lymphocyte transformation and macrophage activation appear to be linked, mitogen and antigen responses in vitro may be suppressed during acute toxoplasmosis (ANDERSON et al. 1979). SUZUKI et al. (1981a, b) and SUZUKI and KOBAYASHI (1983) studied the effects of a *Toxoplasma* infection on the primary antibody response and the initiation of memory cells. A population of plastic-adherent, radiation-resistant cells was found to be responsible for suppresssion of antibody responses against both T-dependent (sheep red blood cells) and T-independent [dinitrophenol (DNP)-Ficoll] antigens. Marked suppression induced by the same cell population was also noticed in the initiation of memory cells to DNP-keyhole limpet haemocyanin (DNP-KLH). However, if memory cells were established prior to a *Toxoplasma* infection, expansion of the anamnestic response was not affected. A possible mechanism for the immunosuppression induced by *Toxoplasma* infection is that macrophages are activated by T-lymphocytes and the activated macrophages non-specifically inhibit the proliferation of T-helper cells, causing suppression of primary and anamnestic responses. DTH also develops slowly during a chronic *Toxoplasma* infection, though whether or not this is a result of immunosuppression has not been firmly established (HANDMAN et al. 1980a).

Supernates from in vitro cultures of sensitised animal or human cells stimulated with antigen can activate murine and human monocyte populations to inhibit *Toxoplasma* proliferation (JONES et al. 1975; BORGES and JOHNSON 1975). SHIRAHATA and SHIMIZU (1979) described a factor released by splenic lymphocytes from *Toxoplasma*-infected mice which inhibited parasite multiplication in non-immune macrophages in vivo. The apparent mw of this lymphokine (called growth inhibitory factor; GIF) was between 38 000 and 55 000. It could be precipitated by 50%–80% ammonium sulphate and was isoelectric at pH 4.9–5.9.

The production of a number of lymphokines has recently been examined both in vivo and in vitro. When stimulated by soluble antigens, immune spleen cells produced detectable amounts of macrophage migration inhibitory factor (MIF), macrophage-activating factor (MAF), interferon (IFN) and GIF. Peak GIF and IFN activity could be detected in sera from immune mice 6 and 24 h respectively after boosting with soluble antigen in vivo, whereas the release of these lymphokines in vitro was simultaneous with peak activities after 36 h of culture. The IFN produced by antigen activation in vivo and in vitro was characterised as γ-IFN by pH2 lability when compared with non-specific stimulation by poly I:C (SAKURAI et al. 1981). The necessity of a fully competent T-lymphocyte response has been supported by studies in which athymic (*nu/nu*) mice were infected with a normally avirulent strain of *Toxoplasma*. Nude mice were unable to produce antibodies and there was an apparent enhanced proliferation of tissue cysts (BUXTON 1980). Normal infections in nude mice can be attained by reconstitution with T-lymphocytes from thymus-bearing litter mates (HOF et al. 1976).

7 Parasite Evasion of Host Immunity

Survival of *Toxoplasma* in host cells has been extensively studied in both animal models and humans, and would appear to be a major factor in the pathogenesis of disease and the evasion of the host's immune response. Human mononuclear phagocytes and activated murine macrophages are able to inhibit intracellular replication or kill *Toxoplasma*, whereas normal (i.e. resident and unelicited) mouse macrophages are not. This may be due in part to the fact that the mode of killing in these cell populations is different. In the activated macrophage, aminophylline and tosyllysine chloromethylketone (TLCK) both inhibit killing. TLCK, aminophylline and inhibitors of DNA and protein synthesis, cytochrome *c*, microtubule aggregation and other intracellular processes did not interfere with monocyte killing of intracellular *Toxoplasma* (MCLEOD and REMINGTON 1980). These observations would imply that *Toxoplasma* has developed a mechanism for survival in normal macrophages, based on the host cell's failure to activate oxidative metablism. This survival mechanism is not effective in monocytes, which are able to kill *Toxoplasma,* and may be responsible for the large number of subclinical infections in immunologically normal individuals.

All cells of monocyte lineage will readily kill antibody-coated parasites. WILSON et al. (1980b) showed that phagocytosis of antibody-coated *Toxoplasma* stimulated oxidative metabolism in normal macrophages, whereas no oxidative respiratory burst was observed during the entry of uncoated *Toxoplasma*. Monoclonal antibodies against defined membrane antigens (HAUSER and REMINGTON 1981) and heat denaturation also enhanced phagocytosis, but formalin-killed *Toxoplasma* did not. The factors leading to intracellular killing would appear to depend, at least partially, on the lack of integrity of the parasite membrane.

Successful growth of *Toxoplasma* inside macrophages depends on the inhibition of lysosome-phagosome fusion (JONES and HIRSCH 1972). Parasite antigens are not detectable on the surface of infected macrophages and are only present in the parasitophorous vacuole (HANDMAN et al. 1980a). When formalin-killed parasites are used to infect cells, antigen can be detected on the macrophage surface. These observations have two important implications. Firstly, antigen presentation by infected or uninfected macrophages would appear to be a result of the uptake of antigens from the milieu, and not a result of intracellular parasitism. It is possible that extracellular *Toxoplasma* organisms are predisposed to killing by either antibody and complement or antibody alone. These dead organisms may then be processed in a manner different from that of viable parasites. Secondly, the absence of antigen presentation on infected cells would explain the ability of *Toxoplasma* to survive the effector mechanisms of both cellular and humoral immunity. In this case, the activated macrophage, with its enhanced capacity to kill intracellular organisms, would have an essential role in the control of infection. In these studies, antigen was detected either by direct or indirect immunofluorescence, using a polyvalent hyperimmune serum (HANDMAN et al. 1980a). The possibility that T-dependent antigen(s) not recognised by antibody may be present in association with class II MHS antigen must not be excluded. Although cytotoxic T-lymphocytes (CTL) have not been observed in vitro, there is evidence that such T-dependent antigens exist on the macrophage surface. Antigen-pulsed non-immune macrophages (i.e. non-antigen presenters) are able to elicit a DTH reaction comparable to that observed by the use of formalin-killed tachyzoites. In the normal course of events, it would be expected that antigen-presenting cells (APC) would be effective in eliciting a CTL response. HANDMAN et al. (1980a) suggest that a cytotoxic response would lead to an ineffective effector mechanism, as only those cells expressing antigen would be killed, and infected cells (i.e. non-APC) would remain intact. These studies also implied that the immunosuppression previsously described during toxoplasmosis is not universal, as the DTH response remains unaffected.

As well as the possibility that the antigens on APC are T dependent, the detection methods used by HANDMAN et al. (1980a) may not have been sufficiently sensitive. VAN KNAPEN (1982) was able to detect *Toxoplasma* antigens on the surface of infected cells, by using the more sensitive peroxidase-anti-peroxidase (PAP) technique. These were shown to be PAS-positive and shared properties ascribed to PEF (Sect. 5.2). However, the role that this antigen may have in eliciting T-lymphocyte responses must remain questionable as it is not known whether it is expressed on APC, as all studies on this secretory antigen have been carried out using an established human cell line (HEp2).

7.1 Parasite Secretions and Immune Escape

Toxoplasma relies on the formation of a parasitophorous vacuole for intracellular survival. Recent studies on the vacuole membrane have indicated that parasite secretions are responsible for at least part of its formation. Nascent tubules can be detected on rare occasions in rhoptries during penetration and are associated with active invasion by *Toxoplasma*. These tubules then appear to be secreted during vacuolar formation around the parasite (NICHOLS et al. 1983) and are a typical feature of the parasitophorous vacuole (SHEFFIELD and MELTON 1968). They then fuse with the intravacuolar membrane, and NICHOLS et al. (1983) suggest that this secreted product is a surface-active component, with digitonin-like properties, as it readily lyses the cholesterol-rich host cell plasmalemma, but not limiting membranes surrounding intracellular organelles. The origin and kinetics of tubule formation remain unknown, but it has been suggested that the rhoptries contain steroids such as cholesterol bound to a surface-active material. Only when secretion is initiated does this complex become active and the presence of tubules in rhoptries represents a reorganisation of their contents (NICHOLS et al. 1983).

Once the parasitophorous vacuole has formed, tubules can be seen inside the phagosome, but distinct from the parasite (SHEFFIELD and MELTON 1968). These tubules may be responsible for the expression of *Toxoplasma* antigen inside phagosomes that has been described by HANDMAN et al. (1980a) in macrophages and VAN KNAPEN (1982) in HEp2 cells. Macrophage activation may have direct effects on the rhoptries, resulting in little or no tubule secretion, thus allowing normal phagocytosis and antigen presentation. NICHOLS et al. (1983) suggest that the secretory tubules act on the phagosome membrane in much the same way as on the plasmalemma, and prevent lysosomal fusion.

Toxoplasma would appear to have two independent mechanisms which ensure its survival inside host cells. These are (a) inhibition of the respiratory burst in macrophages (but not monocytes) and (b) inhibition of lysosomal fusion. The relationship between the latter and defined secretory antigens is not firmly established, so the roles of PEF and secretory (non-PEF) antigen can only be speculated about in the light of present-day knowledge. However, the morphology and distribution of the lysosomal vesicles (NORRBY et al. 1968) and PAS-positive secretions (VAN KNAPEN, 1982) suggest that they are partially identical to the tubules observed by NICHOLS et al. (1983).

7.2 The Role of Tissue Cysts in Immune Evasion

Tissue cysts are formed within the host cell cytoplasm after repeated endodyogeny of the invading organism. They may contain up to several thousand organisms, and the absence of a granulomatous response has resulted in speculation as to the origin of the cyst wall. Carbohydrate moieties are present that bind to a number of lectins, and this initially led to the hypothesis that the cyst wall was of host cell origin (SETHI et al. 1977) due to the paucity of tachyzoite carbohydrate. The recent definition of glycoconjugates on the tachyzoite membrane has indicated that the presence of glycoproteins on the cyst wall may be of parasite rather than host origin.

Cyst formation is not thought to occur as a result of host immunity, though SHIMIDA et al. (1974) reported the induction of cyst formation by the RH strain in vitro by the addition of *Toxoplasma*-specific antibody to the culture medium. The in vivo growth of the RH strain is different to that of an avirulent (i.e. cyst-forming) strain (FURGUSON and HUTCHISON 1981), so must be regarded as a poor model for cyst formation. There is also considerable evidence that immunosuppressed hosts develop cysts more rapidly and in greater numbers than intact hosts (BUXTON 1980; KRAHENBUHL and REMINGTON 1982).

It has not been established whether glycoconjugates detected on tachyzoites are also present on the cyst wall. Most studies on cyst formation have compared cyst wall structure with membranes of tachyzoites from a virulent strain (e.g. RH or BK) of *Toxoplasma*. As virulent strains do not normally form cysts in vivo, the validity of these studies is dubious. However, studies on the invasion of tachyzoites of the avirulent Beverley strain have shown that parasite-derived particles are deposited on the inner membrane of the phagosome, forming a granular layer, which thickened during development to form the cyst wall. Both the granular deposits and the cyst wall reacted with ferritin-labelled antibody (MATSUBAYASHI and AKAO 1966).

7.3 Antigen Capping of *Toxoplasma* Tachyzoites

Antigenic variation per se has not been described in *Toxoplasma*, though capping of antigen has been proposed as a mechanism by which the parasite may evade the lethal effects of humoral antibody. Tachyzoites of RH strain *Toxoplasma* were found to remove antibodies bound to their surface by the accumulation of membrane components at the anterior pole of the parasite (DZBENSKI et al. 1976). Capping is prevented by addition of metabolic inhibitors or low temperature (DZBENSKI and ZIELINSKA 1976) and hence is an active process. There was no internalisation of the resulting complex and capping was found to occur in the minority of parasites.

DZBENSKI et al. suggested that the parasite is able to secrete an antigen which is then exposed on the surface. The existence of such an antigen has been corroborated by JOHNSON et al. (1983a, c), who described a monoclonal antibody (FMC 20) which reacts with a membrane antigen of intracellular origin (Sect. 3). The role that this antigen may have in cap formation though must remain speculative until further analysis of both FMC 20 and cap formation has been carried out. The subclass of the antibodies involved in capping should be elucidated, as only non-complement fixing antibodies would have any significant effect in blocking immune effector mechanisms.

8 Vaccination Against Toxoplasmosis

The primary use of a *Toxoplasma* vaccine would be to prevent abortion and neonatal death in humans and domestic animals. This could be achieved efficiently in two ways, by eradicating toxoplasmosis through vaccination of cats or by a vaccination programme for seronegative women who intend to become pregnant. Clearly, eradication of *Toxoplasma* is an unrealistic aim, due to the vast reservoir of infected wild felids and rodents. Vaccination in farm animals should ideally become a routine precaution in areas where toxoplasmosis is known to be pandemic, as preliminary screening may prove to be uneconomic. Any potential vaccine must afford long-term protection, be safe, cost-effective and easy to deliver.

Vaccines using killed organisms have on the whole been unsuccessful. When guinea pigs were inoculated with formalin-killed RH strain tachyzoites, there was protection from challenge with virulent organisms. However, parasites could be isolated from tissues, indicating that the vaccine did not induce sterilising immunity (CUTCHINS and WARREN 1956). Attempts to induce immunity in more susceptible host species (e.g. rabbits and mice) have also met with limited success (reviewed by KRAHENBUHL and REMINGTON 1982). When avirulent strains have been used as live vaccines, resistance against challenge with virulent strains has been reported. However, organisms in live vaccines persist, as parasites can be isolated from infected animals or their offspring, and can be identified as being of the vaccinating avirulent strain (NAKAYAMA and SATO 1980; KRAHENBUHL and REMINGTON 1982).

The use of attenuated organisms for vaccine preparation has been studied by a number of workers. SEAH and HUCAL (1975) showed that a 10 000 to 20 000 rad dose of X-irradiation abolished the virulence of RH *Toxoplasma*, but 5000 rad did not.

Immunisation with irradiated parasites induced protection, but only for a limited period; if challenge was delayed, then protection was lost. CHHABRA et al. (1979) reported that protection could be prolonged by administering a second (booster) dose of irradiated tachyzoites, and TRAN MANH SUNG (1982) was able to induce immunity using parasites irradiated at a lower (6000-rad) split dose. Survival curves of cell damage following irradiation follow exponential relationships. Therefore, regardless of total dose received by a given cell population, there is the possibility that a few cells still have the capacity to devide and proliferate. Therefore there is a distinct possibility that vaccines using irradiated *Toxoplasma* may have a small number of organisms that are able to survive and serve as a live vaccine. GRIMWOOD (1980) established a minimum fluence of ultraviolet radiation which did not inhibit entry of tachyzoites into cells but did prevent intracellular multiplication. Parasites irradiated at this level (70 J m^{-2}) also delayed mortality of mice challenged with virulent *Toxoplasma*.

Temperature-sensitive mutants of the RH strain of *Toxoplasma* have been described (PFEFFERKORN and PFEFFERKORN 1976), one of which (ts-4) grew normally at 37 °C but persisted for only 40 h at 38° and 39 °C and 20 h at 40 °C. Further studies on ts-4 showed that its pathogenicity in mice was very much reduced when compared with other mutants and the parent RH strain. The lack of persistence in vivo was at first thought to be a result of the febrile illness experienced during acute infection. However, the highest temperature recorded in mice (38.4 °C) was not markedly restrictive and cellular immunity appears to play an essential role in controlling a ts-4 infection (WALDELAND et al. 1983). The ts-4 mutant was further tested for any potential it may have as a vaccine, and elicited protection against 10^6 sporozoites after 3 months and 10^2 sporozoites after 12 months. ts-4 could not be isolated from any of the host tissues tested beyond 2 months following immunisation, and appeared not to confer stage-specific immunity (WALDELAND and FRENKEL 1983). The ts-4 mutant would appear, on the basis of these preliminary investigations, to combine the advantages of both living and killed vaccines, with its unique ability to proliferate for only a short period and then die, yet elicit an effective cellular response. Another mutant strain of *Toxoplasma* has recently been described, which is resistant to monoclonal antibody-mediated killing. It was apparent that this mutant does not express a 22 000 mw membrane protein. Monoclonal antibody-resistant mutants may offer possibilites not only for vaccines, but also for better understanding of the pathogenesis of toxoplasmosis (KASPER et al. 1982).

JOHNSON et al. (1983b) found that certain monoclonal antibodies could induce protection against virulent and highly virulent strains of *Toxoplasma*. The two antibodies that gave the most significant protection (FMC 19 and FMC 22) reacted against 35 000 and 14 000 mw antigens (Sect. 4.2; Table 2). Other monoclonal antibodies which precipitated the same antigens when analysed by SDS-PAGE did not give the same degree of protection, and it was assumed that they were recognising different epitopes on the same antigen. Those monoclonal antibodies that induced protection were of the IgG3 and IgG1 isotype. The two monoclonals that did not afford significant protection were of the IgG1 (FMC 23) and IgG2a (2G11) isotypes. It is possible that the lack of protection induced by FMC 23 when compared with FMC 19 or FMC 22 may be due to different epitopes being recognised. The lack of protection observed when 2G11 was used could also be due to the different functional properties that the IgG2a subclass has compared with IgG1 and IgG3. Both SETHI et al. (1981) and HAUSER and REMINGTON (1981) observed that pretreatment of *Toxoplasma* with membrane-specific monoclonal antibodies enhanced the capacity of resident murine macrophages to kill *Toxoplasma*. These studies, together with the fact that both the successful immunising monoclonal antibodies of JOHNSON et al. (1983b) are unable to fix complement, would imply that antibody coating of parasites and subsequent intracellular killing may be the primary mechanism for the immunity which they describe. HAUSER and REMINGTON (1981) suggested that the enhanced intracellular killing that monoclonals induce when compared with immune serum could have been artificial since the level of specific immunoglobulin in the monoclonal antibody preparation will be in excess of that present in immune serum. It is therefore expedient to determine (a) whether antigens isolated by affinity chromatography using the FMC 19 and FMC 22 antibodies are protective and (b) whether they are able to elicit a humoral response, similar to those obtained after passive immunisation with monoclonal antibody.

There are a number of ways that protective antibodies may act. *Toxoplasma* may be killed by complement-mediated lysis, by macrophages through the action of cytophilic antibody or by T_k cells through ADCC. Antibodies may also affect the metabolism of the parasite directly by membrane perturbation, thus affecting parasite survival inside host cells. The antibodies used by JOHNSON et al. (1983b) are unable to fix complement and IgG1 is not cytophilic for macrophages. Whether killing was mediated by other immune effector cells or was a result of a direct physiological effect on *Toxoplasma* has not been determined. Previous studies (HANDMAN and REMINGTON 1980b) showed that the IgG1 isotype could not be detected during *Toxoplasma* infections, and they proposed that it may be a blocking antibody with parasite-protective effects. The role of IgG1 during a naturally acquired infection is still open to question, despite the role that JOHNSON et al. (1983b) show that it may have in protection. Monoclonal IgG1 antibodies have only been described when mice have been injected with parasite preparations (JOHNSON et al. 1980a) or immunisation has been carried out in vitro (SETHI and BRANDIS 1981). When immunisation has been carried out using active infections, the IgG1 isotype has not been isolated from hybrid cultures (HANDMAN et al. 1980a; HUGHES, unpublished).

ERLICH et al. (1983) established that the 6000 mw glycoprotein antigen further characterised by SHARMA et al. (1983) elicited a high IgM response in adults with acute toxoplasmosis. The reactivity of IgG against this antigen during the latent phase of infection was much reduced, being directed mostly against other antigens of higher molecular weight. The authors suggest that as well as the important role that this antigen may have in diagnosing acute adult infection, it may be an effective component in a vaccine, due to its capacity to stimulate an elevated humoral response.

The majority of studies carried out on vaccination against toxoplasmosis have been concerned with the elicitation of a high humoral response. Antibodies undoubtedly are essential for the clearance of extracellular *Toxoplasma* by complement-mediated killing and enhancement of phagocytosis. However, the initiation of an effective CMI response should be a consideration when developing vaccines. A vaccine which induces a high antibody response which eventually leads to the killing of parasites by complement, ADCC or enhanced macrophage killing relies on there being viable parasites present in the host. A more pragmatic approach may be to include T-dependent antigens in the vaccine which are able to mediate non-specific and/or specific cellular responses as well as B-cell help, either directly or through the action of lymphokines. The release of suppressor factors and the expansion of suppressor T-lymphocytes should also be investigated for any part they may have in the depletion of cellular and humoral immunity and the persistence of infection.

9 Conclusion

The advances made over the past 5 years have meant that the development of inexpensive, sensitive and specific diagnostic tests based on the use of defined soluble antigens can now be approached in a more rational manner. Such tests have important diagnostic and epidemiological implications, and will allow large-scale surveys to be carried out as well as the routine screening of women before pregnancy. In a recent epidemiological report on congenital toxoplasmosis in England, Wales and Northern Ireland, HALL (1983) found that 91 cases had been reported to the Communicable Diseases Surveillance Centre (CDSC) in the period 1975 to 1980. However, she discovered that the reporting criteria varied between *Toxoplasma* reference laboratories and that the decision to report is often subjective. In the same 5-year period, 1478 cases of ocular toxoplasmosis were reported (BANNISTER 1983), and the incidence of eye disease in 5- to 15-year-olds was approaching 2/1000, and in the 15- to 25-year-old age group was over 5/1000. Improved diagnostic capabilities will make possible the follow-up of children that are born of mothers who seroconvert during pregnancy. This will elucidate the relationship of eye disease and other sequelae of asymptomatic congenital toxoplasmosis to maternal infection.

McCARTHY (1983) stresses that standardised reporting of toxoplasmosis by diagnostic laboratories is essential for accurate epidemiological surveys. He also stresses the important role that clinicians have in reporting and confirming both congenital and acquired toxoplasmosis. These measures could be taken almost immediately. The serological and clinical definition of what constitutes acute and congenital toxoplasmosis is essential, and must be carried out using existing serology so that future tests can be accurately assessed.

Present-day serology basically relies on the detection of specific IgM and IgG antibodies reactive in two or more tests. Classical serological interpretation has relied on the assumption that the IFA or dye test measures the antibody response against membrane antigens, whereas that IHA detects antibody directed against cytoplasmic components which are released by complement-mediated lysis. The studies carried out by Johnson and his colleages, with the positive identification of a shared cytoplasmic and membrane antigen, imply that this dogma may be incorrect. The use of such antigens in diagnosis is questionable, but their absence from tests using cytoplasmic antigens may improve the definition of acute, subacute and chronic infections.

Education of the public, particularly during pregnancy, has been suggested as a way of decreasing the incidence of toxoplasmosis (FRENKEL 1981). It can be argued that the timing of such a campaign is critical, as there may be a huge influx of requests for antenatal screening, and it is doubtful whether the reference laboratories in the

United Kingdom could handle such a demand with present methodologies. However, improved diagnosis, with the (longer term) prospect of vaccination, will clearly mean that such measures can be taken. However, FRENKEL (1981) argues that because the modes of acquiring *Toxoplasma* infection are so limited technically simple and common-sense measures can be preventative. He also states that an emphasis on preventing infection in mothers would also increase physicians' awareness of toxoplasmosis as a possible cause of low birth weight and prolonged jaundice. This extra commitment would have immediate implications, as the incidence of congenital infection could be more accurately assessed, without large-scale ante- and postnatal screening.

Acknowledgements. Thanks are due to Dr. L. Hudson for his advice and to Mrs. K. Dorelli for secretarial assistance. This work was supported by MRC grant no. G8222691/T and NFRCD grant no. A/8/F138.

References

Ambroise-Thomas P, Cagnard MY, Roumaintzeff M, Colombert G (1982) Selection de l'immunite toxoplasmique par intradermoreactions a l'aide d'exo-antigens de *Toxoplasma gondii.* Lyon Med 248:79-84

Anderson SE, Remington JS (1974) Effect of normal and activated human macrophages on *Toxoplasma gondii.* J Exp Med 139:1154-1174

Anderson SE, Krahenbuhl JL, Remington JS (1979) Longitudinal studies of lymphocyte response to *Toxoplasma* antigens in humans infected with *T. gondii.* J Clin Lab Immunol 2:293-297

Araujo FG, Remington JS (1980) Antigenaemia in recently acquired acute Toxoplasmosis. J Infect Dis 141:144-150

Araujo FG, Handman E, Remington JS (1980) Use of monoclonal antibodies to detect antigens of *Toxoplasma gondii* in serum and other body fluids. Infect Immun 30:12-16

Balfour AH, Bridges JB, Harford JP (1980) An evaluation of the ToxHA test for the detection of antibodies to *Toxoplasma gondii* in human serum. J Clin Pathol 33:644-647

Balfour AH, Fleck DG, Hughes HPA, Sharp D (1982) Comparative study of three tests (dye test, indirect haemagglutination test, latex agglutination test) for the detection of antibodies to *Toxoplasma gondii* in human sera. J Clin Pathol 35:228-232

Balsari A, Poli G, Molina V, Dovis M, Petruzzelli E, Boniolo A, Rolleri E (1980) ELISA for toxoplasma antibody detection: a comparison with other serodiagnostic tests. J Clin Pathol 33:640-643

Bannister DB (1983) Toxoplasmosis 1976-1980: review of reports to CDSC. CDR 83/06:3-4

Beverley JKA, Beattie CP (1958) Glandular toxoplasmosis. A survey of 30 cases. Lancet 2:379-383

Bjerrum OJ (1977) Immunochemical investigation of membrane proteins. A methodological survey with emphasis placed on immunoprecipitation in gels. Biochim Biophys Acta 472:135-195

Bloomfield MM, Remington JS (1970) Comparison of three strains of *T. gondii* by polyacrylamide-gel electrophoresis. Trop Geogr Med 22:367-370

Borges JS, Johnson WD (1975) Inhibition of multiplication of *Toxoplasma gondii* by human monocytes exposed to T-lymphocyte products. J Exp Med 141:483-496

Buxton D (1980) Experimental infection of athymic mice with *Toxoplasma gondii.* J Med Microbiol 13:307-311

Camargo ME, Leser PG (1976) Diagnostic information from serological tests in human toxoplasmosis II. Evolutive study of antibodies and serological patterns in acquired toxoplasmosis, as detected by haemagglutination, complement fixation, IgG and IgM-immunofluorescence tests. Rev Inst Med Trop Sao Paulo 18:227-238

Camargo ME, Leser PG, Leser WSP (1976) Diagnostic information from serological tests in human toxoplasmosis I - a comparative study of haemagglutination, complement fixation IgG- and IgM-immunofluorescence tests in 3,752 serum samples. Rev Inst Med Trop Sao Paulo 18:215-226

Camargo ME, Ferreira AW, Mineo JR, Takiguti CK, Nakahara CS (1978a) Immunoglobulin G and immunoglobulin M enzyme-linked immunosorbent assays and defined toxoplasmosis serological patterns. Infect Immun 21(1):55–58

Camargo ME, Leser PG, Kiss HB, Neto VA (1978b) Serology in early diagnosis of congenital toxoplasmosis. Rev Inst Med Trop Sao Paulo 20:152–160

Carlier Y, Bout B, Dessaint JP, Capron A, Van Knapen F, Ruitenberg EG, Bergquist R, Huldt G (1980) Evaluation of the enzyme-linked immunosorbent assay (ELISA) and other serological tests for the diagnosis of toxoplasmosis. Bull WHO 58:99–105

Chhabra MB, Mahajan RC, Ganguly NK (1979) Effects of ^{60}Co irradiation of virulent *Toxoplasma gondii* and its use in experimental immunization. Int J Radiat Biol 35:433–440

Chordi A, Walls KA, Kagan IG (1964) Analysis of *Toxoplasma gondii* antigens by agar diffusion methods. J Immunol 93:1034–1044

Cutchins EC, Warren J (1956) Immunity patterns in the guinea pig following *Toxoplasma* infection and vaccination with killed *Toxoplasma*. Am J Trop Med 5:197–209

Desgeorges PT, Billiault X, Ambroise-Thomas P, Bouttaz M (1980) Mise en evidence et cinetique d'apparition d'exo-antigenes produits par *Toxoplasma gondii* en culture in vitro. Lyon Med 243:737–740

Dzbenski TH, Zielinska E (1976) Antibody-induced formation of caps in *Toxoplasma gondii*. Experimentia 32:454–456

Dzbenski TH, Michalak T, Plonka WS (1976) Electron microscopic and radioisotopic studies on cap formation in *T. gondii*. Infect Immun 14 (5):1196–1201

Erlich HA, Rodgers G, Vaillancourt P, Araujo FG, Remington JS (1983) Identification of an antigen-specific immunoglobulin M antibody associated with acute *Toxoplasma* infection. Infect Immun 41:683–690

Feldman JD (1972) Immunological enhancement: a study of blocking antibodies. Adv Immunol 15:167–214

Feldman HA, Lamb GA (1966) A micromodification of the *Toxoplasma* dye test. J Parasitol 52:415

Fleck DG (1961) Serological tests for *Toxoplasma*. Nature 190:1018–1019

Fleck DG, Payne RA (1963) Tests for *Toxoplasma* antibody. Monthly Bulletin of the Ministry of Health and the Public Health Laboratory Service 22:97–102

Frenkel JK (1948) Dermal hypersensitivity to *Toxoplasma* antigens (toxoplasmins). Proc Soc Exp Biol Med 68:634–639

Frenkel JK (1967) Adaptive immunity to intracellular infection. J Immunol 98:1309–1319

Frenkel JK (1981) Congenital toxoplasmosis: prevention or palliation? Am J Obstet Gynecol 141:359–361

Fujita K, Kamei K, Fujita T, Shiori-Nakana K, Tsunematsu Y (1969) Purification of *Toxoplasma* antigen for haemagglutination tests. Am J Trop Med Hyg 18:892–901

Furguson DJP, Hutchison WM (1981) Comparison of the development of avirulent and virulent strains of *Toxoplasma gondii* in the peritoneal exudate of mice. Ann Trop Med Parasitol 75:539–546

Gentry LO, Remington JS (1971) Resistance against cryptococcus conferred by intracellular bacteria and protozoa. J Infect Dis 123:22–31

Gransden WR, Brown PM (1983) Pneumocystic pneumonia and disseminated toxoplasmosis in a male homosexual. Br Med J 286:1614

Grimwood BG (1980) Infective *Toxoplasma gondii* trophozoites attenuated by ultraviolet irradiation. Infect Immun 28:532–535

Hafizi A, Modabber FZ (1978) Effect of cyclophosphamide on *Toxoplasma gondii* infection: reversal of the effect by passive immunization. Clin Exp Immunol 33:389–394

Hall SM (1983) Congenital toxoplasmosis in England, Wales and Northern Ireland. Br Med J 287:453–455

Handman E, Remington JS (1980a) Serological and immunochemical characterization of monoclonal antibodies to *Toxoplasma gondii*. Immunology 40:579–588

Handman E, Remington JS (1980b) Antibody responses to *Toxoplasma* antigens in mice infected with strains of different virulence. Infect Immun 19:215–220

Handman E, Chester PM, Remington JS (1980a) Delayed hypersensitivity to *Toxoplasma* and unrelated antigens in *Toxoplasma* infected mice: induction and elicitation of delayed-type hypersensitivity by antigen pulsed macrophages. Infect Immun 28:524–531

Handman E, Goding JW, Remington JS (1980b) Detection and characterization of membrane antigens of *Toxoplasma gondii*. J Immunol 124:2578–2583

Hauser WE, Remington JS (1981) Effect of monoclonal antibodies on phagocytosis and killing of *Toxoplasma gondii* by normal macrophages. Infect Immun 32:637–640

Hof H (1983) The role of macrophages in acquired cell-mediated immunity to *Toxoplasma gondii*. Med Microbiol Immunol 171:199–202

Hof H, Emmerling P, Hohne K, Seeliger HPR (1976) Infection of congenitally athymic (nude) mice with *Toxoplasma gondii*. Ann Microbiol (Paris) 127B:503–507

Hughes HPA (1981) Characterisation of the circulating antigen of *T. gondii*. Immunol Lett 3:99–101

Hughes HPA (1982) The antigenic structure of *Toxoplasma gondii*. Lyon Med 248:135–139

Hughes HPA, Balfour AH (1981) An investigation of the antigenic structure of *Toxoplasma gondii*. Parasite Immunol 3:235–248

Hughes HPA, Van Knapen F (1982) Characterisation of a secretory antigen from *Toxoplasma gondii* and its role in circulating antigen production. Int J Parasitol 12:433–437

Hughes HPA, Lee DL, Balfour AH (1980) A multiple staining technique for agarose gels. Sci Tools 23:39–41

Hughes HPA, Van Knapen F, Atkinson HJ, Balfour AH, Lee DL (1982) A new soluble antigen preparation of *Toxoplasma gondii* and its use in serological diagnosis. Clin Exp Immunol 49:239–246

Huldt G (1966) Experimental toxoplasmosis. Effect of inoculation of *Toxoplasma* in seropositive rabbits. Acta Pathol Microbiol Scand 68:592–604

Hutchison WM, Dunachie JK, Work K, Siim JC (1971a) The life cycle of the coccidian parasite *Toxoplasma gondii* in the domestic cat. Trans R Soc Med Hyg 65(3):380–399

Hutchison WM, Dunachie JF, Furguson DJ, Gardner IC (1971b) Endogenous development of the coccidian parasite *Toxoplasma gondii*. Trans R Soc Trop Med Hyg 65(4):429–430

Hyde B, Barnett EV, Remington JS (1975) Method for differentiation of non specific from specific *Toxoplasma* IgM fluorescent antibodies in patients with rheumatoid factor. Proc Soc Exp Biol Med 148:1184–1188

Jacobs L, Lunde MN (1957) A haemagglutination test for toxoplasmosis. J Parasitol 43:308–314

Johnson AM, McNamara PJ, Neoh SH, McDonald PJ, Zola H (1981a) Hybridomas secreting monoclonal antibody to *Toxoplasma gondii*. Austr J Exp Biol Med Sci 59:303–306

Johnson AM, McDonald PJ, Neoh SH (1981b) Molecular weight analysis of the major polypeptides and glycoproteins of *Toxoplasma gondii*. Biochem Biophys Res Commun 100:934–943

Johnson AM, Haynes WD, Leppard PJ, McDonald PJ, Neoh SH (1983a) Ultrastructural and biochemical studies on the immunohistochemistry of *Toxoplasma gondii* antigens using monoclonal antibodies. Histochemistry 77:209–215

Johnson AM, McDonald PJ, Neoh SH (1983b) Monoclonal antibodies to *Toxoplasma* cell membrane surface antigens protect mice from toxoplasmosis. J Protozool 30:351–356

Johnson AM, McDonald PJ, Neoh SH (1983c) Molecular weight analysis of *Toxoplasma gondii* soluble antigens. J Parasitol 69:459–464

Jones TC, Hirsch JG (1972) The interaction between *Toxoplasma gondii* and mammalian cells II. The absence of lysosomal fusion with phagocytic vacuoles containing living parasites. J Exp Med 136:1173–1194

Jones TC, Len L, Hirsch JG (1975) Assessment in vitro of immunity against *Toxoplasma gondii*. J Exp Med 141:466–482

Karim KA, Ludlam GB (1975a) Serological diagnosis of congential toxoplasmosis. J Clin Pathol 28:383–387

Karim KA, Ludlam GB (1975b) The relationship and significance of antibody titres as determined by various serological methods in glandular and ocular toxoplasmosis. J Clin Pathol 28:42–49

Kasper LH, Crabb JH, Pfefferkorn ER (1982) Isolation and characterization of a monoclonal antibody-resistant antigenic mutant of *Toxoplasma gondii*. J Immunol 129:1694–1699

Kasper LH, Crabb JH, Pfefferkorn ER (1983) Purification of a major membrane protein of *Toxoplasma gondii* by immunoabsorption with a monoclonal antibody. J Immunol 130:2407–2412

Kohler G, Milstein C (1975) Continuous cultures of fused cells secreting antibody of predefined specificity. Nature 256:495–497

Krahenbuhl JL, Remington JS (1982) Immunology of *Toxoplasma* and toxoplasmosis. In: Cohen S, Warren KS (eds) Immunology of parasitic infections, 2nd Ed. Blackwell, Oxford

Krahenbuhl JL, Ruskin J, Remington JS (1972) The use of killed vaccines in immunization against an intracellular parasite: *Toxoplasma gondii*. J Immunol 108:425–431

Kramer J, Chalupsky J, Hubner J (1970) Relationship of the immunofluorescence reaction and the Sabin-Feldman reaction. J Hyg Epidemiol Microbiol, Immunol 14:129–134

Lin TM, Halbert SP, O'Connor GR (1980) Standardized quantitative enzyme-linked immunoassay for antibodies to *Toxoplasma gondii*. J Clin Microbiol 11:675–681

Luna MA, Lichtiger B (1971) Disseminated toxoplasmosis and cytomegalovirus infection complicating Hodgkin's disease. Am J Clin Pathol 55:499–505

Lunde MN, Jacobs L (1967) Differences in Toxoplasma dye test and haemagglutination antibodies shown by antigen fractionation. Am J Trop Med Hyg 16:26–30

Lycke E, Norrby R (1966) Demonstration of a factor of *Toxoplasma gondii* enhancing the penetration of *Toxoplasma* parasites into cultured host cells. Br J Exp Pathol 47:248–256

Lyng J, Siim JC (1982) The WHO international standard for anti-*Toxoplasma* serum human. Lyon Med 248:107–108

Matsubayashi H, Akao S (1966) Immuno-electron microscopic studies on *Toxoplasma gondii*. Am J Trop Med Hyg 15:486–491

Mauras G, Dodeur M, Leget P, Senet J-M, Bourrillon R (1980) Partial resolution of the sugar content of *Toxoplasma gondii* membranes. Biochem Biophys Res Commun 97:906–912

McCarthy M (1983) Of cats and women. Br Med J 287:445–446

McLeod R, Remington JS (1980) Inhibition or killing of an intracellular pathogen by activated macrophages is abrogated by TLCK or aminophylline. Immunology 39:599–605

Milatovic D, Braveny I (1980) Enzyme-linked immunosorbent assay for the serodiagnosis of toxoplasmosis. J Clin Pathol 33:841–844

Mineo JR, Camargo ME, Ferreira AW (1980) Enzyme-linked immunosorbent assay for antibodies to *Toxoplasma gondii* polysaccharides in human toxoplasmosis. Infect Immun 27:283–287

Mitchell GF, Marchalonis JJ, Smith PM, Nicholas NL, Warner NL (1977) Studies on immune responses to larval cestodes in mice. Immunoglobulins associated with the larvae of *Mesocestoides*. Aust J Exp Biol Med Sci 55:187–211

Nakayama I, Sato K (1980) Augmentation of resistance to *Toxoplasma* in mice by pretreatment with attenuated vaccine. Tokai J Exp Clin Med 5:311–321

Naot Y, Remington JS (1980) An enzyme-linked immunosorbent assay for detection of IgM antibodies to *Toxoplasma gondii*: use for diagnosis of acute aquired toxoplasmosis. J Infect Dis 142:757–766

Naot Y, Remington JS (1981) Use of enzyme linked immunosorbent assays (ELISA) for detection of monoclonal antibodies: experience with antigens of *Toxoplasma gondii*. J Immunol Methods 43: 333–341

Naot Y, Desmonts G, Remington JS (1981) IgM-enzyme linked immunosorbent assay test for the diagnosis of congenital *Toxoplasma* infection. J Pediatr 98:32–36

Naot Y, Guptill DR, Mollenax J, Remington JS (1983) Characterization of *Toxoplasma gondii* antigens that react with human immunoglobulin M and immunoglobulin G antibodies. Infect Immun 41:331–338

Nichols BA, Chiappino ML, O'Connor GR (1983) Secretion from the rhoptries of *Toxoplasma gondii* during host cell invasion. J Ultrastruct Res 83:85–98

Norrby R (1971) Immunological study on the host cell penetration factor of *Toxoplasma gondii*. Infect Immun 3:278–286

Norrby R, Lindholm L, Lycke E (1968) Lysosomes of *Toxoplasma gondii* and their possible relation to the host-cell penetration of *Toxoplasma* parasites. J Bacteriol 96:916–919

Ordonez GA, Newman JT, Stone MJ (1982) Serological diagnosis of *Toxoplasma gondii* infections by rapid separation of serum immunoglobulins M & G with CM Bio-Gel A. J Clin Microbiol 16:751–753

Palmer AA (1974) The diagnosis of lymphomas. Med J Aust 2:96–98

Pande PG, Shukla RR, Sekariah PC (1961) A heteroglycan from *Toxoplasma gondii*. Nature 190:644–645

Pettersen EK (1968) Preparation of *Toxoplasma gondii* antigen for the complement fixation test. Acta Pathol Microbiol Scand 74:35–40

Pfefferkorn ER, Pfefferkorn LC (1976) *Toxoplasma gondii*: Isolation and preliminary characterization of temperature-sensitive mutants. Exp Parasitol 39:365–376

Pokorny J, Fruhbauer Z, Curic B, Zastera M (1972) A Tween-ether preparation of *Toxoplasma gondii* antigen for the complement fixation test. Bull WHO 46:127–130

Pokorny J, Fruhbauer Z, Curic B, Zastera M (1979) Verification of Tween-ether antigen for complement fixation test in the diagnostic of toxoplasmosis. J Hyg Epidemiol Microbiol Immunol 23:353–356

Raizman RE, Neva FA (1975) Detection of circulating antigen in acute experimental infections with *Toxoplasma gondii*. J Infect Dis 132:44–48

Remington JS (1974) Toxoplasmosis in the adult. Bull N Y Acad Med 50:211–227

Remington JS (1982) Toxoplasmosis in homosexuals. Lyon Med 248:133–134

Remington JS, Desmonts G (1976) Toxoplasmosis. In: Remington JS, Klein JO (eds) Infectious diseases of the fetus and newborn infant. Saunders, Philadelphia

Remington JS, Miller MJ, Brownlee I (1968a) IgM antibodies in acute toxoplasmosis: I Diagnostic significance in congenital cases and a method for their rapid demonstration. Pediatrics 41:1082–1091

Remington JS, Miller MJ, Brownlee I (1968b) IgM antibodies in acute toxoplasmosis: II Prevalence and significance in acquired cases. J Lab Clin Med 71:855–866

Remington JS, Krahenbuhl JL, Mendenhall JW (1972) A role for activated macrophages in resistance to infection with *Toxoplasma*. Infect Immun 6:829–834

Rollins DF, Tabbara KF, O'Conner R, Araujo FG, Remington JS (1983) Detection of toxoplasmal antigen and antibody in ocular fluids in experimental ocular toxoplasmosis. Arch Ophthalmol 101:455–457

Ruskin J, Remington JS (1968) Immunity and intracellular infection: resistance to bacteria in mice infected with a protozoan. Science 160:72–74

Ryning FW, Remington JS (1978) Effect of cytochalasin D on *T. gondii* cell entry. Infect Immun 20:739–743

Sabin AB (1941) Toxoplasmic encephalitis in children. JAMA 116:801–807

Sabin AB, Feldman HA (1948) Dyes as microchemical indicators of a new immunity phenomenon affecting a protozoan parasite (*Toxoplasma*). Science 108:660–663

Sakurai H, Takei Y, Omata Y, Suzuki N (1981) Production and properties of *Toxoplasma* growth inhibitory Factor (Toxo-GIF) and interferon (1FN) in the lymphokines and the circulation of *Toxoplasma* immune mice. Zentralbl Bakteriol Hyg [A] 251:134–143

Schreiber RD, Feldman HA (1980) Identification of the activator system for antibody to *Toxoplasma* as the classical complement pathway. J Infect Dis 141:366–369

Seah SKK, Hucal G (1975) The use of irradiated vaccine in immunization against experimental toxoplasmosis. Can J Microbiol 21:1379–1385

Sethi KK, Brandis H (1981) Generation of hybridoma cell lines producing monoclonal antibodies against *Toxoplasma gondii* or rabies virus following fusion of in vitro-immunized spleen cells with myeloma cells. Ann Immunol 132 C:29–41

Sethi KK, Rahman A, Pelster B, Brandis H (1977) Search for the presence of lectin-binding sites on *Toxoplasma gondii*. J Parasitol 63:1076–1080

Sethi KK, Endo T, Brandis H (1980) Hybridomas secreting monoclonal antibody with specificity for *Toxoplasma gondii*. J Parasitol 66:192–196

Sethi KK, Endo T, Brandis H (1981) *Toxoplasma gondii* trophozoites precoated with specific monoclonal antibodies cannot survive within normal murine macrophages. Immunol Lett 2:343–346

Sharma SD, Remington JS (1981) Macrophage activation and resistance of intracellular infection. In: Pick E (ed) Lymphokines, vol 3. Academic, New York

Sharma SD, Mullenax J, Araujo FG, Erlich HA, Remington JS (1983) Western blot analysis of the antigens of *Toxoplasma gondii* recognized by human IgM and IgG antibodies. J Immunol 131:977–983

Sheffield HG, Melton ML (1968) The fine structure and reproduction of *Toxoplasma gondii*. J Parasitol 54:209–226

Shimida K, O'Connor GR, Yoneda C (1974) Cyst formation by *Toxoplasma gondii* (RH strain) in vitro. Arch Ophthalmol 92:496–500

Shirahata T, Shimizu K (1979) Some physicochemical characteristics of an immune lymphocyte product which inhibits the multiplication of *Toxoplasma* within mouse macrophages. Microbiol Immunol 23:17–50

Siegel JP, Remington JS (1983) Circulating immune complexes in toxoplasmosis: detection and clinical correlates. Clin Exp Immunol 52:157–163

Stray-Pedersen B (1980) A prospective study of acquired toxoplasmosis among 8,043 pregnant women in the Oslo area. Am J Obstet Gynecol 136:399–406

Stray-Pedersen B, Lorentzen-Styr AM (1979) The prevalence of *Toxoplasma* antibodies among 11736 pregnant women in Norway. Scand J Infect Dis 11:159–165

Sulzer AJ, Hall EC (1967) Indirect fluorescent antibody tests for parasitic diseases. IV Statistical study of variation in the indirect fluorescent antibody (IFA) test for toxoplasmosis. Am J Epidemiol 86:401–407

Suzuki Y, Kobayashi A (1983) Suppression of unprimed T and B cells in antibody responses by irradiation-resistant and plastic-adherant suppressor cells in *Toxoplasma gondii*-infected mice. Infect Immun 40:1–7

Suzuki Y, Watanabe N, Kobayashi A (1981a) Nonspecific suppression of primary antibody responses and presence of plastic-adherent suppressor cells in *Toxoplasma gondii*-infected mice. Infect Immun 34:30–35

Suzuki Y, Watanabe N, Kobayashi A (1981b) Nonspecific suppression of initiation of memory cells in *Toxoplasma gondii*-infected mice. Infect Immun 34:36–42

Tonjum AM (1962) Soluble antigens produced by *Toxoplasma gondii*. Acta Path Microbiol Scand 54:96–98

Tran Manh Sung R (1982) Les essais de radio-vaccins dans la toxoplasmose murine. Lyon Med 248:101–106

Turunen HJ (1983) Detection of soluble antigens of *Toxoplasma gondii* by a four layer modification of an enzyme immunoassay. J Clin Microbiol 17:768–773

Van Knapen F (1982) Detection and significance of circulating antigens and immune complexes. Lyon Med 248:51–54

Van Knapen F, Panggabean SO (1977) Detection of circulating antigen during acute infections with *Toxoplasma gondii* by enzyme-linked immunosorbent assay. J Clin Microbiol 6:545–547

Van Loon AM, Van der Veen J (1980) Enzyme-linked immunosorbent assay for quantitation of toxoplasma antibodies in human sera. J Clin Pathol 33:635–639

Van Loon AM, Van der Logt JTM, Heesen FWA, Van der Veen J (1983) Enzyme-linked immunosorbent assay that used labelled antigen for detection of immunoglobulin M and A antibodies in toxoplasmosis: comparison with indirect immunofluorescence and double-sandwich enzyme-linked immunosorbent assay. J Clin Microbiol 17:997–1004

Voller A, Bidwell DE, Bartlett A, Fleck DG, Perkins M, Oladehin B (1976) A microplate enzyme-immunoassay for *Toxoplasma* antibody. J Clin Pathol 29:150–153

Waldeland H (1976) Toxoplasmosis in sheep. The reliability of a microtiter system in Sabin and Feldman's dye test. Acta Vet Scand 17:426–431

Waldeland H, Frenkel JK (1983) Live and killed vaccines against toxoplasmosis in mice. J Parasitol 69:60–65

Waldeland H, Pfefferkorn ER, Frenkel JK (1983) Temperature-sensitive mutants of *Toxoplasma gondii*: pathogenicity and persistence in mice. J Parasitol 69:171–175

Walls KW, Bullock SL, English DK (1977) Use of the enzyme-linked immunosorbent assay (ELISA) and its microadaptation for the serodiagnosis of toxoplasmosis. J Clin Microbiol 5:273–277

Walton BC, Benchoff BM, Brooks WH (1966) Comparison of the indirect fluorescent antibody test and the methylene blue dye test for detection of antibodies to *Toxoplasma gondii*. Am J Trop Med Hyg 149–152

Warren J, Sabin AB (1942) The complement fixation reaction in toxoplasmic infection. Proc Soc Exp Biol Med 51:11–14

Welch PC, Masur H, Jones TC, Remington JS (1980) Serologic diagnosis of acute lymphadenopathic toxoplasmosis. J Infect Dis 142:256–264

Werk R, Bommer W (1980) *Toxoplasma gondii*: membrane properties of active energy dependent invasion of host cells. Tropenmed Parasitol 31:417–421

Wielaard F, van Gruijthuijsen H, Duermeyer W, Joss AWL, Skinner L, Williams SH, Van Elven EH (1983) Diagnosis of acute toxoplasmosis by an enzyme immunoassay for specific immunoglobulin M antibodies. J Clin Microbiol 17:981–987

Wilson CB, Desmonts G, Couvreur J, Remington JS (1980a) Lymphocyte transformation in the diagnosis of congenital *Toxoplasma* infection. New Eng J Med 302:785–788

Wilson CB, Tsai V, Remington JS (1980b) Failure to trigger the oxidative metabolic burst by normal macrophages. Possible mechanism for survival of intracellular pathogens. J Exp Med 151:328–346

Wing EJ, Krahenbuhl JL, Remington JS (1979) Studies of macrophage function during *Trichinella spiralis* infection in mice. Immunology 36:479–485

Antigens of African Trypanosomes

M.J. TURNER

1 Introduction

The African trypanosomes present their hosts with the best characterised antigens in parasitology – the variant surface glycoproteins, or VSGs. The structure of the genes encoding VSGs and the mechanism by which the expression of these genes is regulated have been studied intensively (reviewed in STEINERT and PAYS 1985). Research in this area is at the forefront of molecular biology, and the most modern techniques of recombinant DNA analyses have been applied. A number of the steps in VSG biosynthesis have been well characterised (reviewed in TURNER et al. 1985) and the VSGs have been purified and studied extensively (reviewed in CROSS 1984; TURNER 1984a). The sequence of several of the proteins has been determined, one directly, and the remainder indirectly by DNA sequencing (reviewed in CROSS 1984; TURNER 1984a). These data have been used as the basis for calculations of the secondary structure potential, in an attempt to predict VSG structure (LALOR et al. 1984). Recently, the three-dimensional structure of part of a VSG molecule has been solved to 6 Å resolution (FREYMANN et al. 1984), and a structure at higher resolution is expected shortly. In spite of these attentions, new methods of controlling trypanosomiasis, both in man and his domestic animals, still seem remote. Trypanosomiasis (known as sleeping sickness in man and nagana in cattle) is endemic throughout central Africa. It is estimated that 35 million people and 25 million cattle are exposed to the risk of infection in an area about the size of the continental United States, posing a huge obstacle to the economic development of Africa (WHO 1979). The secret of the success of this remarkable parasite does indeed lie in the structure of its surface antigens, the VSGs, which are the products of a process called "antigenic variation" (reviewed in TURNER 1984b). The purpose of this article is to illustrate how the VSGs are essential to para-

Merck Sharp & Dohme, Research Laboratories, 126 E. Lincoln Avenue, P.O.Box 2000, Rahway, NJ 07065, USA

site evasion of the immune response; how they can be used for diagnosis of disease; why attempts to use them to induce protective immunity have provided only limited success; and how they may yet provide targets for new chemotherapeutic reagents.

2 Escape: Antigenic Variation

Trypanosomes are cyclically transmitted by the bite of the tsetse fly or, outside the tsetse belt of central Africa, non-cyclically by other biting diptera, and, in the case of *T. equiperdum*, venereally. Once in the blood stream, they multiply rapidly until antibodies capable of both agglutinating and lysing the trypanosomes appear in the serum. Parasite numbers fall, until a new population emerges, antigenically distinct from the first, which in turn acts as the progenitor of a third population before it too is overwhelmed, and so on. A single trypanosome can produce the same fluctuating parasitaemia, indicating that variation is a property of the individual, not the population. Antibody acts selectively rather than inductively – in an in vitro culture system antigenic variation occurs at about the same rate as it does in vivo – between 1 in 10^4 and 1 in 10^5 trypanosomes are switching at any one time (DOYLE et al. 1980). The limits of antigenic variation have not been precisely defined, but molecular genetic analysis suggests that there may be up to 1000 genes encoding different surface antigens (VAN DER PLOEG et al. 1982).

Antigenic variation is achieved by changing the composition of the surface coat of the trypanosome, which can be visualised in transmission electron microscopy as a dense layer some 12–15 nm thick (VICKERMANN 1969). As shown in Fig. 1, the surface coat of *T.b.brucei* is an adaption to life in the blood stream of the mammalian host. It is found on the infective metacyclic trypanosome present in the salivary glands and is in all stages in the blood stream, but is lost when trypanosomes are ingested by the tsetse fly, and does not reappear until metacyclic trypomastigotes develop in the salivary glands some 3 weeks later. The surface coat on any one trypanosome is made up of a matrix of about 10^7 identical molecules (CROSS 1975), the variant surface glycoproteins which have received so much attention in recent years. By switching to the expression of a different VSG gene, the trypanosome can modulate its immuno-chemical profile.

The biology of antigenic variation is complex, largely because the expression of different VSGs is non-random (reviewed in TURNER 1984b). An early model for antigenic variation (VICKERMAN 1976), based on observations made by GRAY (1965a, b), postulated that all variants assume a common antigenicity in the fly midgut on the loss of the surface coat and that in the salivary glands all metacyclic trypomastigotes acquired the same surface coat to become the first variant in a preprogrammed order of expression. Although this is now known to be an oversimplification, some elements of this model remain. Thus, it is certainly true that all variants assume a common antigenicity in the fly midgut (SEED 1964). Secondly, although the metacyclic trypanosome population is antigenically heterogeneous (LE RAY et al. 1978) there does seem to be a restriction on the number of VSG genes which can be expressed in the salivary glands. Thus, in the most detailed study of its sort to date, a set of 12 different variant specific monoclonal antibodies was prepared which by immuno-

Fig. 1. Life cycle of *Trypanosoma brucei.* Trypanosomes infective to the mammalian host appear in the salivary glands of the tsetse fly as non-replicative metacyclic forms, which possess a surface coat. When injected into the mammal, these rapidly differentiate into long, slender replicative trypomastigotes which can persist in the blood stream because of their ability to undergo antigenic variation. Some trypanosomes differentiate into a form which does not replicate in the blood stream, the short stumpy from. On ingestion by the tsetse fly, short stumpies differentiate to long, slender dividing trypomastigotes, which lack the surface coat. This step, which can readily be brought about in vitro, is the first in a sequence of morphological and physiological changes which take place in the fly, leading after about 3 weeks to the development of epimastigotes and then infective metacyclic trypomastigotes in the salivary glands to complete the cycle

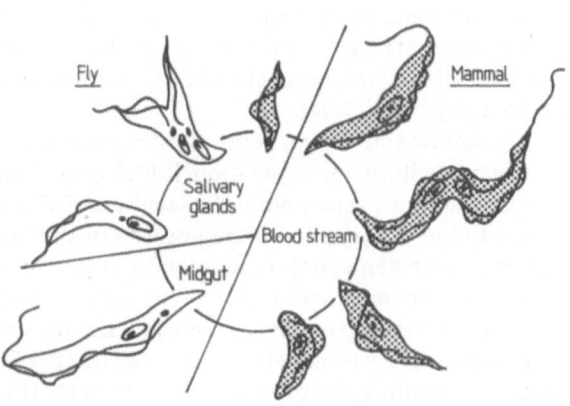

fluorescence accounted for all the metacyclic trypomastigotes present in the saliva of flies infected with a strain of *T. congolense* (CROWE et al. 1983). Furthermore, the same set of monoclonal antibodies neutralised up to 10^6 trypanosomes produced from the same strain of *T. congolense* in an in vitro culture system. The percentage of the population accounted for by each monoclonal antibody varied from 1% to 50%. The situation may be slightly more complex in *T. b. rhodesiense,* where nine different variant-specific antibodies accounted for about 80% of the metacyclic population (BARRY et al. 1983), but nevertheless a picture is emerging of perhaps 0.1% of the VSG gene repertoire being capable of expression in the environment of the salivary glands. What it is about the expression of these genes, or about the product of those genes, that makes them successful in that environment is not known.

The third feature of the original model which is only partially true is that there is an ordered sequence of appearance of variants in the blood stream of the mammalian host. Certain variants, termed predominant variants, do tend to appear early in the course of both chronic and acute infections, whereas others tend to be detected later, but there is nothing absolute about the order (GRAY 1965a, b; VAN MEIRVENNE et al. 1975; MILLER and TURNER 1981). The existence of predominant antigenic types has important consequences for serodiagnosis (see below). The basis for predominance is obscure. There is hope that analysis of the gene rearrangements associated with VSG gene expression may provide the answer. There are, however, complications. It has been calculated that random generation of variants followed by selection operating through growth rates would not be capable of producing the degree of variant ordering reported in the literature (KOSINSKI 1980), but there is evidence that growth rates of individual clones of trypanosomes may not correlate with the growth rates observed in mixtures (SEED 1978; MILLER and TURNER 1981). That is, there may be competition effects in populations of trypanosomes which may relate to predominance. Also, trypanosomes are not just confined to the vasculature, and although one variant may predominate in the blood stream at any one time, quite a different pattern may exist simultaneously in the lymph or the brain (reviewed in SEED et al. 1984). Thus, predominance may be a reflection of a very complex population dynamic, which is not amenable to a simple biochemical explanation.

A further complication is that although the variant ingested by the tsetse fly is "lost", the same variant then tends to appear early in infection following cyclical transmission (HAJDUK and VICKERMAN 1981) – that is, there is some form of anamnestic response, the basis of which is not understood but may be explained at the molecular genetic level.

As mentioned above, gene counting implies the existence of large numbers of VSG genes within any population. It is also true that in most species of salivarian trypanosomes, with the important exception of *T. b.gambiense*, different geographical isolates express very different repertoires of VSG genes, thus enormously amplifying the size of the gene pool. A population of trypanosomes, all of whom are capable of expressing the same repertoire of VSG genes, constitutes a serodeme. The number of serodemes in circulation is not known. It has become clear that VSG gene repertoires can evolve rapidly by novel processes, such as partial gene conversion (PAYS 1983a, b), further expanding the diversity of the system. It is therefore obvious that immunisation against sleeping sickness is likely to be, at best, formidably difficult.

3   Diagnosis: Predominant Antigens and Cross-reactive Determinants

Two species of trypanosome are infective for human beings – *Trypanosome brucei gambiense* and *Trypanosoma brucei rhodesiense*. Both are lethal unless treated, although *T.b.gambiense* gives a more chronic form of the disease. If infected individuals are to be treated successfully, it is essential that on-the-spot diagnosis should be feasible, so that cases can be isolated for immediate treatment. Two types of test are currently in use, parasitological and serological. Both have serious disadvantages. Parasitological diagnosis is based on detection of the parasite by light microscopy in blood films or gland punctures. The same method is also used in detection of *T.b.brucei*, *T.congolense* or *T.vivax* infections in domestic cattle. The method suffers from lack of sensitivity, especially in *T.b.gambiense* infections in which fluctuations in the parasitaemia are such that parasites are frequently not detectable by this method.

A number of serodiagnostic tests have been developed over the past 20 years but, in the main, their use has been limited because they involve the use of expensive equipment and expensive and labile reagents, and they require skilled personnel. The literature is reviewed in more detail elsewhere (VOLLER and DE SAVIGNY 1981), but the following examples may be taken as representative. The indirect fluorescent antibody test (WÉRY et al. 1970) uses air-dried films of *T.b.brucei* or *T.b.gambiense* as a source of antigen for the detection of antitrypanosomal antibody (which may be anti-VSG or antibody to unidefined common components). Use of such fluorescence tests is not feasible in the field. An ELISA assay developed by VOLLER et al. (1975)

uses soluble antigen prepared from *T.b.brucei* by sonication and centrifugation to coat the walls of a microtitre plate. Circulating antibody in patient serum is detected following incubation on the plate and subsequent development with alkaline phosphatase-labelled antiglobulin. Again, the antigens detected are not defined. The capillary-passive haemagglutination test (BONÉ and CHARLIER 1975) uses a saline extract of *T.b.gambiense* coupled by glutaraldehyde treatment to type O negative human erythrocytes. Treated erythrocytes can be stored frozen in capillaries. In the test, diluted patient plasma is drawn into the capillary and the contents are allowed to mix. The erythrocytes are sedimented by centrifugation, and the capillaries are then held at 45°, pellet uppermost. In the absence of antibody, erythrocytes sediment down the tube; in the presence of antibody the pellet remains intact. This test has been modified for use in haemagglutination trays, and has been commercially developed as Cellognost (Behringwerke) and as Testryp (RIT, Smith-Kline). Once again, the antigens detected are not defined. Elevated IgM levels, considered diagnostic for trypanosomiasis, are detected in the immunoprecipitin test of CUNNINGHAM et al. (1967). The capillary flocculation test (Ross 1971) uses a particulate preparation from *T.b.brucei* or *T.b.rhodesiense* as a source of antigen to detect antibodies in animals infected with *T.b.brucei*, *T.b.gambiense*, *T.b.rhodesiense*, *T.evansi*, *T.congolense*, *T.vivax* and *T.simiae*. The particulate fraction is prepared following disruption of trypanosomes by sonication, centrifugation and washing of the pellet. The rate at which flocculation occurs in capillaries on mixing with test serum is a function of antibody titre. Although the antigen is not defined, it cannot be VSG, because of the mode of preparation, and obviously includes pan-species determinants. All the tests measure circulating antibody, not antigen. None of them used defined trypanosome antigens. All are, to greater or lesser extents, inconvenient for use in the field. Any serodiagnostic test which is to have real value in the field, must be rapid and simple to perform, and should use whole blood or saliva. The reagents must be stable for use in the tropics. The test must be portable and should not require elaborate equipment, and it should be inexpensive to produce and cheap to perform (WHO 1981). These problems have been solved in a most elegant and appealing way in the Card Agglutination of Trypanosomes Test (CATT) (MAGNUS et al. 1978), which is now undergoing extensive trials for its efficiency in detecting *T.b.gambiense* infections. The test takes advantage of the fact that there are very few serodemes of *T.b.gambiense* in circulation, perhaps only one or two (GRAY 1972, 1974). These serodemes share the same set of predominant antigens (GRAY 1974); that is, in all *T.b.gambiense* infections, the same variant antigen types (VATs) are expressed early in the course of infection. The corollary to this is that all infected individuals also produce antibody to the same group of VATs. In the CATT test, trypanosomes expressing predominant VATs are isolated and produced in bulk in rodents, and are then fixed, stained with Coomassie Brilliant Blue and stabilised such that they can be freeze dried. In the field, the freeze-dried preparations are reconstituted, and one drop of the suspension is mixed with one drop of whole blood from the patient in a depression in a small plastic card. If antibody to the predominant VATs is present, the fixed trypanosomes agglutinate, and this can easily be seen as a blue ring around the edge of the droplet. Such individuals can then be isolated for further testing. The CATT test thus fulfils all the requirements for field assay for trypanosomiasis – it is simple, reliable and cheap. It does have two drawbacks. The first is that it does not seem suitable for diagnosis of *T.b.rhodesiense*, or indeed of the cattle trypanosomiases. This is because *T.b. gambiense* is unique in the restricted repertoire of predominant VATs that it expresses. In all other species, very many serodemes seem to be in circulation and there is no evidence for sharing of predominant antigen types amongst these serodemes. Secondly, because the test is based on the presence of antitrypanosome antibodies, it does not distinguish between cured persons and those with active infections. At this stage, this is not really a serious problem – it is better to generate a low level of false positives in this way than to produce false negatives. In the long term, however, the development of a sensitive test for circulating free antigens, or for immune complexes, is desirable.

If a test for circulating antigen is to be developed, then a suitable antigen has to be identified. There are two possibilities. One is to test for circulating VSG. The other is to test for some non-variant-specific antigen. A test based on detection of VSG has the dual advantage that VSGs are the most abundant proteins in trypanosomes, and it is known that death of trypanosomes leads to release of VSG (reviewed in TURNER 1982). The obvious drawback is that all VSGs are immunochemically distinct, and it would be necessary to have an antibody which cross-reacted with all of them. As it happens, all VSGs do share a common antigenic determinant. VSGs are held onto the plasma membrane by a glycolipid which is attached to the C-terminal amino acid (HOLDER 1983; FERGUSON and CROSS 1984). VSGs are released from trypanosomes by a phospholipase which cleaves the hydrophobic component from the glycolipid, producing a water-soluble VSG (CARDOSO DE ALMEIDA and TURNER 1983). In certain species, most notably in rabbits, the carbohydrate component of the glycolipid is immunogenic, and the antibody produced cross-reacts with all VSGs (BARBET and MCGUIRE 1978; HOLDER and CROSS 1981). This carbohydrate is therefore known as the cross-reacting determinant, or CRD. Anti-CRD has no immunoprophylactic value for two reasons: firstly, because the organisation of VSGs within the surface coat is such as to make the CRD inaccessible to antibody (CARDOSO DE ALMEIDA 1983) and secondly, because anti-CRD only binds very weakly to VSG which still contains the complete glycolipid moiety (CARDOSO DE ALMEIDA and TURNER 1983) probably because of steric hinderance by the fatty acid-containing component. It may be possible, however, to devise a test for circulating VSG based on detection of the CRD. At present, very little is known about levels of VSG in the serum of chronically infected individuals. It is known that VSGs do not continually synthesise and release VSGs, but that VSG release is generally associated with senescence and cell death (BLACK et al. 1982). In hyperacute infections, it is possible to detect 28–320 µg VSG/ml plamsa in rats containing 4×10^8–2×10^9 trypanosomes/ml blood (DIFFLEY et al. 1980). These parasitaemias are totally unrelated to any levels found in field infections – if they were, diagnosis would be trivial. The methods used were insufficiently sensitive to measure VSG release in chronic low-level infections, where production of antibody may mean that released VSG is actually present in the form of immune complexes, which may be cleared very rapidly from the circulation. However, assuming that amounts detectable in acute infections are an accurate reflection of the amount that would be released in a chronic infection, then at a parasitaemia of 10^4/ml of blood, about 1 ng/ml VSG might be present. Given the limitations upon a serodiagnostic test imposed by the requirements for a field assay described above, it is going to be formidably difficult to achieve this kind of sensitivity in the near future, and therefore tests for circulating antibody are likely to be used for some time yet. The same limitations apply to a test based on non-VSG antigens. It is well known that all VATs share common antigenic determinants. These include membrane-bound antigens (SEED 1964), which are presumably transport proteins and the like, and soluble antigens which could be "housekeeping" enzymes (LE RAY et al. 1971; AFCHAIN et al. 1975; LE RAY 1975). None of them are present at the same abundance as VSGs, which implies that they are going to be even more difficult to detect in low-grade, chronic parasitae-mias. Most of the published work has used hyperimmune serum for immunochemi-cal comparison, but it is possible of course that some of them may be sufficiently immunogenic to use in a variation of the CATT test for detection of circulating anti-body generated in chronic infections. With the exception of the CATT test, all the serodiagnostic tests described above which use trypanosome antigens detect cir-culating antibody which recognises common (though not necessarily species-speci-fic) antigens. The implication is that the detection of antibodies specific for defined antigens should be feasible. This would be particularly useful for *T.b.rhodesiense* and, indeed, if species-specific antigens could be identified, it could also be extremely use-ful in diagnosis of different cattle trypanosomiases. Since the regime of chemothera-py adopted may differ for different species, this is an important practical considera-tion. Problems of producing sufficient antigen could be overcome by recombinant DNA technology, and the polypeptides produced could be used, for example, to coat coloured latex beads to use as the basis for a simple agglutination test. Unfortunately, there is very little information currently available on the specificity of the antibodies produced during a chronic infection, or on the antigens that they recognise. In recognition of this problem, the UNDP/World Bank/WHO Special Programme for Research and Training in Tropical Diseases has arranged to distribute human sera from patients with *T.b.gambiense* and *T.b.rhodesiense* infections to researchers who are attempting to identify common parasite antigens, although by the nature of the material, amounts available are strictly limited (WHO 1981).

Although field-based diagnostic tests must be a priority, there is still a place for lab-based tests for identification and classification of different strains and species col-lected in the field. Such information is very valuable in epidemiological studies. More sophisticated technologies can be considered for lab tests of this kind, and several methods have been used. These include analysis of interspecific and isoenzy-me polymorphism (GIBSON et al. 1980; GODFREY and KILGOUR 1978), and restriction endonuclease analysis of kinetoplast DNA (STEINERT et al. 1973, 1976; BORST et al. 1980, 1981). Since these tests are not antigen based, they will not be discussed further in this review.

Probes for VSG genes have also been used to discriminate species of trypanoso-mes, and may prove useful in the identification of different serodemes. Thus, PAYS et al. (1981, 1983c) used cDNA clones specific for two VSGs from the ANTAR-1 serode-me of *T.b.brucei*, AnTat 1.1 and AnTat 1.8, to look at the extent to which the genes for these two VSGs were conserved in different serodemes and in different species. The AnTat 1.8 probe recognised a number of bands on Southern blots of restriction enzy-me digests of DNA from *T.b.brucei*, *T.b.rhodesiense* and *T.evansi*. These different bands are assumed to be characteristic of different members of the AnTat 1.8 gene "family". The AnTat 1.8 gene family appeared to be much more highly conserved in the 14 stocks of *T.b.gambiense* examined than in stocks of *T.b.brucei*, *T.b.rhodesiense* and *T.evansi* and the pattern of bands was such that it was considered diagnostic of

T.b.gambiense. There were minor differences between some of the *T.b.gambiense* stocks which could in itself indicate that the technique could provide a powerful approach to distinguishing between stocks of *T.b.gambiense* which apparently express the same repertoire of VSGs. In contrast, the AnTat 1.1 gene was invariably absent from the stocks of *T.b.gambiense*, and could therefore apparently be used to distinguish between *gambiense* and non-*gambiense* species. By extension it could be possible to isolate VSG-specific DNA clones specific for each species to use as the basis for a comparatively simple assay for laboratory-based typing.

   Conservation of the AnTat 1.8 gene sequence in the genome of *T.b.brucei, T.b. gambiense, T.b.rhodesiense* and *T.evansi* and analysis of variation in kDNA sequences (BORST et al. 1980) has led to the suggestion that these represent variants of the same species, rather than different species. However, in another study (MASSAMBA and WILLIAMS 1984), cDNAs specific for three VSGs of the ILTAR-1 serodeme hybridised to DNA from five different strains of *T.b.brucei* and five different strains of *T.b. rhodesiense*, yet failed to hybridise to any of the strains of *T.b.gambiense, T.evansi, T.congolense* or *T.vivax* used. Interestingly, the one *T.b.brucei* genome DNA to which the VSGs did *not* hybridise was AnTat 1.8. However, a 2.2-kb genomic DNA clone prepared from ILTAR-1 DNA, which mapped to a region about 10 kb 3' to the ILTat 1.3 VSG coding sequence, hybridised to DNA from *all* strains of *T.b.brucei* tested, including AnTat 1.8, and to DNA from *T.b.rhodesiense*, but not from *T.b.gambiense, T.evansi, T.congolense* or *T.vivax*. This study suggest a closer genetic relationship between *T.b.brucei* and *T.b.rhodesiense* than any of the other species or subspecies. MASSAMBA and WILLIAMS (1984) also showed that the hybridisation analysis could be carried out in the form of a "dotblot", in which either purified trypanosomes or infected blood could be lysed on nitrocellulose filters to release DNA, which could then be fixed and detected with radiolabelled probes in the conventional way. The sensitivity was such that an adequate signal was obtained from 10^4 purified trypanosomes, or 10^5 trypanosomes in infected blood. This shows a way in which the use of such probes might eventually be used in a fairly simple assay for species discrimination, although a substantial increase in sensitivity coupled with a non-isotope detection system is still needed. Such studies are still in their infancy, but there is a clear potential for use of probes based on VSG-related sequences, in epidemiological analyses of trypanosomiasis.

4 Immunity

Man possesses a potent innate mechanism of humoral immunity which protects him from most species of salivarian trypanosomes. Normal human serum lyses all species except *T.b.gambiense* and *T.b.rhodesiense*. This phenomenon forms the basis of the blood incubation infectivity test (BIIT, RICKMAN and ROBSON 1970a,b), which remains the best way to distinguish between the three morphologically indistinguishable members of the *brucei* group: *T.brucei brucei* (lysed by normal human serum), *T.b.gambiense* and *T.b.rhodesiense* (not lysed). The active principle in serum is believed to be high-density lipoprotein (HDL) (RIFKIN 1978a,b), but the mechanism by which it acts is obscure. Calcium ions are apparently necessary, and α_2-macroglobin acts synergistically with HDL (D'HONDT et al. 1979). It is thought that exchange of lipids between HDLs and the trypanosome plasma membrane might create a lesion leading to an ultimate failure of the permeability barrier (D'HONDT and KONDO 1980). The surface coat could form a barrier inhibiting contact between the HDL and the plasma membrane, and experiments designed to decrease lateral diffusion of VSGs, and thus to minimise the chance of close contact, seem to support this hypothesis (RIFKIN 1983). It is worth noting that neither uncoated procyclic trypomastigotes nor coated senescent short stumpy forms are lysed by HDL (RIFKIN 1978b), indicating that a fundamental property of the plasma membrane of long slender blood stream trypomastigotes may be implicated. A complete understanding of this phenomenon could have profound implications for future work. For example, if it correlates with expression of a particular human apoprotein, then gene therapy of cattle, to allow them to express similar levels of immunity, could have a great economic impact. Different breeds of cattle all do get infected, but they show different levels of susceptibility to trypanosomiasis (reviewed in ROELANTS and PINDER 1983). It is by no means certain that so-called trypanotolerant cattle have developed a similar mechanism of non-specific immunity as that used by man, since parasitaemia usually develops in such animals. Reviewing all the published data, ROELANTS and PINDER (1983) propose that increased resistance in cattle does not correlate with acquired humoral immunity, and that innate mechanisms are of importance. What such mechanisms are is not known at all.

Studies on the pathogenicity of *T.b.brucei* in mice suggest that those mice which respond rapidly to produce an anti-VSG response to the infecting inoculum have the greatest success in controlling the initial parasitaemia (reviewed in BLACK et al. 1985). Furthermore, anti-VSG responses in such mice are said to develop only in response to the appearance in the blood stream of the non-replicative senescent, short stumpy trypomastigotes, and that the rate at which this differentiation step proceeds is under host control. That is, those mice which have more success in controlling the initial parasitaemia do so because they induce a switch to the short stumpy forms at an earlier stage, and then are the essential immunising species for the development of an anti-VSG response. The nature of the inducement provided by some strains of mice to promote accelerated switching to short stumpy forms is not understood, nor is it known if this mechanism is employed in trypanotolerant cattle.

Many attempts have been made to immunise against trypanosomiasis, but protection can only be induced against the immunising VAT. Such attempts include the use of whole trypanosomes (HERBERT and LUMSDEN 1968; WELLDE et al. 1975), soluble extracts (HERBERT and LUMSDEN 1968), and purified VSGs (CROSS 1975; BALTZ et al. 1977). From our current knowledge about the rate of antigenic variation, and about the antigenic heterogeneity of parasite populations in the blood stream, we can see that attempts to immunise with clones or with sVSG produced from clones of trypanosomes are doomed to failure. Current hopes for a trypanosome vaccine are centred on induction of antibody cross-reactive with all VSGs, and on induction of immunity against the full repertoire of metacyclic VATs.

Variant surface glycoproteins show quite remarkable amino acid sequence diversity. Out of the 450–500 amino acids in the sequence, the N-terminal 350–400 are hypervariable, with very little detectable homology other than the placement of cysteines (VSG biochemistry is reviewed in detail in TURNER 1982; 1984a,b; CROSS 1984).

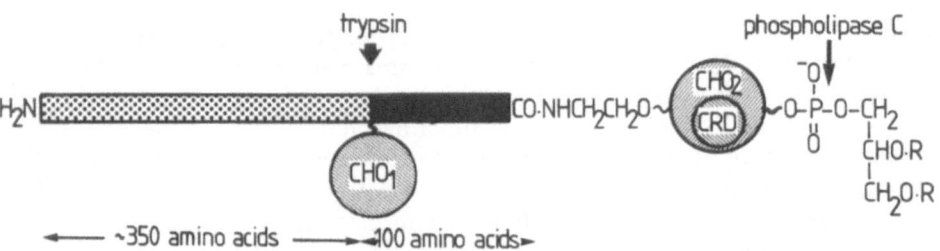

Fig. 2. Anatomy of a VSG. VSGs contain 450–500 amino acids, of which the N-terminal 350 or so are hypervariable (indicated by the *dotted area*) and the remainder from a more conserved region (*black area*). Proteolysis with trypsin or other enzymes cleaves the molecule into two domains. Homologies in the conserved domain allow division of VSGs into two sequence classes. Class I VSGs have a C-terminal aspartate, class II VSGs C-terminal serine, and in both classes the C-terminal amino acid is glycerated (CHO_2). The linkage is uncertain, but the carbohydrate is coupled through an ethanolamine which is in amide bond with the α-carboxyl group of the C-terminal amino acid. The carbohydrate forms part of a glycolipid, the exact structure of which is unknown, but which includes a phosphoglyceride containing two moles of myristic acid (R = C14:0 myristate) and which anchors the VSG to the plasma membrane. VSGs are released from trypanosomes through the action of a membrane-bound phospholipase, probably a phospholipase C, which cleaves off the diglyceride. CHO_2 contains a conserved antigenic determinant, the cross-reacting determinant or CRD. VSGs also contain at least on N-linked oligosaccharide (CHO_1). The exact number and location of the N-linked oligosaccharide differs in the two sequence classes, and between VSGs of the same sequence class, but it is usually found in the C-terminal third of the molecule, and is not available for lectin binding on living trypanosomes. The VSG exists as a dimer in solution

The anatomy of a VSG is shown in Fig. 2. Homologies in the C-terminal 100 amino acids, which seem to form a distinct domain which can be cleaved from an N-terminal domain by trypsin, allow the division of VSGs into two sequence classes, but as yet there have been no reports that such homologies can form the basis for the development of sequence-class specific antibodies, far less for the induction of cross-reactive immunity. Furthermore, the surface coat appears to be arranged on the trypanosome so that the N-terminal domain is exposed to the immune system, whereas the C-terminal domain is "masked". Since peptide immunogens have been used to stimulate the production of antibody which binds to sites in viral surface antigens formerly considered non-immunogenic when presented on the intact virus (reviewed in LERNER 1982) to the immune system, the production of neutralising, cross-reactive, class-specific antibodies is at least a formal possibility. It seems unlikely, however, since plenty of variant-specific monoclonal antibodies have been described which do not bind to living trypanosomes, but only to purified VSG or air-dried or acetone-fixed preparations of trypanosomes (MILLER et al. 1984; HALL and ESSER 1984). Such antibodies see "cryptic", or buried, variant-specific antigenic determinants, and they are incapable of neutralising infections produced with the homologous clones of trypanosomes (HALL and ESSER 1984).

As described in Sect. 3, anticarbohydrate antibodies can be prepared which recognise all VSGs by binding to the cross-reacting determinant which forms part of the glycolipid attached to the C-terminal amino acid. The reasons why such antibodies are of no immunoprophylactic value have already been discussed. Again, it remains at least a paper possibility that if anti-CRD could be prepared having a high affinity for the form of the VSG found on the plasma membrane of living trypanosomes, then such antibodies could be protective. Such a possibility must be considered to be remote.

It has proved possible to generate serodeme-specific immunity against *T. congolense* by immunising mice by repeated challenge with infected tsetse flies, followed by a curative dose of Berenil (NANTULYA et al. 1980). The immunity is short lived and metacyclic specific. As described in Sect. 2, within a serodeme of *T. congolense*, 12 monoclonal antibodies accounted for all the VATs expressed in a metacyclic population, and a cocktail of the 12 neutralised up to 10^6 metacyclic trypanosomes (CROWE et al. 1983). Production of a cocktail vaccine, using either peptide immunogens or polypeptides produced from recombinant DNA, should be feasible. There are two serious drawbacks. Firstly, each serodeme expresses a different repertoire of metacyclic VATs, and there may be very many serodemes in circulation. Secondly, at least in *T. b. rhodesiense,* the repertoire within a serodeme showed significant changes over a 20-year period (BARRY et al. 1983), which would imply that new cocktails would have to be produced periodically. No information is available as yet on the repertoire of metacyclic VATs in *T. b. gambiense*, in which only one or two serodemes are circulating, and the VSG gene repertoire seems to be much more stable. It is in this species that hopes for a successful vaccine based on VSGs expressed on metacyclic trypanosomes are most likely to be realised.

Recently, a novel approach to the control of trypanosomiasis has been reported (MAUDLIN et al. 1984). Rabbits were immunised with a membrane fraction produced from procyclic (uncoated) trypomastigotes of *T. congolense*. When tsetse flies were allowed to feed on such rabbits, the numbers of metacyclic trypomastigotes which developed in the flies were greatly reduced, presumably because ingested antibody blocket the development of procyclic trypomastigotes into epimastigotes. Such immunisation would not protect an individual, but might knock down the transmission rate. Whether such a scheme could be used in the field, where there is a large reservoir of infection in game animals, is by no means certain. Again, it may have some value in *T. b. gambiense* infections, where man himself seems to provide the principle reservoir. The nature of the antigens involved in the response is unknown.

5 Cure: VSGs as Targets for Directed Chemotherapy

Variant surface glycoproteins should provide excellent targets for chemotherapy, because of their essential role in the survival of the trypanosome in the blood stream of the mammalian host. Any agent which inhibits VSG synthesis, induces incorrect expression at the cell surface, or causes release of the surface coat in blood stream trypomastigotes is a potential trypanocide, because underlying the surface coat are common antigenic determinants (BARRY 1979) and structures which activate complement by the alternate pathway (TETLEY et al. 1981). One possibility is to inhibit VSG gene transcription. The expression of VSGs is controlled by a complex set of gene rearrangements, discussion of which is beyound the scope of this article (see STEINERT and PAYS 1985). A key point, however, is that transcription is discontinuous. The first 25 nucleotides of the mRNA are encoded in a so-called mini-exon (VAN DER PLOEG et al. 1982a; BOOTHROYD and CROSS 1982) and the remainder in one large exon

(VAN DER PLOEG et al. 1982a; BORST et al. 1981). Transcripts of 140 nucleotides containing the mini-exon are produced at a high level and donate the mini-exon either by priming transcription of the main exon or by bimolecular splicing to transcripts containing the main exon (CAMPBELL et al. 1984). Recent experiments show that transcription of the two exons is controlled by different polymerases showing different sensitivities to inhibition by α-amanitin. Most surprisingly, transcription of the main exon was not inhibited by levels of α-amanitin of up to 1000 µg/ml and inhibition of transcription of the mini-exon required 200 µg/ml (KOOTER and BORST 1984). In contrast, transcription of these other proteins was inhibited at 5 µg/ml, the level expected for a typical RNA polymerase II. Further characterisation of the polymerases involved in VSG transcription might prove a fruitful approach to chemotherapy.

A second possibility is to interfere with translation of VSG mRNA or with post-translational modification. It is known that VSG precursors have N-terminal signal peptides (BOOTHROYD et al. 1981; MCCONNELL et al. 1982), and although these are unusually long, they can be accurately processed by the signal peptidase found in stripped dog pancreas microsomal vesicles. The trypanosome signal peptidase is therefore unlikely to be a target. Similarly, VSGs are N-glycosylated by a pathway that seems broadly similar to that found in the mammalian host, in that N-glycosylation can be inhibited by tunicamycin, and precursor oligosaccharide can be accurately transferred to nascent VSG by the enzymes found in the dog microsome system (MCCONNELL et al. 1982). There may be differences in the processing of transferred oligosaccharides in parasite and host, based on the difference in the structure of the dolichol-bound precursor found in the mammalian host, and in the insect flagellate *Crithidia fasciculata*, which like the trypanosomes is a member of the order *Kinetoplastida*. In *C. fasciculata*, this intermediate has the structure: Man $_7$GlcNac$_2$, whereas in rat liver it has the structure: Glu$_3$Man$_9$GlcNac$_2$ (PARODI et al. 1981). This has not been studied in the trypanosome, but might merit closer attention.

The trypanosome mRNA encodes a sequence of 17–22 amino acids at the C-terminus which are absent from the mature VSG. Most of these amino acids are hydrophobic and would appear to form a suitable membrane-anchoring peptide (BOOTHROYD et al. 1980, 1981). However, they are not used for this purpose in the mature VSG, although they may be used to hold a precursor to the membrane of the endoplasmic reticulum. Instead, they are removed and replaced by the C-terminal glycolipid discussed earlier. The nature and timing of this unusual post-translational modification is not yet fully understood, nor is it clear why it should be necessary to replace the hydrophopic tail by a glycolipid (reviewed in TURNER et al. 1985). Presumably, the two events (removal of the tail and addition of the lipid) are tightly coupled to ensure that the VSG is not released from the membrane of the e.r. or Golgi body. The structure of the glycolipid is not yet fully understood. The α-carboxyl group of the C-terminal amino acid of the mature VSG is in amide linkage to ethanolamine (HOLDER 1983), and coupled to the ethanolamine, through uncharacterised linkages, are carbohydrate (*N*-acetylglucosamine) and galactose, with occasionally small amounts of mannose (HOLDER and CROSS 1981), phosphate (CARDOSO DE ALMEIDA and TURNER 1983), glycerol and myristic acid (FERGUSON and CROSS 1984; FERGUSON et al. 1985). Determination of the complete structure is urgently needed, and the pathway through which it is synthesised and transferred to nascent VSG merits detailed investigation. This problem is discussed in more detail elsewhere.

Lastly, activation of VSG release offers a potential approach to chemotherapy. Trypanosomes possess a membrane-bound enzyme (CARDOSO DE ALMEIDA et al. 1985), probably a phospholipase C (JACKSON 1983; FERGUSON et al. 1985), which is normally quiescent, but which is activated by perturbation of the plasma membrane to clip off part of the C-terminal glycolipid (probably a glyceride) and release soluble VSG. In the lab, activation of this pathway is brought about by breaking trypanosomes. This is the key to the purification of VSGs for biochemical analysis. The big question is whether this pathway is active under physiological conditions, i.e. can VSG release be activated without busting the plasma membrane. To answer this question, we need to know more about the role of VSG release during the life cycle of the trypanosome. The available evidence suggests that VSGs are not continuously shed from dividing blood stream trypomastigotes, but that they are shed from senescent short-stumpy forms, probably in consequence of cell death (BLACK et al. 1982). It is not known whether VSG release is an essential component of the antigenic variation process itself, simply because the event is so rare that it has not yet been possible to monitor it at the level of the individual trypanosome. VSGs are certainly released in soluble form during the transformation to procyclic trypomastigotes in vitro, an event which is equivalent to the transformation seen in the fly midgut. In this in vitro synthesis, repression of VSG synthesis release is detectable between 6 and 8 h after induction of transformation. VSG is formed in the culture supernatants as a mixture of the soluble form and an N-terminal proteolytic fragment (OVERATH et al. 1983). This release is slow compared with the rate of release which can be generated in broken trypanosomes (complete VSG release within 1 h; CARDOSO DE ALMEIDA and TURNER 1983).

If VSG release is a physiological event, there must be a pathway to activate the enzyme, which is normally inactive. Uptake of calcium ions (VOORHEIS et al. 1982) and treatment of trypanosomes with local anaesthetics such as benzyl alcohol (JACKSON 1983) are both said to stimulate VSG release, and both these pathways also activate adenylate cyclase (VOORHEIS and MARTIN 1980, 1981, 1982). Whether cyclase activation and VSG release are truly coupled events has yet to be established. Further research into the physiological stimulus, if any, controlling VSG release, may yet be the most fruitful new approach towards the controls of trypanosomiasis.

6 Conclusions

The reasons for the tremendous interest in antigenic variation in the African trypanosomes are obvious, quite apart from the fact that the parasite is of enormous social and economic importance. The trypanosome offers a fascinating insight into the way in which a simple eucaryote controls gene expression. Large amounts of comparatively well-defined material are available, both for molecular genetic analysis and for biochemical characterisation of the protein. The hypervariability in the protein sequence of members of this multigene family provides an approach to the relationship between sequence and structure and function. The unusual mode of attachment and release of VSGs is of great interest to membrane biochemists – VSGs may be the first-characterised examples of a new class of membrane proteins. Yet it is important to relate this research to some of the problems in the field, some of which will not be solved by elaborate technology. A simple and reliable field test for trypanosomiasis, both in man and in cattle, is needed – the CATT test beautifully exemplifies the kind of technology needed. On the other hand, development of new drugs or of vaccines may be stimulated by the basic research conducted on antigenic variation. A clear difficulty lies in the translation from the identification of sites of potential chemotherapeutic interest into marketable drugs active against such sites. Although the first part of the process may be accomplished by academic scientists, the second part must of necessity involve the pharmaceutical industry, and it is a hard fact that the production of trypanocides is not seen as commercially viable. The same strictures apply to vaccine production. The best hope must be for sponsorship of development costs as well as research by the World Bank/UNDP/WHO Special Programmes for Research and Training in Tropical Diseases, for at the end of the day the goal of research on these antigens must be the control of a parasite which still provides a major obstacle to the economic development of an area of Africa larger than the United States.

References

Afchain D, Le Ray D, Van Meirvenne N, Fruit J, Capron A (1975) Comparative immunoelectrophoretic analysis of culture and bloodstream forms of *Trypanosoma (Trypanozoon) brucei gambiense*. Demonstration of stage and antigenic type specific antigens. Ann Immunol 126C:45–50

Baltz T, Baltz D, Pautrizel R, Richet C, Lamblin G, Degand P (1977) Chemical and immunological characterisation of specific glycoproteins from *Trypanosoma equiperdum* variants. FEBS Lett 82:93–96

Barbet AI, McGuire TC (1978) Cross-reacting determinants in variant-specific antigens of African trypanosomes. Proc Nat Acad Sci USA 75:1989–1993

Barry JD (1979) Capping of variable antigen of *Trypanosoma brucei* and its immunological and biological significance. J Cell Sci 37:287–302

Barry JD, Crowe JS, Vickerman K (1983) Instability of the *Trypanosoma brucei rhodesiense* metacyclic variable antigen repertoire. Nature 306:699–701

Black SJ, Hewett RS, Sendashonga CN (1982) *Trypanosoma brucei* variable surface coat is released by degenerating parasites, but not by actively dividing parasites. Parasite Immunol 4:233–244

Black SJ, Sendashonga CN, O'Brien C, Borowy NK, Naessens M (1985) Regulation of parasitaemia in mice infected with *Trypanosoma brucei*. Curr Top Microbiol Immunol 117:93–118

Boné GJ, Charlier J (1975) L'Hémagglutination indirecte en capillaires: une methode de diagnostic de la trypanosomiase applicable sur le terain. Ann Soc Belge Med Trop 55:559–569

Boothroyd JC, Cross GAM (1982) Transcripts coding for different variant surface glycoproteins of *Trypanosoma brucei* have a short identical exon at their 5' end. Gene 20:279–287

Boothroyd JC, Cross GAM, Hoeijmakers JHJ, Borst P (1980) A variant surface glycoprotein of *Trypanosoma brucei* synthesised with a C-terminal hydrophobic "tail" absent from purified glycoprotein. Nature 288:624–626

Boothroyd JC, Paynter CA, Cross GAM, Bernards A, Borst P (1981) Variant surface glycoproteins of *Trypanosoma brucei* are synthesised with cleavable hydrophopic sequences at the carboxy and amino termini. Nucleic Acids Res 9:4743–4745

Borst P, Fase-Fowler F, Hoeijmaker JHJ, Frasch ACC (1980) Variation in maxi-circle and mini-circle sequences in kinetoplast DNAs from different *Trypanosoma brucei* strains. Biochem Biophys Acta 610:197–210

Borst P, Frasch ACC, Bernards A, Van der Ploeg LHT, Hoeijmakers JHJ, Arnberg AC, Cross GAM (1981) DNA rearrangement involving the genes for variant antigens in *Trypanosoma brucei*. Cold Spring Harbour Symp Quant Biol 45:935–943

Campbell DA, Thornton DA, Boothroyd JC (1984) Apparent discontinuous transcription of *Trypanosoma brucei* variant surface antigen genes. Nature 311:350–352

Cardoso de Almeida ML (1983) The mode of attachment of VSGs to the plasma membrane of *T. brucei*. Ph.D. Thesis, University of Cambridge.

Cardoso de Almeida ML, Turner MJ (1983) The membrane form of variant surface glycoproteins of *Trypanosoma brucei*. Nature 302:349–352

Cross GAM (1975) Identification, purification and properties of clone-specific glycoprotein antigens constituting the surface coat of *Trypanosoma brucei*. Parasitology 71:393–417

Cross GAM (1984) Structure of the variant glycoproteins and surface coat of *Trypanosoma brucei*. Philos Trans R Soc Lond [Biol] 307:3–12

Crowe JS, Barry JD, Luckins AG, Ross CA, Vickerman K (1983) All metacyclic variable antigen types of *Trypanosoma congolense* identified using monoclonal antibodies. Nature 306:389–391

Cunningham MP, Bailey NM, Kimber CD (1967) The estimation of IgM immunoglobulin in dried blood, for use as a screening test in the diagnosis of human trypanosomiasis in Africa. Trans R Soc Trop Med Hyg 61:688–700

D'Hondt J, Kondo M (1980) Carbohydrate alters the trypanocidal activity of normal human serum with *Trypanosoma brucei*. Mol Biochem Parasitol 2:113–123

D'Hondt J, Van Meirvenne N, Moens L, Kondo M (1979) Ca^{2+} is essential cofactor for trypanocidal activity of normal human serum. Nature 282:613–615

Diffley P, Strickler JE, Patton CCL, Waksman BH (1980) Detection and quantification of variant specific antigen in the plasma of rats and mice infected with *Trypanosoma brucei brucei*. J Parasitol 66:185–191

Doyle JJ, Hirumi H, Hirumi K, Lupton EN, Cross GAM (1980) Antigenic variation in clones of animal infective *Trypanosoma brucei* derived and cloned in vitro. Parasitology 80:354–370

Ferguson MAJ, Cross GAM (1984) Myristylation of the membrane form of a *Trypanosoma brucei* variant surface glycoprotein. J Biol Chem 259:3011–3015

Ferguson MAJ, Holder K, Cross GAM (1985) *Trypanosoma brucei* variant surface glycoprotein has a *sn*-1,2-dimyristyl glycerol membrane anchor at its COOH-terminus. J Biol Chem 260:4963–4968

Freymann DM, Metcalf P, Turner MJ, Wiley DC (1984) 6 Å-resolution X-ray structure of a variable surface glycoprotein from *Trypanosoma brucei*. Nature 311:167–169

Gibson WC, Marshall TF, Godfrey DG (1980) Numerical analysis of enzyme polymorphism: new approach to the epidemiology and taxonomy of trypanosomes of the subgenus *Trypanozoon*. Adv Parasitol 18:171–247

Godfrey DG, Kilgour V (1978) Enzyme electrophoresis in characterising the causative organism of Gambian trypanosomiasis. Trans R Soc Trop Med Hyg 70:219–224

Gray AR (1965a) Antigenic variation in clones of *Trypanosoma brucei*. I. Immunological relationships of the clones. Ann Trop Med Parasitol 59:27–36

Gray AR (1965b) Antigenic variation in a strain of *Trypanosoma brucei* transmitted by *Glossina morsitans* and *G. palpalis*. J Gen Microbiol 41:195–214

Gray AR (1972) Variable agglutinogenic antigens of *Trypanosoma gambiense* and their distribution among isolates of the trypanosomes collected in different places in Nigeria. Trans R Soc Trop Med Hyg 66:283–284

Gray AR (1974) Antigenic similarity among isolates of *Trypanosoma gambiense* from different countries in Africa. Trans R Soc Trop Med Hyg 68:150–151

Hajduk SL, Vickerman K (1981) Antigenic variation in cyclically transmitted *Trypanosoma brucei*. Variable antigen type composition of metacyclic trypanosome populations from the salivary glands of *Glossina morsitans*. Parasitology 83:596–607

Hall T, Esser K (1984) Topological mapping of protective and non-protective epitopes on the variant surface glycoproteins of the WRATat 1 clone of *Trypanosoma brucei rhodesiense*. J Immunol 132:2059–2063

Herbert WJ, Lumsden WHR (1968) Single dose vaccination of mice against experimental infection with *Trypanosoma (Trypanozoon) brucei*. J Med Microbiol 1:23–32

Holder AA (1983) Carbohydrate is linked through ethanolamine to the C-terminal amino acid of *Trypanosoma brucei* variant surface glycoprotein. Biochem J 209:261–262

Holder AA, Cross GAM (1981) Glycopeptides from variant surface glycoproteins of *Trypanosoma brucei*. C-terminal location of antigenically cross-reacting carbohydrate moieties. Mol Biochem Parasitol 2:135–150

Jackson DG (1983) Studies on the mechanism of release of the surface coat glycoproteins from *Trypanosoma brucei*. Ph.D. Thesis, University of Dublin

Kooter JM, Borst P (1984) α-Amanatin insensitive transcription of variant surface glycoprotein genes provides further evidence for discontinuous transcription in trypanosomes. Nucleic Acids Res 12:9457-9472

Kooter JM, De Lange T, Borst P (1984) Discontinuous synthesis of mRNA in trypanosomes. EMBO J 3:2387-2392

Kosinski RJ (1980) Antigenic variation in trypanosomes: a computer analysis of variant order. Parasitology 80:343-357

Lalor TM, Kjeldgaard M, Shimamoto GT, Strickler JE, Konigsberg WH, Richards FF (1984) Trypanosome variant-specific glycoproteins: a polygene protein family with multiple folding patterns. Proc Nat Acad Sci USA 81:998-1002

Le Ray D (1975) Antigenic structure of *Trypanosoma brucei (Protozoa, Kinetoplastida)*. Immunoelectrophoretic analysis and comparative study. Ann Soc Belge Med Trop 55:132-311

Le Ray D, Afchain D, Jadin J, Capron A, Famerae L (1971) Immuno-taxonomic interrelationships between *T. brucei, T. rhodesiense* and *T. gambiense*. Ann Parasit Hum Comp 46:523-531

Le Ray D, Barry JD, Vickerman K (1978) Antigenic heterogeneity of metacyclic forms of *Trypanosoma brucei*. Nature 273:300-302

Lerner RA (1982) Tapping the immunological repertoire to produce antibodies of predetermined specificity. Nature 299:592-596

Magnus E, Vervoort T, Van Meirvenne N (1978) A card agglutination test with stained trypanosomes (C.A.T.T.) for the serological diagnosis of *T. b. gambiense* trypanosomiasis. Ann Soc Belge Med Trop 55:169-176

Massamba NN, Williams RO (1984) Distinction of African trypanosome species using nucleic acid hybridisation. Parasitology 88:55-65

Maudlin I, Turner MJ, Dukes P, Miller N (1984) Maintenance of *Glossina morsitans morsitans* on antiserum to procyclic trypanosomes reduces infection rates with homologous and heterologous *Trypanosoma congolense* stocks. Acta Trop (Basel) 41:253-258

McConnell J, Gurnett AM, Cordingley JS, Walkier JE, Turner MJ (1981) Biosynethesis of *Trypanosoma brucei* variant surface glycoprotein. I. Synthesis and processing of an N-terminal signal peptide. Mol Biochem Parasitol 4:225-42

McConnell J, Cordingley JS, Turner MJ (1982) The biosynthesis of *Trypanosoma brucei* variant surface glycoproteins - in vitro processing of signal peptide and glycosylation using heterologous rough endoplasmic reticulum vesicles. Mol Biochem Parasitol 6:161-174

Miller EN, Turner MJ (1981) Analysis of antigenic types appearing in the first relapse populations of clones of *Trypanosoma brucei*. Parasitology 82:63-80

Miller EN, Allan LM, Turner MJ (1984) Topological analysis of antigenic determinants on a variant surface glycoprotein of *Trypanosoma brucei*. Mol Biochem Parasitol 13:67-81

Nantulya VM, Doyle JJ, Jenni L (1980) Studies on *Trypanosoma (Nannomonas) congolense*. II. Antigenic variation in three cyclically transmitted stocks. Parasitology 80:123-131

Overath P, Ozichos J, Stock V, Nonnengaesser C (1983) Repression of glycoprotein synthesis and release of surface coat during transformation of *Trypanosoma brucei*. EMBO J 2:1721-1728

Parodi AJ, Allue LAQ, Cazzulo JJ (1981) Pathway of protein glycosylation in the trypanosomatid *Crithidia fasciculata*. Proc Nat Acad Sci USA 78:6201-6205

Parsons M, Nelson RG, Watkins KP, Agabian N (1983) Sequences homologous to the variant antigen mRNA spliced leader are located in tandem repeats and variable orphons in *Trypanosoma brucei*. Cell 34:901-909

Pays E, Lheureux M, Vervoort T, Steinert M (1981) Conservation of variant specific surface antigen gene in different trypanosome species and subspecies. Mol. Biochem Parasitol 4:349-367

Pays E, Van Assel S, Laurent M, Darville M, Vervoort T, Van Meirvenne N, Steinert M (1983a) Gene conversion as a mechanism for antigenic variation in trypanosomes. Cell 34:371-381

Pays E, Delauw MF, Van Assel S, Laurent M, Vervoort T, Van Meirvenne N, Steinert M (1983b) Modifications of a *Trypanosoma b. brucei* antigen gene repertoire by different DNA recombinational mechanisms. Cell 35:721-731

Pays E, Dereck P, Van Assel S, Babiker EA, Le Ray D, Van Meirvenne N, Steinert M (1983c) Comparative analysis of a *Trypanosoma brucei gambiense* antigen gene family and its potential use in epidemiology of sleeping sickness. Mol Biochem Parasitol 7:63-74

Rickman LR, Robson J (1970a) The blood incubation infectivity test: a simple test which may serve to distinguish *Trypanosoma brucei* from *T. rhodesiense*. Bull WHO 42:650

Rickman LR, Robson J (1970b) The testing of proven *Trypanosoma brucei* and *T. rhodesiense* by the blood incubation infectivity test. Bull WHO 42:911

Rifkin MR (1978a) Identification of the trypanocidal factor in normal human serum: high density lipoprotein. Proc Nat Acad Sci USA 75:3450-3454

Rifkin MR (1978b) *Trypanosoma brucei*: some properties of the cytotoxic reaction induced by normal human serum. Expl Parasitol 46:189-206

Rifkin MR (1983) Interaction of high density lipoprotein with *Trypanosoma brucei*: effect of membrane stabilisers. J Cell Biochem 23:53-70

Roelants GE, Pinder M (1983) The immunobiology of African trypanosomiasis. Curr Top Immunobiol 12:225-274

Ross JPJ (1971) The detection of circulating trypanosomal antibodies by capillary-tube agglutination test. Ann Trop Med Parasitol 65:327-333

Seed JR (1978) Competition among serologically different clones of *Trypanosoma brucei gambiense in vivo*. J Protozool 25:526-529

Seed JR (1984) Antigenic similarity among culture forms of the *brucei* group of trypanosomes. Parasitology 54:593-596

Seed JR, Edwards R, Sechelski J (1984) The ecology of antigenic variation. J Protozool 31:48-53

Steinert M, Pays E (1985) Genetic control of antigenic variation. Br Med Bull 41:149-155

Steinert M, Van Assel S, Borst P, Mol JNM, Kleisen CM, Newton BA (1973) Specific detection of kinetoplast kDNA in cytological preparations of trypanosomes by hybridisation with complementary RNA. Exp Cell Res 76:175-185

Steinert M, Van Assel S, Borst P, Newton BA (1976) Evolution of kinetoplast DNA. In: Saccone C, Kroon AM (eds) The genetic function of mitochondrial DNA North Holland, Amsterdam pp 71-81

Tetley L, Vickerman K, Moloo SK (1981) Absence of a surface coat from metacyclic *Trypanosoma vivax*: possible implications for a vaccination against *vivax* trypanosomiasis. Trans R Soc Trop Med Hyg 75:409-414

Turner MJ (1982) Biochemistry of the variant surface glycoproteins of salivarian trypanosomes. Adv Parasitol 21:69-153

Turner MJ (1984a) The biochemistry of variant surface glycoproteins of the African trypanosomes. Biochem Soc Symp 49:169-181

Turner MJ (1984b) Antigenic variation in its biological context. Philos Trans R Soc Lond [Biol] 307:27-40

Turner MJ, Cardoso de Almeida ML, Gurnett AM, Raper J, Ward J (1985) Biosynthesis, attachment and release of variant surface glycoproteins of the African trypanosomes. Curr Top Microbiol Immunol 117:23-55

Van Meirvenne N, Magnus E, Janssens PG (1975) Antigenic variation in syringe-passaged populations of *Trypanosoma (Trypanozoon) brucei*. I. Rationalisation of the experimental approach. Ann Soc Belge Med Trop 55:1-23

Van der Ploeg LHT, Liu AYC, Michels PAM, De Lange T, Borst P, Majumber HK, Weber H, Veeneman GH, Van Boom J (1982a) RNA splicing is required to make the messenger RNA for a variant surface antigen in trypanosomes. Nucleic Acids Res 10:3591-3604

Van der Ploeg LHT, Valerio D, De Lange T, Bernards A, Borst P, Grosveld FG (1982b) An analysis of cosmid cloned of nuclear DNA from *Trypanosoma brucei* shows that the genes for VSGs are clustered in the genome. Nucleic Acids Res 10:5905-5923

Vickerman K (1969) On the surface coat and flagellar adhesion in trypanosomes. J Cell Sci 5:163-193

Vickerman K (1976) Antigenic variation in African trypanosomes. In: Porter K, Knight J (eds) Parasites in the immunised host. Mechanisms of survival. Ciba Federation Symposium, vol 25 (new series). Associated Scientific, Amsterdam pp 58-80

Voller A, De Savigny D (1981) Diagnostic serology of tropical parasitic diseases. J Immunol Methods 46:1-29

Voller A, Bidwell A, Bartlett A (1975) A serological study on human *Trypanosoma rhodesiense* infections using a micro-scale enzyme-linked immunosorbent assay. Tropenmed Parasitol 26:247-251

Voorheis HP, Martin BR (1980) "Swell dialysis" demonstrated that adenylate cyclase in *Trypanosoma brucei* is regulated by calcium ions. Eur J Biochem 113:223-227

Voorheis HP, Martin BR (1981) Characterisation of the calcium-mediated mechanism activating cyclase in *Trypanosoma brucei*. Eur J Biochem 123:371-376

Voorheis HP, Martin BR (1982) Local anaesthetics including benzyl alcohol activate the adenylate cyclase in *Trypanosoma brucei* by a calcium-dependent mechanism. Eur J Biochem 123:471–476

Voorheis HP, Bowles DJ, Smith GA (1982) Characterisation of the release of the surface coat protein from bloodstream forms of *Trypanosoma brucei*. J Biol Chem 257:2300–2304

Wellde BT, Schoenbeckler MJ, Diggs CL, Langbelin HR, Sadun EH (1975) *Trypanosoma rhodesiense:* variant specificity of immunity induced by irradiated parasites. Exp Parasitol 37:125–129

Wéry M, Van Wettere P, Wéry-Paskoff S, Van Meirvenne N, Mesatena M (1970) The diagnosis of human African trypanosomiasis (*T.gambiense*) by the use of the fluorescent antibody test. 2. First results of field applications. Ann Soc Belge Med Trop 50:711–730

WHO (1979) The African trypanosomes. WHO Technical Report series, no. 635, WHO Geneva

WHO (1981) Scientific working group on African trypanosomes. Application of modern technoloy in the development of serodiagnostic reagents and studies on antigenic variation in African trypanosomiasis. WHO, Geneva

Antigens of Taeniid Cestodes in Protection, Diagnosis and Escape

L.J.S. HARRISON[1] and R.M.E. PARKHOUSE[2]

1 Introduction

Parasites of the family Taeniidae (the taeniids) have a worldwide distribution but tend to have a higher prevalence in the developing countries. This is probably the most important cestode family as it contains several zoonotic species which cause health hazards to man and expense due to medical treatment. Economic losses to agricultural and related industries are incurred due to infection of domestic food animals and subsequent treatment or condemnation of infected carcasses. Perhaps more important is the loss of potential export markets and constraints placed on the development of livestock industries in those developing countries with high instances of infection amongst their domestic livestock.

Taeniids have two mammalian hosts in their life cycles, the adult being an obligate intestinal parasite producing eggs which are voided with the faeces and form the only free-living stage in the life cycle. If they are ingested by the intermediate host, the eggs hatch and activate in the intestines. The whole process of hatching and activation of these invasive larvae or oncospheres and their subsequent penetration of the gut wall is of obvious importance, as they are prime targets for immune attack; their biology was reviewed by LETHBRIDGE (1980). Once through the intestinal barrier, the larvae pass round the body in either the blood or lymph and then develop into the mature larval or metacestode stage in the tissues, the exact location being species dependent. The definitive host becomes infected with an adult tapeworm after ingesting the metacestode.

Members of this family can be divided into two main groups, as some species such as *Echinococcus granulosus, E. multilocularis* and *Taenia multiceps* undergo asexual development in the intermediate host whereas others do not replicate at this stage. The life cycles of *E. granulosus* and *T. multiceps* usually involve dogs

1 University of Edinburgh, Centre for Tropical Veterinary Medicine, Easter Bush, Roslin, Midlothian EH25 9RG, Scotland, United Kingdom
2 National Institute for Medical Research, The Ridgeway, Mill Hill, London NW7 1AA, United Kingdom

as the definitive host and domestic ruminants as the intermediate host but unfortunately man can also act as an intermediate host, especially in the case of *E. granulosus*, with potentially serious consequences. The fox is the usual definitive host for *E. multilocularis*, but dogs and cats can be involved. A variety of small rodents normally act as intermediate hosts, but again man can also become infected with the larvae. This parasite in man characteristically damages vital organs and infection can be fatal.

Parasites such as *T. solium, T. saginata, T. ovis* and *T. hydatigena* do not multiply in the intermediate host. Man is the definitive host for the first two parasites but infection is not considered to be a serious human health problem (PAWLOWSKI and SCHULTZ 1972). Unfortunately man can also act as intermediate host for *T. solium* with, in some cases, severe disturbances such as neurological disorders. In some rarer cases infections can also be fatal (VELASCO-SUAREZ et al. 1982). *T. saginata* and *T. solium* cause economic problems to agricultural and related industries due to infection of the intermediate hosts, cattle and swine respectively, with the larvae. Infected carcasses are either treated and downgraded in value or condemned as unfit for human consumption (GRINDLE 1978; ACEVEDO-HERNANDEZ 1982; DE ALUJA 1982). *T. ovis* and *T. hydatigena* have a low infectivity to man. Their usual definitive host is the domestic dog and intermediate hosts are sheep and goats; so that these parasites can also cause economic loss to agriculture if infections become prevalent. The definitive host of *T. pisiformis* is the domestic dog and the intermediate host is the rabbit. Therefore, this species can cause problems if rabbits for the domestic market become infected.

Taenia crassiceps and *T. taeniaeformis*, which have mice and rats as intermediate hosts, do not normally infect either man or domestic food animals, but can be easily maintained in the laboratory, making them very useful models of larval taeniid infections. *T. pisiformis, T. ovis* and *T. hydatigena* are also suitable models for laboratory study, the latter two especially for veterinary work as the intermediate hosts are small domestic ruminants. Control of those taeniids of veterinary or medical importance is desirable if the health hazard to man and/or economic losses are unacceptable. The various perspectives and options on this subject have been reviewed (GEMMEL 1978). One aspect is the development of an immunological approach to the diagnosis of infection and the use of vaccination to prevent infection. However, as pointed out by URQUHART (1980), before an effective vaccine is produced many other factors such as potential markets and competing products such as drugs must be taken into account.

There are many reports in the literature on attempts to use serological techniques to diagnose taeniid cestode infections in both man and domestic animals. In addition there is a great deal of evidence to suggest that it is possible to vaccinate animals against challenge infection with these parasites. The parasite antigens involved have been the subject of some recent reviews (PERCY and LAUFFAU 1979; MITCHELL and ANDERS 1982). Antigens of significance are those which induce protection in the host to challenge infection and those antigens which could be used in specific immunodiagnostic tests. There is, however, very little known about their physicochemical characteristics or indeed about their location in the tissues, surfaces or excretions/secretions of these parasites. In addition the species which have been studied in most detail in this respect are not usually those of great economic or veterinary/medical importance but model systems, such as *T. taeniaeformis* and *T. pisiformis* (for review see WILLIAMS and SANDEMAN 1982). The effector mechanisms of host resistance to parasite infection and the genetic and molecular basis of this resistance are also not fully understood. Neither are the mechanisms by which the parasite, in an appropriate host, manages to evade the host's immune response and the genetic and molecular basis of this evasion (MITCHELL 1982a, b). A rational approach to all of these questions can only be precisely designed when the chemical nature of the target antigens is known.

The purpose of this review is to attempt to summarize the known parasitology and biology of these taeniid cestodes and the extent to which the modern immunochemical/molecular biological approach to their diagnosis, cure and prevention has been applied, especially in areas where the antigens involved are defined biochemical entities. As a corollary, since much remains to be done, the article will also serve as a reference point of entry for the modern molecular approaches which must surely soon be attempted.

2 Protective Antigens

An important consideration in the vaccination of animals against helminth infections is the type of antigen to be used. In helminths this has usually meant the use of one of three alternatives; irradiation-attenuated live helminths, somatic extracts or excretory/secretory antigens collected after culture of various stages of larval or adult helminths in vitro (CLEGG and SMITH 1978; TAYLOR and MULLER 1980). Other factors are the use of various adjuvants as potentiators of the immune response, the effects of different routes of immunization on the development of protective immunity and the possible use of heterologous, instead of homologous, antigens (LLOYD 1981) as immunogens. Innate resistance may also play an important role by affecting the response of a host to infection. Factors such as age, strain, sex and general health of the host and, in addition, the infectivity of different strains of the same parasite, should be taken into account (RICKARD and WILLIAMS 1982).

Many attempts have been made to immunize animals against cestode infections, but despite the success which has been achieved, relatively little information is available on the nature and location of the protective antigens. This section attempts to review what is known about their physicochemical nature.

It has frequently proved difficult to obtain sufficient quantities of homologous helminth material for use in immunization and diagnostic studies and it is often impossible or impracticable to maintain certain species of helminths in the laboratory. In the absence of suitable immunogens manufactured by genetic engineering, heterologous antigens from antigenically closely related species of taeniids, which can be maintained in the laboratory, have often been used. These studies take advantage of the extensive cross-reactions between taeniid antigens (HEATH et al. 1979; LLOYD 1979; RICKARD and BRUMLEY 1981).

Evidence suggests that oncospheres or oncospheral products are a potent source of protective antigens (GALLIE and SEWELL 1981; RICKARD 1982). Information available on the metacestode stage of *T. taeniaeformis* in rodents indicates that detergent-solubilized protective oncospheral antigens were present in a high-pressure liquid chromatography fraction of oncospheres (MW > 200 000) (RAJASKARIAH et al. 1982). More recent investigations were performed by BOWTELL et al. (1983), who used a combination of radiolabelling techniques, immune coprecipitation and sodium dodecyl sulphate polyacrylamide gel electrophoresis (SDS-PAGE) under reducing conditions. These authors identified two major antigens in somatic preparations of 28-day-old *T. taeniaeformis* larvae (MW 40 000 and 200 000) which were also present in oncospheres. A further two major oncospheral antigens (MW 55 000 and 60 000) were also present in the excretions/secretions of 28-day-old larvae. Unfortunately none of these *T. taeniaeformis* antigens proved to be protective when isolated by SDS-PAGE under reducing conditions (LIGHTHOWLERS et al. 1984). In contrast antigen activity survived extraction and subsequent electrophoresis in sodium deoxycholate providing non-reducing conditions were employed. As yet the structural basis for these differences remains obscure.

Excretory/secretory products of *T. pisiformis* oncospheres cultured for 6–9 days, as well as extracts of whole dead larvae, could be used successfully to immunize rabbits against infection. Six components were detected by disc electrophoresis of the tissue culture medium; two of these were glycoproteins (HEATH 1973, 1976). Protective immunogens (MW > 300 000) were also detected in excretions of *T. pisiformis* oncospheres (RICKARD and KATIYAR 1976).

When the antibody response of resistant and susceptible groups of calves to challenge infection with *T. saginata* eggs was analysed by SDS-PAGE, under reducing conditions, followed by immunoelectrotransfer blot techniques (Western blotting), two antigens (MW 26 000 and 35 000) were identified to which antibodies were present in the sera of resistant but not susceptible cattle (HARRISON et al. 1984). These oncospheral antigens may be of importance in protection, and since immunized Balb/c mice responded to the same antigens, the production of monoclonal antibodies (HARRISON 1984) against these determinants is feasible.

Protective antigens have also been extracted and partially characterized from metacestodes of the rat strain of *T. taeniaeformis* and from their excretory/secretory antigens. A highly antigenic protein (MW 140 000) could be obtained from both sources (KWA and LIEW 1977). Extracts of entire *T. taeniaeformis* metacestode surface membranes prepared by incubating the metacestodes in Triton X100 were effective in immunizing rats against a challenge infection with *T. taeniaeformis* eggs. The active component was not precipitated by 10% w/v trichloroacetic acid (TCA) (WILLIAMS and SANDEMAN 1982). Immunogens with a similar solubility in 10% w/v TCA were also released from *T. taeniaeformis* metacestodes maintained in vitro overnight and such crude in vitro products were effective in immunizing rats against challenge infection with this parasite (AYUYA and WILLIAMS 1979).

In general, however, less success has been reported for vaccination with metacestode material as opposed to oncospheral products. These findings are supported by results obtained in systems such as *T. saginata, T. hydatigena* and *T. ovis* (RICKARD and ADOLPH 1977; LLOYD 1979; RICKARD and BRUMLEY 1981).

Protective antigens may also be present in somatic extracts derived from adult *T. saginata* strobilate material. Such extracts were capable of inducing complete protection against infection with *T. saginata* in cattle (GALLIE and SEWELL 1976).

3 Diagnostic Antigens

The two major criteria for a reliable diagnostic test are sensitivity and specificity. Most of the modern techniques are potentially of an adequate sensitivity, but are lacking in acceptable levels of specificity. This may be primarily a failure in the quality of the antigens used. In principle it is possible to design tests for circulating antibodies to the parasite or alternatively for antigens released by the parasite. If they are to be effective, however, detailed information is first needed on the presence or absence of cross-reactions, in the chosen system, between different parasite families and genera. The necessary strict specificity demanded for reliable diagnosis is not, of course, a prerequisite of protective antigens, where effectiveness through cross-reactions against a group of parasites would hardly be a problem. Epidemiological information is also needed on the occurrence of the various parasites in different areas and, in particular, which parasites display cross-reactions and hence yield false-positive results. The current state of diagnostic serology for tropical parasitic diseases and some often-neglected criteria, such as antigen purity and validation of immunodiagnostic methods by detailed studies on assay predictive value, sensitivity, specificity, cross-reactivity and precision, have been outlined (VOLLER and DE SAVIGNY 1981).

Unfortunately, it is still common to encounter the term "parasite antigen" in the literature when the author refers to a total, saline-soluble, extract of the parasite. In these cases it would be preferable if the more accurate term "saline extract" were used in preference to "antigen". Knowledge of the total complexity of the antigens in a parasite extract, and of the subsets of these which cross-react with other taeniids and/or other parasites, is obviously essential if specific diagnostic antigens are to be identified (PARKHOUSE et al. 1984).

Diagnosis of the adult parasite is usually a much more simple matter since the appearance of proglottids or eggs in the faeces is an immediate indication of infection. Therefore most work has concentrated on the more urgent requirement for reliable diagnosis of metacestode infections.

BIGUET et al. (1965) and CAPRON et al. (1968) carried out comparative antigen analyses of several members of the family Taeniidae. Hyperimmune sera against *T. saginata* were prepared and 23 antigens in this parasite were identified by immunoelectrophoresis. Up to 16 of these were present in six other cestode genera and between two and five of the antigenic components of *T. saginata* were detected in various nematode and trematode species. In addition these authors identified a major antigen of *E. granulosus* (antigen 5) which is now also known to occur in *T. hydatigena, E. multilocularis, T. ovis* and *T. solium* (VARELA-DIAZ et al. 1977a, b; YARZABAL et al. 1977a; YOUNG and HEATH 1979; CONDER et al. 1980; SCHANTZ et al. 1980).

Antigen 5 is probably the most highly characterized substance from these taeniids but the terminology used to describe this and other major antigens of these cestodes is very confusing. There has been little comparison of antigens between laboratories and no standard descriptions or nomenclature. However, there appear to be two main lipoprotein antigens in *E. granulosus*. The first, designated antigen A (ORIOL et al. 1971; WILLIAMS et al. 1971), probably corresponds to antigen 5 described by CAPRON et al. (1968) and the second has been designated antigen B (ORIOL et al. 1971). This complex situation was reviewed by WILLIAMS and SANDEMAN (1982).

The physicochemical characteristics of these hydatid antigens have been studied in detail (BOUT et al. 1974; ORIOL and ORIOL 1975; DOTTORINI and TASSI 1977; PIANTELLI et al. 1977). Antigen A (or antigen 5) is a lipoprotein with α- and β-carboxylesterase activity and a MW of about 400 000. It is considered to be composed of non-covalently linked subunits. These subunits migrate as a single component (MW 67 000) on SDS-PAGE under non-reducing conditions, but dissociate into two smaller subunits (MW 47 000 and 20 000) on reducing gels.

Antigen B is heat stable at 100 °C and has a MW of 150 000. It gives three components in the 10 000–20 000 MW range on SDS-PAGE with or without reducing conditions. These subunits may, however, become associated through non-covalent linkages to produce aggregates in a range of sizes.

These relatively well characterized antigens of *E. granulosus* (antigens A and B) also represent one of the few instances in which the location of the antigen within the parasites' tissues is known. The lack of information on the location or characteristics of antigens of the taeniids and on the location of host immunoglobulin hinders the understanding of the factors governing the immune response of the host and of the mechanisms which have evolved to evade these responses. Factors such as the presence or absence of immune complexes on the parasite's surface and their fate if present are important considerations if the way in which parasite antigens are presented to the host is to be elucidated.

Echinococcus granulosus antigen A (antigen 5) is mainly localized in the brood capsule wall, the germinal membranes and the subtegumental cells of the protoscolices, where it is thought to be synthesized. Smaller amounts are also found in the walls of the collecting ducts and flame cells. Some also finds its way into the cyst fluid and is thought to arrive there after being liberated from the areas of manufacture. Antigen B is most densely located in the tegument of the protoscolices, especially anterior to the suckers. It is also located in the calcareous corpuscles. The subtegumental cells and the brood capsule parenchymal cells are thought to be where this antigen is synthesized (YARZABAL et al. 1976, 1977b; RICKARD et al. 1977; DAVIES et al. 1978).

Recently it has been shown that of the 20–22 antigens in *T. saginata* proglottids identifiable by immunoelectrophoresis, there may be at least 7 components which do not share identity with the host and which are absent from other closely related helminth parasites (GEERTS et al. 1979). Three of these antigens may be of some value in the diagnosis of *T. saginata* metacestode infection in cattle but there has been little success so far in exploiting them. Evidence suggests that the techniques at present in use may not be sufficiently sensitive to detect infection in cattle with cyst burdens of less than 100 (GEERTS et al. 1981). Little or no information is available on the location of these antigens.

One of the electrophoretically identifiable antigens of *T. solium*, designated antigen B by FLISSER et al. (1980), is recognized by antibodies from a large proportion of cysticercosis patients who show a detectable serological response. However, serodiagnosis in such cases of human cysticercosis is confounded by the very large number of patients who show a limited humoral response to infection or whose antibody response is below that detectable by immunoelectrophoresis.

There are up to 11 antigenic components in aqueous extracts of *T. solium* metacestodes, 8 of which may react with antibodies from infected human sera. The *T. solium* antigen B has an isoelectric point at pH 8.6 and is distinguished from the other main antigens, designated A, C and E. At this pH antigens A and E migrate to the cathode while antigen C migrates to the anode. Again, there is little or no information available on location of these antigens within the tissues of the metacestode.

There is some evidence to suggest the existence of stage-specific antigens in the development of taeniid larvae from the invasive oncosphere to the fully developed metacestode (CRAIG and RICKARD 1981a). A difference in the time course of antibody production in *T. ovis*- and *T. hydatigena*-infected lambs was detected when the response to oncospheral antigens was compared with that against metacestode antigens. HARRISON (1984), by the use of monoclonal antibodies prepared against. *T. saginata* oncospheres and metacestodes, demonstrated the presence of stage-specific antigenic determinants both on the surface of the oncospheres and fully developed metacestodes. As yet there is very little information on the physicochemical characteristics and location within the parasite of stage-specific antigens in the taeniids. However, immunodiagnostic tests specific for the stage of the parasites' development might indicate the time course of the infection in the host.

Until cross-reacting components have been identified and discarded, any attempts to apply sensitive diagnostic techniques in the field will be confounded by cross-reactions with other parasite infections (CRAIG and RICKARD 1980; HARRISON and SEWELL 1981). However, attempts to produce immunodiagnostic reagents by immunoaffinity purification of antigens did not result in absolute specificity and some reduction in sensitivity was recorded (CRAIG and RICKARD 1981b). Further work, using selected hybridoma antibodies, did not totally eliminate cross-reactions (CRAIG et al. 1980, 1981). However, the antigens involved were not characterized at the biochemical level, nor was any information available on their location within the parasite.

Mammalian hosts infected with cestodes commonly develop immediate-type hypersensitivity reactions to cestode antigens. The most reactive of the allergens of *T. saginata* is carbohydrate in nature (MACHNIKA-ROGUSKA and ZWIERZ 1970). The major allergenic component of *T. taeniaeformis* is a protein of 50 000 MW, present in both the adult and metacestode (LEID and WILLIAMS 1974), which migrates cathodally on electrophoresis at pH 8.4. Conversely a component of *T. solium*, designated antigen H, which migrates anodally at pH 8.6, is reported to have a tendency to provoke an IgE response in man (FLISSER et al. 1980). The heat-stable lipoprotein, antigen B from *E. granulosus*, has been used in the assay of human reagins (DESSAINT et al. 1975).

This small amount of evidence on the physicochemical characteristics of these major allergenic components of the taeniids suggests that they have little structurally in common. In addition, tests for reaginic antibodies (FROYD 1963; GEERTS and KUMAR 1977) generally exhibit poor specificity.

A major problem gap in speciating samples of adult taeniids or speciating taeniid eggs in sewage or pasture is the absence of specific tests for the eggs. Identification on morphological grounds is not possible as the various species yield eggs of very similar structure. In areas of high transmission of, for example, *Echinococcus sp.,* a reliable specific test is an urgent priority. Given the highly conserved morphology of the embryophoral blocks, it seems likely that this reflects a biochemical homogeneity. Hence, specific identification of these eggs might be more fruitfully pursued at the level of the oncospheral membrane or possibly the cell membrane of the hatched and activated oncosphere (CRAIG 1983).

4 Escape

For successful transmission it is essential that the developing metacestodes can survive in the host despite the immune response and resistance to reinfection that the presence of the larvae provoke in the host. The main mechanisms which these and other parasites have evolved to evade the hosts' immune response have been reviewed (OGILVIE and WILSON 1976; BLOOM 1979; WILLIAMS 1979; MITCHELL 1982a, b). The most detailed work on how members of the family Taeniidae evade the immune mechanism has involved the study of the rodent model systems, *T. taeniaeformis* and *T. crassiceps* and also, to some extent, experimental infections of mice with *E. granulosus* and *E. multilocularis* larvae.

One method by which evasion may be achieved is through genetic variation within the host population (WAKELIN 1984). Certain individuals may have a genetically predetermined low responsiveness to infection which enhances the parasite's chances of survival. In the murine model systems, different inbred strains of mice show such variation.

The *T. taeniaeformis* system has been reviewed by MITCHELL et al. (1982). It should be noted that all the strains of mice studied exhibited strong concomitant immunity so that after having been infected they become resistant to further challenge. This is a common feature of larval cestode infections. However, different strains of mice were found to vary greatly in their response to primary oral infection with *T. taeniaeformis* eggs. Initially this difference lay mainly in the rapidity with which the various strains developed an antibody response to infection. In the strain C57BL/6, which showed a rapid response to infection, few strobilocerci developed or survived and those that did were small compared with the larger number which developed in a susceptible strain such as C3H/He.

In a study comparing primary and secondary infections of BALB/c, BDF1 and CH3 mice with *T. crassiceps* metacestodes (SIEBERT and GOOD 1980), CH3 mice showed a stronger humoral response and a slower cellular encapsulation response to secondary infection than BALB/c or BDF1 mice. Early immune damage was less pronounced in the latter two strains. However, resistance to reinfection was similar, suggesting that genetically determined differences may alter the state of equilibrium between the different effector mechanisms without necessarily producing variation in host resistance to reinfection. A similar situation may exist in the closely related cestode family, the Mesocestoidea, since, when six strains of mice were compared for susceptibility to infection with *Mesocestoides corti*, C57BL/6 were most resistant and CBA/H the most susceptible (WHITE et al. 1982).

It has been established that antibody levels play an important role in protective immunity against infection with taeniid larvae. In the early stages following infection the larvae are very susceptible to antibody attack. As they develop they become more resistant (MITCHELL et al. 1982). Indeed it is possible passively to protect animals against infection simply by inoculating them with serum from immunized hosts (SOULSBY and LLOYD 1982). Presumably the antigen targets for these antibodies are the protective antigens present in the oncospheres.

Although cell/parasite interactions monitored in vitro do not necessarily mirror the situation in vivo, they may give some indication of the kind of interactions which occur in the host. When *T. taeniaeformis* strobilocerci were cultured in vitro in the presence of various combinations of peritoneal cells and serum from infected or uninfected rats, cells adhered to the tegument of the larvae. There was evidence of extensive damage to the tegument and profound changes in the adherent cells after only 1 h. Both cells and serum, which had not been heat treated to destroy complement, were needed for the damage to occur but there was no difference between serum from infected or uninfected rats. These results suggested that potent non-specific effector mechanisms can rapidly damage the tegument in vitro, contrasting with the failure of recognition and rejection by host defences in vivo. Established strobilocerci are not always invulnerable to cellular attack, but the balance of the host-parasite relationship in vivo must favour their survival (ENGEL-KIRK et al. 1981). Nothing is known about the targets of this attack.

Living *E. granulosus* protoscolices maintained in vitro stimulate blastic transformation in putatively normal murine lymph node cells (DIXON et al. 1982). Stimulation apparently occurs on contact, is not host derived, is not secreted, is not diminished by maintaining the parasite in vitro and is fully active only on the living organism. This mitogenic response is T cell dependent. It has been postulated that these blastic responses may be the in vitro representation of an immunosuppressive mechanism favouring the survival of the parasite in the host. The physicochemical nature of the "stimulator" has yet to be determined. ALLAN et al. (1981) used mesenteric lymph node cells from *E. granulosus equinus*-infected mice for adoptive cell transfer into syngeneic normal responder recipient mice. The cell transfer inocula were shown to have a depleted T-cell population but to be highly suppressive to the antibody response of recipient mice to sheep red blood cells.

The relationship between immunodepression and parasite survival was also investigated in the *T. crassiceps* mouse model system (GOOD et al. 1982). Excretory/secretory antigens taken from *T. crassiceps* metacestodes cultured in vitro can elicit resistance to reinfection. At the same time this antigenic material can provoke a non-specific immunodepression and eosinophilia to a degree comparable to that produced by injection of living larvae. Depression of the host's antibody response to sheep erythrocytes occurs during established intraperitoneal infections with *T. crassiceps* metacestodes, but not in subcutaneous infections or in individual mice which have overcome their intraperitoneal infections. The degree of immunodepression was apparently related to the high parasite burden in intraperitoneal infections. Larvae in subcutaneous sites were surrounded by a host-derived cyst, and the consequent sequestration of the larval antigens may explain the lack of immunodepression. Intraperitoneal larvae, in contrast, are not encapsulated. Again little is known about the target antigens.

Host immunoglobulin has been demonstrated, both on the surface of several larval cestode species (WILLMS and ARCOS 1977; KWA and LIEW 1978; ALI-KHAN and SIBOO 1981; SIEBERT et al. 1981) and within the cyst fluid of the metacestodes (HUSTEAD and WILLIAMS 1977). Immunoglobulin on the surface may benefit the parasite, as *T. taeniaeformis* larvae coated with normal or immune sera survived better than uncoated controls when transplanted into normal and immune hosts (KWA and LIEW 1978). However, this resistance is not complete, since a proportion of *T. crassiceps* larvae from long-term infected mice are subject to immune damage, both in vitro and in vivo (SIEBERT et al. 1981). These larvae have been shown to shed surface-bound antibody at room temperature. This shedding of immuno-globulins or antigen-immunoglobulin complexes from the surface of the larvae may have a functional role in protecting the parasite from attack from host antibody in vivo. This mechanism has been shown to exist in other parasite systems, such as the sloughing of the glycocalyx by *Fasciola hepatica* (HANNA 1980).

The Taeniidae may also share antigenic components such as blood group antigens (RUSSI et al. 1974) with the host, a factor which can complicate sero-diagnosis (BEN-ISMAIL et al. 1980). These larval cestodes may also produce substances such as spontaneous agglutinins (ORIHARA 1973) and materials which interact with the complement fixation system (HAMMERBERG et al. 1977; HAMMERBERG and WILLIAMS 1978a; PERRICONE et al. 1980). Certain of these factors, which inhibit coagulation and complement cascades, have however, been partially characterized. The active components are thought to be composed of macromolecular poly-sulphated polysaccharide chains, possibly linked to a negatively charged protein (HAMMERBERG and WILLIAMS 1978b; HAMMERBERG et al. 1980). The production of such factors by larval cestodes may contribute to their survival in the immuno-competent host (KASSIS and TANNER 1976; RICKARD and WILLIAMS 1982), and recent circumstantial evidence suggests that the presence of sulphated mucopoly-saccharides on the bladders, as opposed to the scolices, of *T. taeniaeformis* meta-cestodes may be associated with regional resistance to C3 deposition and leucocyte attachment (LETONJA and HAMMERBERG 1983).

5 Conclusions

Members of the family Taeniidae are, therefore, obligate parasites of considerable biological interest. They are very widely disseminated, exhibit comparatively good host adaptability, but despite causing considerable public health, medical, veterinary and economic problems, remain relatively poorly studied.

References

Acevedo-Hernandez A (1982) Economic impact of porcine cysticercosis. In: Flisser A, Willms K, Laclette JP, Larralde C, Ridaura C, Beltran F (eds) Cysticercosis: present state of knowledge and perspectives. Academic, London, pp 63–67
Ali-Khan Z, Siboo R (1981) *Echinococcus mulilocularis*: distribution and persistence of specific host immunoglobulins on cyst membranes. Exp Parasitol 51:159–168

Allan D, Jenkins P, Connor RJ, Dixon JB (1981) A study of immunoregulation of Balb/c mice by *Echinococcus granulosus equinus* during prolonged infection. Parasite Immunol 4:137–142

Ayuya JM, Williams JF (1979) The immunological response of the rat to infection with *Taenia taeniaeformis*. VII. Immunization by oral and parenteral administration of antigens. Immunology 36:825–834

Ben-Ismail R, Carme B, Niel G, Gentillini M (1980) Non-specific serological reactions with *Echinococus granulosus* antigens: role of anti-P1 antibodies. Am J Trop Med Hyg 29:239–245

Biguet J, Rose F, Capron A, Tran Van Ky P (1965) Contribution de l'analyse immunoelectrophoretique a la connaissance des antigenes vermineux. Incidence practiques sur leur standardisation, leur purfication et le diagnostic helminthiases par immunoelectrophorese. Rev Immunol Therapie Anti-microbienne 29:5–23

Bloom BR (1979) Games parasites play: how parasites evade immune surveillance. Nature 279:21–26

Bout D, Fruit J, Capron A (1974) Purification d'un antigene specifique de liquide hydatique. Ann Immunol 125C:755–788

Bowtell DDL, Mitchell GF, Anders RF, Lightowlers MW, Rickard MD (1983) *Taenia taeniaeformis*: immunoprecipitation analysis of the protein antigens of oncospheres and larvae. Exp Parasitol 56:416–427

Capron A, Biguet J, Vernes A, Afchain D (1968) Structure antigenique des helminthes. Aspects immunologiques des relations hote-parasite. Pathol Biol 16:121–138

Clegg JA, Smith MA (1978) Prospects for the development of dead vaccines against helminths. Adv Parasitol 16:165–218

Conder GA, Andersen FL. Schantz PM (1980) Immunodiagnostic tests for hydatidosis in sheep: an evaluation of double diffusion, immunoelectrophoresis, indirect haemagglutination and intradermal tests. J Parasitol 66:577–584

Craig PS (1983) Immunodifferentiation between eggs of *Taenia hydatigena* and *T. pisiformis*. Ann Trop Med Parasitol 77:537–538

Craig PS, Rickard MD (1980) Evaluation of "crude" antigen prepared from *Traenia saginata* for the serological diagnosis of *T. saginata* cysticercosis in cattle using the enzyme-linked immunosorbent assay (ELISA). Z Parasitenkd 61:287–297

Craig PS, Rickard MD (1981a) Anti-oncospheral antibodies in the serum of lambs experimentally infected with either *Taenia ovis* or *Taenia hydatigena*. Z Parasitenkd 64:169–177

Craig PS, Rickard MD (1981b) Studies on the specific immunodiagnosis of larval cestode infections of cattle and sheep using antigens purified by immunoaffinity chromatography in an enzyme-linked immunosorbent assay. Int J Parasitol 11:441–449

Craig PS, Mitchell GF, Cruise KM, Rickard MD (1980) Hybridoma antibody immunoassays for the detection of parasitic infection: attempts to produce an immunodiagnostic reagent for a larval taeniid cestode infection. Aust J Exp Biol Med Sci 58:339–350

Craig PS, Hocking RE, Mitchell GF, Rickard MD (1981) Murine hybridoma derived antibodies in the processing of antigens for the immunodiagnosis of hydatid (*Echinococcus granulosus*) infections in sheep. Parasitology 83:303–317

Davies C, Rickard MD, Bout DT, Smyth JD (1978) Ultrastructural immunocytochemical localization of two hydatid fluid antigens (antigen 5 and antigen B) in the brood capsules protoscoleces of ovine and equine *Echinococcus granulosus* and *E. multilocularis*. Parasitology 77:143–152

De Aluja AS (1982) Frequency of porcine cysticercosis in Mexico. In: Flisser A, Willms K, Laclette JP, Larralde C, Ridaura C, Beltran F (eds) Cysticercosis: present state of knowledge and perspectives. Academic, London, pp 53–62

Dessaint JP, Bout D, Wattre P, Capron A (1975) Quantitative determination of specific IgE antibodies to *Echinococcus granulosus* and IgE levels in sera from patients with hydatid disease. Immunology 29:813–823

Dixon JB, Jenkins P, Allan D (1982) Immune recognition of *Echinococcus granulosus* I. Parasite-activated, primary transformation by normal murine lymph node cells. Parasite Immunol 4:33–45

Dottorini S, Tassi C (1977) *Echinococcus granulosus*: characterization of the main antigenic component (arc 5) of hydatid fluid. Exp Parasitol 43:307–314

Engelkirk PG, Williams JF, Signs MM (1981) Interactions between *Taenia taeniaeformis* and host cells in vitro: rapid adherence of peritoneal cells to strobilocerci. Int J Parasitol 11:463–474

Flisser A, Woodhouse E, Larralde C (1980) Human cysticercosis: antigens, antibodies and non-responders. Clin Exp Immunol 39:27–37

Froyd D (1963) Intradermal tests in the diagnosis of bovine cysticercosis. Bull Epizootic Dis Afr 11:303–306

Gallie GJ, Sewell MMH (1976) Experimental immunization of 6-month old calves against infection with the cysticercus stage of *Taenia saginata*. Trop Anim Health Prod 8:233–242

Gallie GJ, Sewell MMH (1981) Inoculation of calves and adult cattle with oncospheres of *T. saginata* and their resistance to challenge infection. Trop Anim Health Prod 13:147–154

Geerts S, Kumar V (1977) In-vivo diagnosis of bovine cysticercosis. Vet Bull 47:653–664

Geerts S, Kumar V, Aertz N (1979) Antigenic components of *Taenia saginata* and their relevance to the diagnosis of bovine cysticercosis by immunoelectrophoresis. J Helminthol 53:293–299

Geerts S, Kumar V, Aertz N, Ceulemans F (1981) Comparative evaluation of immunoelectrophoresis, counterimmunoelectrophoresis and enzyme linked immunosorbent assay for the diagnosis of *Taenia saginata* cysticercosis. Vet Parasitol 8:299–307

Gemmel MA (1978) Perspectives on options for hydatidosis and cysticercosis control. Vet Med Rev 78:3–48

Good AH, Siebert AE Jr, Robbins P, Zaun S (1982) Modulation of the host immune response by larvae of *Taenia crassiceps*. In: Flisser A, Willms K, Laclette JP, Larralde C, Ridaura C, Beltran F (eds) Cysticercosis: present state of knowledge and perspectives. Academic, London, pp 593–610

Grindle RJ (1978) Economic losses resulting from bovine cysticercosis with special reference to Botswana and Kenya. Trop Anim Health Prod 10:127–140

Hammerberg B, Williams JF (1978a) Interaction between *Taenia taeniaeformis* and the complement system. J Immunol 120:1033–1038

Hammerberg B, Williams JF (1978b) Physicochemical characterization of complement-interacting factors from *Taeniae taeniaeformis*. J Immunol 120:1039–1045

Hammerberg B, Musoke AJ, Williams JF (1977) Activation of complement by hydatid cyst fluid of *Echinococcus granulosus*. J Parasitol 63:327–331

Hammerberg B, Dangler C, Williams JF (1980) *Taenia taeniaeformis*: chemical composition of parasite factors affecting coagulation and complement cascades. J Parasitol 66:569–576

Hanna REB (1980) *Fasciola hepatica:* gylcocalyx replacement in the juvenile as a possible mechanism for protection against host immunity. Exp Parasitol 50:103–114

Harrison LJS (1984) Surface and secreted antigens of *Taenia saginata*. Ann Trop Med Parasitol 78:235

Harrison LJS, Sewell MMH (1981) Antibody levels in cattle naturally infected with *Taenia saginata* metacestodes in Britian. Res Vet Sci 31:62–64

Harrison LJS, Parkhouse RME, Sewell MMH (1984) Variation in 'target' antigens of appropriate and in-appropriate hosts of *Taenia saginata* metacestodes. Parasitology 88:569–663

Heath DD (1973) Resistance to *Taenia pisiformis* larvae in rabbits-I. Examination of the antigenically protective phase of larval development. Int J Parasitol 3:485–489

Heath DD (1976) Resistance to *Taenia pisiformis* larvae in rabbits: immunization against infection using non-living antigens from in vitro culture. Int J Parasitol 6:19–24

Heath DD, Lawrence SB, Yong WK (1979) Cross-protection between the cysts of *Echinococcus granulosus, Taenia hydatigena* and *T. ovis* in lambs. Res Vet Sci 27:210–212

Hustead ST, Williams JF (1977) Permeability studies on taeniid metacestodes: I. Uptake of proteins by larval stages of *Taenia taeniaeformis, T. crassiceps* and *Echinococcus granulosus*. J Parasitol 63:314–321

Kassis AI, Tanner CE (1976) The role of complement in hydatid disease: In vitro studies. Int J Parasitol 6:25–35

Kwa BH, Liew FY (1977) Immunity in taeniasis-cysticercosis. I. Vaccination against *Taenia taeniaeformis* in rats using purified antigens. J Exp Med 146:118–131

Kwa BH, Liew FY (1978) Studies on the mechanism of long-term survival of *Taenia taeniaeformis* in rats. J Helminthol 52:1–6

Leid RW, Williams JF (1974) The immunological response of the rat to infection with *Taenia taeniaeformis*. II. Characterization of reaginic antibody and an allergen associated with the larval stage. Immunology 27:209–225

Lethbridge RC (1980) The biology of the oncosphere of cyclophyllidean cestodes. Helminth Abstr 49:59–72

Letonja T, Hammerberg B (1983) Third component of complement, immunoglobulin deposition, and leucocyte attachment related to surface sulphate on larval. *T. taeniaeformis*. J Parasitol 69:637–644

Lightowlers MW, Mitchell GF, Bowtell DDL, Anders RF, Rickard MD (1984) Immunization against *Taenia taeniaeformis* in mice: studies on the characterization of antigens from oncospheres. Int J Parasitol 14:321–333

Lloyd S (1979) Homologous and heterologous immunization against the metacestodes of *Taenia saginata* and *Taenia taeniaeformis* in cattle and mice. Z Parasitenkd 60:87–96

Lloyd S (1981) Progress in immunization against parasitic helminthes. Parasitology 83:225–242

Machnika-Roguska B, Zwierz C (1970) Intradermal test with antigenic fractions of *Taenia saginata*. Acta Parasitol 17:293–329

Mitchell GF (1982a) Effector mechanisms of host-protective immunity to parasites and evasion by parasites. In: Mettrick DF, Dresser SS (eds) Parasites – their world and ours. Elsevier Biomedical, Amsterdam, pp 24–33

Mitchell GF (1982b) Genetic variation in resistance of mice to *Taenia taeniaeformis:* analysis of host-protective immunity and immune evasion. In: Flisser A, Willms K, Laclette JP, Larralde C, Ridaura C, Beltran F (eds) Cysticercosis: present state of knowledge and perspectives. Academic, London, pp 575–584

Mitchell GF, Anders (1982) Parasite antigens and their immunogenicity infected hosts. In: Sela M (ed) The antigens Academic, London, pp 69–179

Mitchell GF, Anders RF, Brown GV, Handeman E, Roberts-Thompson IC, Chapman CB, Forsyth KP, Kahl LP, Cruise KM (1982) Analysis of infection characteristics and antiparasite immune responses in resistant compared with susceptible hosts. Immunol Rev 61:137–188

Ogilvie BM, Wilson RJM (1976) Evasion of the immune response by parasites. Br Med Bull 32:177–181

Orihara M (1973) Studies on serologic diagnosis of multilocular echinococcosis, especially on the haemagglutination test using fractionated antigens and utilization of *Cysticercus fasciolaris* antigens. Jpn J Vet Res 21:93–94

Oriol C, Oriol R (1975) Physicochemical characteristics of a lipoprotein antigen of *Echinococcus granulosus*. Am J Trop Med Hyg 24:96–100

Oriol R, Williams JF, Pierez-Escandi MV, Oriol C (1971) Purification of lipoprotein antigens of *Echinococcus granulosus* from sheep hydatid fluid. Am J Trop Med Hyg 20:569–574

Parkhouse RME, Harrison LJS, Ortega-Pierres G, Clark NWT (1984) Diagnostic antigens for bovine and human cysticercosis. Ann Trop Med Parasitol 78:234

Pawlowski Z, Schultz M (1972) Taeniasis and cysticercosis (*Taenia saginata*). Adv Parasitol 10:69–343

Percy P, Lauffau G (1979) Antigens of helminths. In: Sela M (ed) The antigens. Academic, London, pp 83–172

Perricone R, Fontana L, De Carolis C, Ottaviani P (1980) Activation of alternative complement pathway by fluid from hydatid cysts. N Engl J Med 302:808–809

Piantelli M, Pozzuoli R, Arru E, Musiani P (1977) *Echinococcus granulosus:* identification of subunits of the major antigens. J Immunol 119:1382–1386

Rajaskariah GR, Rickard MD, Mitchell GF, Anders RF (1982) Immunization of mice against *Taenia taeniaeformis* using solubilized oncospheral antigens. Int J Parasitol 12:111–116

Rickard MD (1982) Immunization against infection with larval taeniid cestodes using oncospheral antigens. In: Flisser A, Willms K, Laclette JP, Larralde C, Ridaura C, Beltran F (eds) Cysticercosis: present stage of knowledge and perspectives. Academic, London, pp 633–646

Rickard MD, Adolph AJ (1977) Vaccination of lambs against infection with *Taenia ovis* using antigens collected during short-term in vitro incubation of activated *T. ovis* oncospheres. Parasitology 75:183–188

Rickard MD, Brumley JL (1981) Immunisation of calves against *Taenia sagiata* infection using antigens collected by in vitro incubation of *T. saginata* oncospheres or ultrasonic disintegration of *T. saginata* and *T. hydatigena* oncospheres. Res Vet Sci 30:99–103

Rickard MD, Katiyar JC (1976) Partial purification of antigens collected during the in vitro cultivation of the larval stages of *Taenia pisiformis*. Parasitology 72:269–279

Rickard MD, Williams JF (1982) Hydatidosis/cysticercosis: immune mechanisms and immunization against infection. Adv Parasitol 21:230–296

Rickard MD, Davies C, Bout DT, Smyth JD (1977) Immunohistological localization of two hydatid antigens (antigen 5 and antigen B) in the cyst wall, brood capsules and protoscolices of *Echinococcus granulosus* (ovine and equine) and *E. multilocularis* using immunoperoxidase methods. J Helminthol 51:359–364

Russi S, Siracusano A, Vicari G (1974) Isolation and characterization of a blood P1 active carbohydrate antigen of *Echinococcus granulosus* cyst membrane. J Immunol 112:1061–1069

Schantz PM, Shanks D, Wilson (1980) Serologic cross-reactions with sera form patients with echinococcus and cysticercosis. Am J Trop Med Hyg 29:609–612

Siebert AE, Good AH (1980) *Taenia crassiceps*: immunity to metacestodes in BALB/c and BDF1 mice. Exp Parasitol 50:437–446

Siebert AE, Blitz RR, Morita CT, Good AH (1981) *Taenia crassiceps:* serum and surface immunoglobulins in metacestode infections of mice. Exp Parasitol 51:418–430

Soulsby LEJ, Lloyd S (1982) Passive immunization in cysticercosis: characterization of antibodies concerned. In: Flisser A, Willms K, Laclette JP, Larralde C, Ridaura C, Beltran F (eds) Cysticercosis: present state of knowledge and perspectives. Academic, London, 539–548

Taylor AER, Muller R (1980) Vaccines against parasites. In: Taylor, Muller (eds) Symposia of the British Society of Parasitology, vol 18. Blackwell Scientific, Oxford

Urquhart G (1980) Immunity to cestodes. In: Taylor AER, Muller R (eds) Vaccines against parasites, Symposia of the British Society of Parasitology, vol 18, Blackwell Scientific, Oxford, pp 107–114

Varela-Diaz VM, Coltorti EA, Rickard MD, Torres JM (1977a) Comparative antigenic characterisation of *Echinococcus granulosus* and *Taenia hydatigena* cyst fluids by immunoelectrophoresis. Res Vet Sci 23:213–216

Varela-Diaz VM, Eckert J, Rausch RL, Coltorti EA, Hess U (1977b) Detection of the *E. granulosus* arc 5 in sera of patients with surgically confirmed E. multilocularis infection. Z Parasitenkd 53:183–188

Velasco-Suarez M, Brave-Becherelle MA, Quirasco F (1982) Human cysticercosis: medical-social implications and economic impact. In: Flisser A, Willms K, Laclette JP, Larralde C, Ridaura C, Beltran F (eds) Cysticercosis: present state of knowledge and perspectives. Academic, London, pp 47–51

Voller A, De Savigny D (1981) Diagnostic serology of tropical parasitic diseases. J Immunol Methods 46:1–29

Wakelin D (1984) Evasion of the immune response: survival within low responder individuals of the host population. Parasitology 88:639–657

White TR, Thompson RCA, Penhale WJ (1982) A comparative study of the susceptibility of inbred strains of mice to infection with *Mesocestoides corti*. Int J Parasitol 12:29–33

Williams JF (1979) Recent advances in the immunology of cestode infections. J Parasitol 65:337–349

Williams JF, Sandeman RM (1982) Antigens of taeniid cestodes. In: Flisser A, Willms K, Laclette JP, Larralde C, Ridaura C, Beltran F (eds) Cysticercosis: present state of knowledge and perspectives. Academic, London

Williams JF, Perez-Escandi MV, Oriol R (1971) Evaluation of purified lipoprotein antigens of *Echinococcus granulosus* in the immunodiagnosis of human infections. Am J Trop Med Hyg 20:575–579

Willms K, Arcos L (1977) *Taenia solium:* host serum proteins on the cysticercus surface identified by ultrastructural immunoenzyme techniques. Exp Parasitol 43:396–406

Yarzabal L, Dupas H, Bout D, Capron A (1976) *Echinococcus granulosus:* distribution of hydatid fluid antigens in the tissues of the larval stage. I. Localization of the specific antigen of hydatid fluid (antigen 5). Exp Parasitol 40:391–396

Yarzabal LA, Bout DT, Naquira FR, Capron AR (1977a) Further observation on the specificity of antigen 5 of *Echinococcus granulosus*. J Parasitol 63:495–499

Yarzabal LA, Dupas H, Bout D, Naquira F, Capron A (1977b) *Echinococcus granulosus:* the distribution of hydatid fluid antigens in the tissues of the larval stage. II. Localization of the thermostable lipoprotein of parasitic origin (antigen B). Exp Parasitol 42:115–120

Yong WK, Heath DD (1979) "Arc5" antibodies in sera of sheep infected with *Echinococcus granulosus, Taenia hydatigena* and *Taenia ovis*. Parasite Immunol 1:27–38

Nematode Antigens

N.M. ALMOND and R.M.E. PARKHOUSE

1 Introduction

Nematodes form a large diverse phylum living in a range of habitats both free living and parasitic more varied than any other animals except arthropods. Their life cycle is divided into five stages separated by four moults. There are no gross structural alterations between the newborn L1 and mature adult L5 and so all preadult stages can be described as larval (CHITWOOD and CHITWOOD 1974). Most nematodes are bisexual; the female fertilises her eggs with sperm deposited within her by the males. The female may then release fertile ova or hatched larvae depending on the species. Nematodes may be parasitic for all or part of their life cycle. Some, like filarial worms, pass through two hosts during their lives. The study of nematodes is important because within their numbers are species which cause some of the most widespread and debilitating diseases of man. In addition, economic losses arising from parasitosis of crops and domestic animals make nematodes a major drain of human resources, especially in developing countries.

How these relatively large metazoan parasites survive in hosts in the face of a battery of specific and non-specific defence mechanisms is still not fully understood. Many early studies failed or produced contradictory results though the inability of workers to analyse host responses to individual components of the antigen mosaic presented by the worm. The recent application of biochemical methods to separate the constituent molecules has allowed a more rational approach to the investigation of the subtle interactions of host and parasite. Excellent reviews of earlier work on nematode antigens have been written by CLEGG and SMITH (1978) and PERY and LUF-FAU (1979). Here we will attempt to review the advances in our knowledge of nema-

Division of Immunology, National Institute for Medical Research, The Ridgeway, Mill Hill,
London NW7 1AA, United Kingdom

Current Topics in Microbiology and Immunology, Vol. 120
© Springer-Verlag Berlin · Heidelberg 1985

tode antigens at a molecular level and how these have increased our understanding not only of host-parasite interactions, but also of the physiology of nematodes.

2 Methods of Characterising Parasite Antigens

Parasite antigens may be classified in a number of ways (ANDERS et al. 1982). Nematode antigens were originally defined by parasitologists and so the criteria used were the stage of the life cycle and the source. For the convenience of classification, most animals can be divided into three simple and operational compartments: the surface, the excretory-secretory (E/S) compartment and the residual somatic antigens. Until recently little attention was paid to the surface and the other two compartments were used without any serious purification procedures. A major restriction is this regard was, of course, the paucity of available material. Now with the use of metabolic or enzymic radiochemical labelling procedures it is possible to identify the various components of these compartments. Once the surface, for example, is radiolabelled, it is possible to monitor the distribution of its components through various purification procedures such as electrophoresis, chromatography and immunopurification. Finally one important advance which must not be overlooked is the development of hybridoma technology, allowing the production of monoclonal antibodies which can be used as tools for the identification and purification of antigens.

3 Allergens of Nematodes

The immune response to nematodes is noteworthy for being characterised by a pronounced elevation of serum IgE, not only of antiparasite antibody but also of nonspecific immunoglobulin (JARRET and BAZIN 1974). There has been a rigorous pursuit to isolate and purify the allergens of nematodes. Most of this work was initiated to determine the structure of these molecules in the belief that there was a chemical feature common to all molecules which induced anaphylactic antibodies (BERRENS 1971). As a result nematodes such as *Nippostrongylus brasiliensis* and *Ascaris* are associated with research into the control of IgE production as much as the control of parasitosis. Many allergens are still the most highly defined molecules, biochemically speaking, of many nematodes.

HUSSIAN et al. (1973) isolated an allergen (Asc-1) from adults of *Ascaris suum* by gel filtration and polyvinyl chloride block electrophoresis. This dimeric glycoprotein, with a molecular weight of 19 kd, induced an IgE response in rats either following immunisation with the adjuvant *Bordetella pertussis* or during the course of a normal infection. Interestingly, embryonated eggs of *Ascaris*, which are unable to develop into adults in rats, induce reagins in their hosts which bind *Asc-1* (BRADBURY et al. 1974), implying a common allergen of adult worms and larval stages.

Another allergen of *A. suum* was characterised by AMBLER et al. (1974). Called *allergen A*, this was a secreted protein of adult worms. Although of a similar molecular weight to *Asc-1* (in the range of 15–20 kd), it had a smaller proportion of carbohydrates. DANDEU and LUX (1978) purified two allergens from *Ascaris*. One was possibly *allergen A* since it had a similar composition of amino acids. Although these two allergens were chemically, physically and antigenically distinct, they were cross-reactive as reagins.

Another approach to the study of allergenic components has been followed by O'DONNELL and MITCHELL (1978). The components of the body fluid of *Ascaris* were separated by various biochemical methods. They subsequently analysed the ability of serum from patients infected with *Ascaris* to bind to these partially purified components as determined by a radioallergosorbent test (RAST). This test detects the presence of specific IgE antibodies to the various components. The patients' sera were found to bind to a number of the components of the nematode's body fluid which varied in size and isoelectric point. Surprisingly, when these various purified components were used to try and induce reaginic antibody responses in mice by intranasal immunisation with an adjuvant, only the largest component, an aggregate of 360 kd, induced antibodies detectable by passive cutaneous anaphylaxis (PCA). This result is in marked contrast to the finding of HOGARTH-SCOTT (1967), who reported that the allergenic components detectable by PCA tests of a number of nematodes including *Ascaris* were all found to have molecular weights in the 10–50 kd range.

A range of allergenic components have been found in extracts of *Dirofilaria immitis* (FUJITA et al. 1979; FUJITA and TSUKIDATE 1981) and *N. brasiliensis* (PETIT et al. 1980). Interestingly however their molecular weight heterogeneity has been attributed to complexing or aggregation of a small molecular weight unit which contains the reaginic epitope. The extremely labile high molecular weight allergen of *A. lumbricoides* isolated by O'DONNEL and MITCHELL (1978) also demonstrated a similar subunit structure which may account for the failure of previous workers to detect this allergen. Likewise, the results obtained by DANDEU and LUX (1978) could also be explained by a similar model of aggregation.

In conclusion, although exhaustive biochemical analyses have been performed on certain allergenic molecules isolated from nematodes, the reason why they elicit an IgE antibody response is still unknown. The theory, as proposed by BERRENS (1971), that the chemical nature of the epitope determines this response is probably naive. IgE-secreting cells develop from precursors which originally expressed µ chain on the cell surface (MANNING et al. 1976). The isotype of antibodies produced by a cell is determined by nucleic acid rearrangements. In principle, this mechanism allows any given antibody specificity to be associated with all the major heavy chain classes (PARKHOUSE and COOPER 1977). The nature of the signal which controls the isotype switches of B-lymphocytes and perhaps antibody-secreting cells to a particular isotype is yet to be fully elucidated.

A number of factors acting together probably determine the isotype of an antibody response. First the presentation of antigens and epitopes is important. Unlike the IgE responses induced to intact nematodes, purified allergens isolated from these worms will only elicit reaginic antibodies when specific immunisation protocols are followed, i.e. microgram doses of protein in conjunction with adjuvants such as *Bordetella pertussis* administered by certain routes (DANDEU and LUX 1978; FUJITA et al. 1979; O'DONNEL and MITCHELL 1978). In this respect these molecules are no more allergenic, when isolated, than ovalbumin, another "good" allergen (JARRETT 1978). Therefore the IgE response during normal infection depends on an adjuvant effect of the worm in the host. One of the adjuvant effects of the intact worm may be to induce specific T-cell factors. Infection of rats with *N. brasiliensis* induces the development of T cells which selectively enhance the IgE response (SUEMARA and ISHIZAKA 1979). In addition, factors capable of binding IgE have been found in supernatants of mitogen-stimulated cultures of T cells taken from the mesenteric lymph nodes of *N. brasiliensis*-infested rats (SUEMARA et al. 1980). These and other factors have been shown to modulate the IgE response (reviewed by ISHIZAKA 1982). In this way the role of T cells in controlling the IgE response mirrors a similar demonstrable role for T cells to control the development of IgA antibody-secreting cells in the mucosal-associated lymphoid tissue (KIYONO et al. 1982; KAWANISHI et al. 1982), either by stimulating the switch of B cells to a specific isotype or else by expanding the B-cell clones displaying a certain heavy chain class.

As host responses to helminths unfailingly display enhanced IgE responses, one would expect that antibodies of this isotype must be able to mediate a potent anti-helminth effector mechanism. Such a mechanism may arise through the interaction of IgE antibodies and mast cells present in mucosal and connective tissues. Mast cells have granules which contain a number of pharmacological agents such as histamine as well as factors which attract eosinophils and neutrophils. The release of these agents may be detrimental to invading parasites either by a direct effect (ROTHWELL et al. 1974) or by attracting other defence cells to the site of invasion. Mast cells are intimately associated with IgE antibodies because of the presence of high-affinity, IgE-specific receptors on their surface. There is now good evidence that the cross-linking of IgE bound to these mast cell receptors results in the degranulation of the cells (ISHIZAKA and ISHIZAKA 1975). Thus the IgE-mast-cell axis would provide a potential mechanism of recruiting host defences against large extracellular parasites. Furthermore if IgE also interacts directly with eosinophils and macrophages in the killing of nematodes as has been demonstrated to be the case for schistosomes (CAPRON and DESSAINT 1981), this would only accentuate the vital role of IgE.

Apart from its role in host protection, the IgE response associated with nematode infections may contribute to improving diagnostic tests. Thus antibody diagnosis confined to the IgE class may be more filaria specific than when the entire immunoglobulin class spectrum is examined (OTTESEN et al. 1979; WEISS et al. 1982b).

4 The Dynamic Structure of the Surface of Nematodes

The rest of this review will concentrate on nematode antigens and the responses they elicit in the host. First, however, it is pertinent to discuss recent advances in our understanding of nematode structure and physiology.

For many years the cuticle of nematodes was thought to be a relatively inert acellular exoskeleton (LUMSDEN 1975) surrounding layers of muscle. Its composition is not fully elucidated (LUMSDEN 1975; BIRD 1980, LEE 1972) but it is known to consist of two or three layers of varying thickness. The external surface is delineated by a triple-layered structure which bears resemblance to a mammalian cell membrane when visualised by electron microscopy. However, freeze fracture studies of the surface of two very different parasitic nematodes, namely microfilariae of *Onchocerca volvulus* (MARTINEZ-PALOMO 1978) and adults of *Trichinella spiralis* (LEE et al. 1984), have clearly demonstrated that the outer surface of nematodes is not delimited by any membrane, like the plasmalemma which surrounds eucaryotic cells.

Despite the very different structure as compared with mammalian cell surfaces, the cuticle of nematodes is, in contrast to earlier hypotheses, a very active structure. Enzymes and haemoglobin have been located in it (LEE 1972), and adult stages of *Brugia pahangi* have been shown to be able to take up nutrients via their surfaces (CHEN and HOWELLS 1979; HOWELLS and CHEN 1981). More direct evidence of this dynamic structure was obtained first from analyses of the radiolabelled surface proteins of *Trichinella spiralis*.

Trichinella spiralis is a convenient nematode for laboratory investigation because it completes its life cycle in one host within a relatively short period. Moreover, as it completes this cycle, the worm passes through stages in the circulation, gastrointestinal tract and also within intracellular niches. The life cycle of *T. spiralis* is as follows: the infective stage is an intracellular parasite which lives as a relatively quiescent form within striated muscle cells. It is released by digestion in the stomach, following ingestion of raw muscle. The released worms undergo four moults in the small intestine, maturing within 5 days into adult worms. These intestinal stages of worm development are, in fact, also intracellular parasites because they live and move through the cells of intestinal villi (WRIGHT 1979). The adult worms reproduce sexually and the females release many newborn larvae. Following penetration of the host's gut wall the larvae circulate in the lymph and blood before finally invading striated muscle. Each larva penetrates a muscle cell and rapidly grows without further moults into an infective muscle larva. During this phase of growth the parasitised muscle cell responds to infection by undergoing a number of changes. The cross-striations, characteristic of voluntary muscles, disappear, and there is a proliferation of membranes believed to be derived from smooth endoplasmic reticulum (DES-POMMIER 1975). The resulting cell containing an infective larva and sourronded by a collagenous capsule is called a "nurse cell" (PURKERSON and DESPOMMIER 1974), because the changes which occur are thought to allow the growth and survival of the larva within this unusual intracellular niche.

Evidence had accumulated that the surface of *T. spiralis* exposed antigens which were stage specific. In the most compelling experiment, stage specific antibodies were demonstrated by absorption of infected sera with living worms. Thus, absorption with infective larvae, for example, removed antibodies mediating eosinophil adherence to that, but no other, stage (MACKENZIE et al. 1978). PHILIPP et al. (1980) looked for stage specificity by radiolabelling the surface proteins of each stage of *T. spiralis* using standard cell surface radiolabelling methods. The worms were then solubilised in detergent and the detergent-soluble proteins separated by polyacrylamide gel electrophoresis in the presence of sodium dodecyl sulphate (SDS-PAGE). The profiles obtained by autoradiography demonstrated that the surface-labelled proteins were stage specific in their expression (Fig. 1). In addition, a comparison of the surface-labelled proteins of preadult and adult worms removed from the gastrointestinal tract from day 2 to 6 postinfection demonstrated qualitative differences occurring at this time of rapid growth and moulting. Moulting, however, does not necessarily account for all the observed changes through the shedding of the cuticle. Later work revealed that the newborn larva changed its surface 6–8 h after being shed

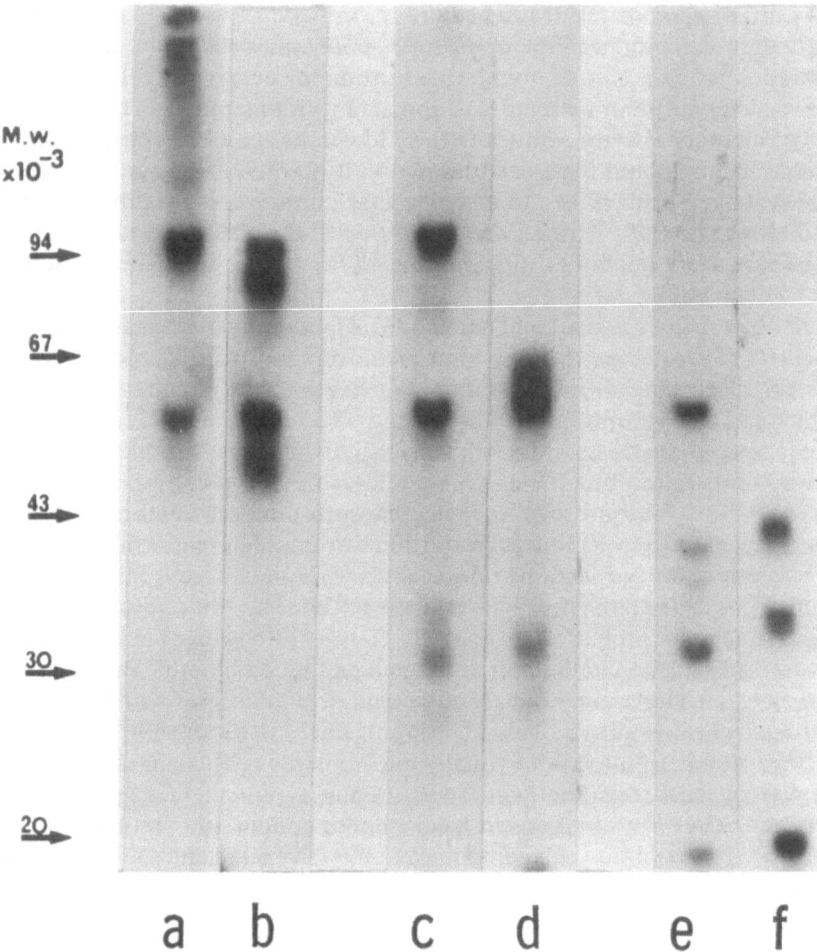

Fig. 1. Surface antigens of three stages of *Trichinella spiralis*. Analysis by SDS-PAGE. Infective larvae (*a, b*), newborn larvae (*c, d*) and 6-day-old-adults (*e, f*) were labelled with ^{125}I solubilised and electrophoresed on gel slabs under non-reducing (*a, c, e*) or reducing (*b, d, f*) conditions. The position of the molecular weight markers run in parallel is indicated on the figure. (CLARK et al. 1982)

by the adult female, at a time when there is no moult (JUNGERY et al. 1983). Thus, changes in surface proteins occur both between defined developmental stages and within one stage.

There is good evidence that the proteins labelled were on the surface of the worms. The number of bands revealed for each stage was highly restricted, not a sign of widespread labelling throughout the worm. In addition, the surface location of the radioactivity was directly demonstrated by autoradiography of sections of the labelled worms (PARKHOUSE et al. 1981). Finally, MAIZELS et al. (1983a) have used the same technique to label surface proteins of *Nippostrongylus brasiliensis* and failed to find any labelling of nematode-derived haemoglobin, a protein known to be located in the cuticle, close to the surface.

A further observation by PHILIPP et al. (1980) was the release of surface-labelled proteins of infective larvae and adults when worms were cultured in vitro. This process was accelerated in the presence of serum, both normal and immune, and especially by immune serum in the presence of neutrophils. The possible role of this phenomenon will be discussed later.

Subsequent papers (PARKHOUSE et al. 1981; CLARK et al. 1982) have provided further information on the organisation of the surface proteins of the intact worm. The SDS-PAGE profile of the surface-labelled proteins of infective larvae of *Trichinella* electrophoresed under reducing conditions consists of two doublets of apparent molecular weights 47 and 55 kd, and 90 and 105 kd (Fig. 1b). The lower molecular weight component of each pair is a lentil lectin-adherent glycoprotein and the closely related tryptic and chymotryptic peptide maps demonstrated that the 90-kd glycoprotein is a dimer of the 47-kd molecule. The two protein components (105 + 55 kd) are similarly related to each, and only display slight differences from the glycoproteins. This, together with the isolation of a monoclonal antibody (NIM-M1) which precipitates all four surface-labelled molecules of infective larvae (ORTEGA-PIERRES et al. 1984a), suggests that all four proteins share common polypeptide sequences. An attractive hypothesis is that all the surface iodinatable proteins of the infective larvae originate from one gene, single or amplified. Its product would then be incorporated into the worm's surface in monomeric or dimeric form and variably glycosylated to yield the lectin-adherent and non-adherent fractions.

Analysis of radiolabelled proteins of all three stages by SDS-PAGE under reducing and non-reducing conditions revealed that some interactions of surface-labelled proteins could be destroyed in the presence of SDS and 2-mercaptoethanol (2ME) but only partially by SDS and iodoacetamide (Fig. 1). (CLARK et al. 1982). Based upon this, the associations can be characterised as non-covalent, but disulphide bond dependent. It was suggested that the aggregates arise from strong non-covalent hydrophobic interaction sites which are stabilised by intrachain disulphide bonds in the molecules concerned. These observations may account for the high levels of disulphide bonding detected histochemically in the outer cuticle by ANYA (1966).

As mentioned above, qualitative changes during development are not restricted to ecdysis. JUNGERY et al. (1983) found that the gel profile of detergent-soluble surface-iodinated proteins of newborn larvae of *T. spiralis* changed 6–8 h after release from the adult female. Only a 64-kd protein could be labelled on newly released larvae. Later, however, three new proteins (58, 34 and 32 kd) became available for labelling. This reorganisation of the surface of newborn larvae can also be demonstrated in two other assays. The time of 6–8 h corresponds with the ability of newborn larvae to activate complement in vitro (MACKENZIE et al. 1980). In addition, a monoclonal antibody raised against newborn larvae only binds newly released larvae, as detected by immunofluorescence (ORTEGA-PIERRES et al. 1984b). Once again, this pattern changes after 6 h when the proportion of newborn larvae stained decreases, presumably because the reorganisation of the surface covers the binding site of the monoclonal antibody. Therefore the availability of the additional proteins to labelling with iodine correlates with the disappearance as well as the appearance of surface determinants.

Finally, the surface of *T. spiralis* is not a homogeneous covering. A monoclonal antibody has been raised which uniquely binds to the surface of the adult male copulatory bell (ORTEGA-PIERRES et al. 1984c). This localised region of available antigenic determinants has a physical correlate. A recent study of the surface of adult worms by scanning electron microscopy has revealed knob-like protuberances found only on the adult male copulatory bell (LICHTENFELS et al. 1984).

The accumulated biochemical evidence demonstrating a restricted set of exposed aggregating proteins on a dynamic cuticle is not unique to *T. spiralis*. Similar results have been obtained for *N. brasiliensis* (MAIZELS et al. 1983a), *Toxocara canis* (MAIZELS et al. 1984a), *Litomosoides carinii* (PHILIPP et al. 1984b) and Brugian filarial worms (MAIZELS et al. 1983b). In contrast, Forsyth and her co-workers (FORSYTH et al. 1981a) found that at least 30 distinct proteins were radioiodinated by similar labelling techniques when applied to microfilariae of *Onchocerca gibsoni* collected in vitro from adult females. A similar result was obtained with radiolabelled microfilariae of *Litomosoides carinii* shed in vitro. Attempts to resolve the surface-labelled detergent-soluble proteins by SDS-PAGE in one dimension failed, resulting in a smear of proteins (PHILIPP et al. 1984b). Whether this is due to the greater penetration of labelling compounds into the microfilariae is not known. However, an autoradiograph showed that reduced silver grains were restricted to the periphery of sections of labelled microfilariae of *O. gibsoni* (FORSYTH et al. 1981a). The occurrence of a highly restricted set of surface antigens as observed for *T. spiralis* is not always rigidly adhered to in all nematodes (MAIZELS et al. 1983a, b, c; PHILIPP et al. 1984b). Nonetheless the concept that the cuticle of nematodes is a dynamic interface between the worm and its environment remains unchanged.

As one might expect, the exposure of stage-specific surface proteins reflects changes in internal metabolism. The same picture of events is obtained when one analyses the composition of metabolically ($[^{35}S]$ methionine) labelled excretory-secretory products of *T. spiralis* (Fig. 2). Here again, each defined stage produced a unique, stage-specific profile of $[^{35}S]$ methionine-labelled components on SDS-PAGE (PARKHOUSE and CLARK 1983). One cannot address the question of stage specificity of metabolically labelled somatic proteins by SDS-PAGE in one dimension, as it yields a smear of unresolved proteins. This is simply a result of the complex, multicellular nature of the worms. However, total glycoproteins of freshly isolated worms resolved by two-dimensional gel electrophoresis and visualised by binding of labelled concanavalin A produce stage-specific profiles (PARKHOUSE and CLARK 1983).

In conclusion, stage-specific surface and secreted proteins are a reflection of changes in the worm's metabolism as their needs and environment change.

5 Nematode Antigens in Protection and Evasion

Our previous discussions have concentrated on the parasite. Immunity, however, depends not only on what is exposed to the host but also on the host's response to those antigens. It was originally thought that the persistence of nematodes in an immunocompetent host simply arose from having a resistant, antigenically inert cuticle (LUMSDEN 1975). We now know, however, that the majority of parasites elicit antibody responses to their surface proteins (OGILVIE et al. 1980) and secreted

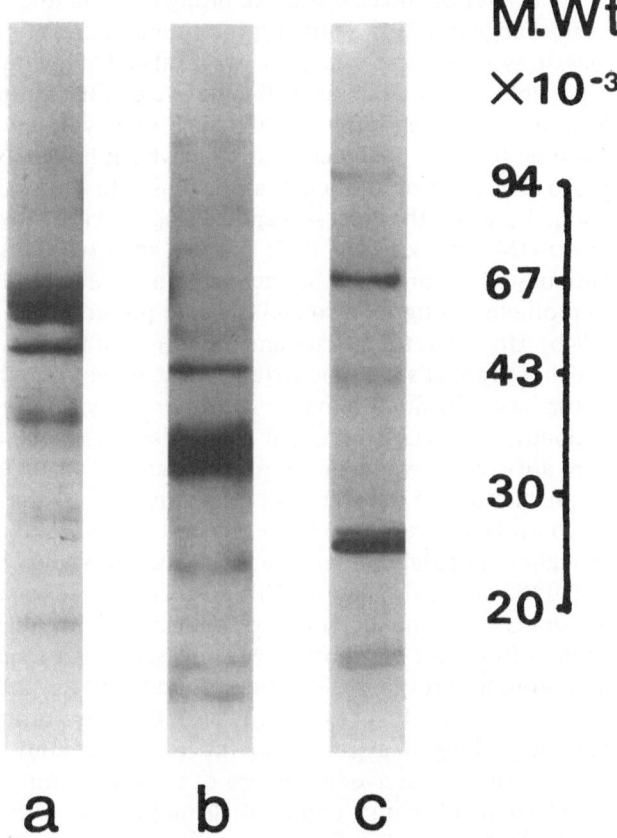

Fig. 2. Secreted antigens of three stages of *Trichinella spiralis*. Infective larvae (*a*), 6-day adults (*b*) and newborn larvae (*c*) were internally labelled in vitro with [^{35}S] methionine (0.6 mCi ml^{-1}) over 48 h at 37°C in RPMl 1640 tissue culture medium containing 3%, V/V, fetal calf serum. Radiolabelled proteins were recovered from the culture supernatants by precipitation with ice-cold ethanol (80%, V/V, final concentration) and analysed by SDS-PAGE under reducing conditions

material can also induce protective immunity in certain systems. Most notable is the identification of a protective secreted glycoprotein of 48 kd from infective larvae of *Trichinella*. It is a product of the stichocytes, cells which secrete many antigenic components, and has been demonstrated to give protection by vaccination with microgram quantities (SILBERSTEIN and DESPOMMIER 1984). As the surface, along with excretory-secretory (E/S) components, form the only interface between viable parasites and the host, the host response to these compartments is critical in determining the outcome of infection. Recent studies utilising biochemically defined antigens have begun to reveal the subtle and active interplay between parasite and host. From the viewpoint of the parasite, a survival strategy should prevent the exposure of any antigen which stimulates the development of protective immunity, preventing the completion of its life cycle. Numerous mechanisms have been postulated; some have been demonstrated to occur.

One potential mechanism exploiting the dynamic nature of the nematode would be to change totally the surface proteins before an effective immune response develops, so as to remain one step (i.e. stage) ahead of the immune response. Whether this is possible clearly depends on the life cycle of the worm. In the case of a primary infection with *T. spiralis* the only stage of development exposed to the systemic immune response is the circulating newborn larvae. Not only are they directly exposed to circulating antibodies as they migrate from gastrointestinal tract to muscle, but they are also the most susceptible stage to killing by leucocytes and macrophages in vitro (MACKENZIE et al. 1978; KAZURA and MESHNICK 1984). Indeed, a monoclonal antibody which binds to a surface protein of newborn larvae has been demonstrated to promote killing in vitro and mediate protection in vivo (ORTEGA-PIERRES et al. 1984b). However, the surface and secretions of the newborn larvae are stage specific and thus the host's immune system will not encounter these antigens until the release of the larvae by adult females. A further 4 days are required before any detectable antibody response is mounted and antibodies mediating eosinophil adherence in vitro are even slower to develop (JUNGERY and OGILVIE 1982). These antibodies will be unable to alter the course of infection since by the time they appear most of the newborn larvae will have been released (DESPOMMIER et al. 1977) and, as they are thought to circulate for less than 3 h (MOLONEY and DENHAM 1979), will have penetrated the host musculature. This mechanism of escape, however, may be unique to *T. spiralis* due to the short duration of the migratory phase and it is only applicable to a primary infection. Newborn larvae released by subsequent infections will encounter a vigorous host response including antibodies to the potentially host-protective 64-kd protein (ORTEGA-PIERRES et al. 1984b). Likewise other parasites with stages which have a prolonged existence in the blood or body cavities must employ other mechanisms to survive in the face of the host's immune system.

The dynamic nature of the cuticle may act in another way to aid the parasite's survival. The release of surface antigens, as described for *T. spiralis* (PHILIPP et al. 1980) and *Toxocara canis* (MAIZELS et al. 1984a), may "divert the attention" of the immune response. The release of surface components may expose new antigenic determinants on the released molecule which are normally buried under the surface of the cuticle in the intact worm. For example, ORTEGA-PIERRES et al. (1984a) raised monoclonal antibodies to surface antigens of *T. spiralis* by fusing myeloma cells with lymphocytes from a mouse which had been exposed to a normal infection. Although all the monoclonals raised precipitated detergent-solubilised surface-labelled proteins of *Trichinella* in an immunocoprecipitation test, only some of the antibodies would interact with the surface of living worms. Presumably these monoclonal antibodies which failed to bind to intact, living parasites recognised epitopes of the surface proteins which are unexposed in the organised surface structure of the parasite.

Another potential mechanism for escape may involve the secretion of components which then function as an immunosorbent. For example, the second-stage larvae of *Toxocara canis* can live for very long periods. They release very large amounts of antigenic E/S components, calculated at about 200 pg protein released/larva per day (MAIZELS et al. 1984a). Most of this material is structurally related to surface proteins. As a result, if one attempts to demonstrate the presence of antibodies to the larval surface by immunofluorescence, no staining of the larval surface will occur unless antimetabolites are added, or the assay is performed in the cold, in order to prevent turnover and shedding of surface materials or the release of secretions which "block" the binding of antisurface antibodies (SMITH et al. 1981).

In contrast, however, one must not forget that the release of material by the parasite is essential for the initiation of host immunity to large extracellular parasites. An immune response will only develop if antigen is presented to lymphoid cells in the context of the host's major histocompatibility complex (MHC) gene products. Therefore only molecules released by the parasite will be suitable processed by antigen-presenting cells. This was demonstrated for *Trichinella* by ORTEGA-PIERRES et al. (1984a). Monoclonal antibodies to determinants not available in the intact worm were raised only with surface-labelled proteins which were released during in vitro culture of labelled viable worms. It is reasonable to propose that their release enables the necessary macrophage processing and presentation to occur. Whilst the release of antigen is necessary for host responses to develop, the newly available determinants not exposed on the living parasite surface will distract the host's immune response from the production of antibodies which will bind to the intact worms' surface and thus may delay the initiation of host-protective immunity.

Apart from the turnover and release of surface components, nematodes, like other parasites, exploit a range of other methods to escape the host's immune system. One strategy of long-lived stages is to mask parasite antigens with host-derived components. For example, infective larvae of *T. spiralis* live within modified muscle cells called "nurse cells" (PURKERSON and DESPOMMIER 1974). The nurse cell membrane and capsule are not seen as foreign by the host immune system, thus protecting the larva within. As an alternative to an intracellular existence parasite antigens may be hidden beneath host proteins bound to the surface of the worm. This phenomenon has been well documented with regard to schistosomes (reviewed by McLAREN 1984). It is only recently, however, that this has been demonstrated to occur on the surface of microfilariae of certain filarial worms. For years the long-term resistance of circulating microfilariae has intrigued immunoparasitologists. The apparent absence of rejection is not simply because the host fails to recognise microfilarial antigens. GREENE et al. (1981) demonstrated that microfilariae of *Onchocerca volvulus* taken from skin of patients were killed in vitro with serum and granulocytes taken from the same patient. This occurred despite the fact that the patients had suffered prolonged microfilaraemia and, in addition, the recovered microfilarae were apparently healthy. This contradictory situation has been resolved by comparing the SDS-PAGE profiles of surface-labelled proteins of microfilariae taken from the host with the profiles of microfilariae released by adult females in vitro. Whilst the SDS-PAGE profiles of surface-labelled uterine microfilariae of both *Litomosoides carinii* (PHILIPP et al. 1984b) and *O. gibsoni* (FORSYTH et al. 1984) were complex, the surface-labelled host-derived microfilariae only yielded one component of 67 kd. By the combined use of electrophoresis and immunochemical procedures the 67-kd band was identified as host serum albumin in both cases.

This observation explains some earlier reports on comparisons between microfilariae obtained from different sources. WEGERHOF and WENK (1979) found that uterine microfilariae of *L. carinii* were more immunogenic than blood-derived microfilariae. Similarly blood-group-like determinants have been described associated with circulating microfilariae of *Loa loa* and *W. bancrofti* (RIDLEY and HEDGE 1977). MAIZELS et al. (1984b) found that a major component of the surface of blood microfilariae of *W. bancrofti* was, in fact, human serum albumin. In addition to having a purely physical blocking capacity serum albumin may play a role as a scavenger of free radicals generated by defence cells. For example, such free radicals generated in vitro have been demonstrated to kill newborn larvae of *T. spiralis* (BUYS et al. 1981).

Heterogeneity between microfilariae of *Brugia pahangi* has also been described. FURMAN and ASH (1983) found differences in the lectin adherence patterns of uterine- and blood-derived microfilariae. In addition functional differences have been described between these microfilariae of *Brugia pahangi*. Complement-mediated cell adherence reactions occur only with blood-derived, and not in vitro shed, microfilariae (JOHNSON et al. 1981). A recent advance in this research has been the development of some of the necessary specific reagents required to analyse host immunoglobulin and lymphocyte subpopulations in the *B. pahangi* cat model. PRE-MARATNE et al. (1984) have raised monoclonal antibodies to cat immunoglobulin classes. These reagents have provided unequivocal proof that host immunoglobulin is on the surface of circulating microfilariae (PREMARATNE et al. 1984). Interestingly the density of the host immunoglobulin on microfilariae is variable and thus indicates that not only do blood-borne microfilariae differ from uterine-derived microfilariae, but heterogeneity exists amongst blood-borne microfilariae themselves. The binding of host-derived components to the surface of microfilariae in the host provides an obvious escape mechanisms through masking of potentially protective nematode antigens. Even host Ig, if bound in a non-specific manner, may prevent further antiparasite antibodies from binding by steric hindrance, and if Fc region determinants were bound to the worm then the immunoglobulin would be unable to mediate cellular adherence.

A feature of many parasitic diseases is the suppression of host immune responses to both parasite-derived and other unrelated antigens. For example, Svet-Moldavsky and co-workers (SVET-MOLDAVSKY et al. 1970) demonstrated non-specific immuno-depression during infection with *T. spiralis*. This suppression is greatest at the time of the migration of newborn larvae and has been found to be mediated by material released by the newborn larvae (FAUBERT 1976). Moreover, this depression of immunity facilitates the survival of the parasite in a primary infection. FAUBERT (1982) demonstrated that suppression was mediated by cyclophosphamide-sensitive suppressor T cells. The injection of cyclophosphamide at specific times during infection was found to decrease the muscle larvae burden in mice.

Another nematode thought to mediate active suppression is the adult stage of *Nematospiroides dubius*. This is a long-lived gastrointestinal-dwelling nematode. Protective immunity can only be elicited in mice by repeated infections accompanied by drug treatment needed to reduce the worm burden (BEHNKE and PARISH 1980). In addition the presence of adults of *N. dubius* markedly reduces the inflammatory responses generated to other gastrointestinal helminths such as *Trichuris muris* (JENKIN and BEHNKE 1977). More striking is the demonstration that the presence of adult worms inhibits the development of the immunity normally elicited in mice when immunised with irradiated larva of *N. dubius,* which can develop to the L4 stage only (BEHNKE et al. 1983). There have been many attempts to try and determine the immunosuppressive mechanism used by *N. dubius*. It is non-specific (ALI and BEHNKE 1983, 1984) and although a role for suppressor T cells has been invoked (PRIT-CHARD et al. 1984), the exact mechanism has yet to be elucidated.

Immunosuppression is also a feature of infection with filarial nematodes. Many workers have reported that despite the chronic persistence of microfilariae, little anti-microfilariae antibody is detectable until infections become occult (WONG and GUEST 1976; WONG and SUTER 1979; PONUNDURAI et al. 1974). Others however have reported that low levels of antimicrofilarial antibodies are detectable during patent infections with *Wuchereria bancrofti* (GROVE and DAVIS 1978) and *Brugia timori* (MAIZELS et al. 1983c) and also in experimental infections of cats with *B. pahangi* (AU et al. 1982). The discrepancy in results may be due to the relative efficiency of the antibody detection methods used. Whilst microfilariae are present most of the antibody produced will be bound either to the parasites themselves or to their immunogenic E/S components, resulting in an underestimation of the amount of antibody being produced.

Not only are humoral responses reduced during infection with filarial worms, but depression of cell-mediated responses has also been documented in these patients, both in response to homologous parasite antigens and also to unrelated substances. Hyporesponsiveness to parasite antigens has been demonstrated most clearly in standard proliferative assays conducted in vitro using crude antigenic extracts from microfilariae or adult worms (OTTESEN et al. 1977; PIESSENS et al. 1980a, b). Most striking was the correlation between patent microfilaraemia and markedly reduced proliferative responses to microfilarial antigens, whilst responses to unrelated antigens remained intact (PIESSENS et al. 1980c).

Immunosuppression has been described as being both parasite specific and non-specific and due to plastic adherent and non-adherent cells as well as to suppressive factors in the serum (PIESSENS et al. 1980c, 1982). Patients with patent *Brugia malayi* infections also displayed reduced ratios of cells expressing helper versus suppressor surface markers (PIESSENS et al. 1982). However, overnight culture in vitro or active removal of cells expressing the OKT8$^+$ phenotype restored the capacity of the patients' peripheral blood leucocytes to proliferate in response to microfilarial antigens (PIESSENS et al. 1982). Conversely, suppression could be induced in normal cells by incubation of cells in the presence of microfilarial antigen. The addition of these cells suppressed the proliferative response of normal cells in culture (PIESSENS et al. 1982). Microfilaraemic patients' serum has been found to have a similar suppressive effect (PIESSENS et al. 1980c).

Recently an immunohistochemical analysis of the surface phenotype of cells within the granuloma-containing adult worms of *Onchocerca volvulus* has been undertaken (PARKHOUSE et al. 1985). This study failed to demonstrate a bias towards T cells of suppressor surface phenotype (T8) within the onchocercoma. The same nodules, however, contained many cells of macrophage and dendritic morphology and surface marker phenotype. The macrophages were defined histochemically by a positive acid phosphatase reaction and consisted of approximately equal numbers of activated (DR+ve) and non-activated (DR−ve) cells. The HLA-DR-positive, acid phosphatase negative dendritic cells observed are classical antigen-presenting cells; these forms were also abundant, indicating that the immunohistological analysis does not support the possibility of immune depression due to a lack of antigen-processing. All of these different cell types, however, are known to secrete a variety of soluble factors which exert both positive and negative effects upon the different cellular components of the immune system (UNANUE 1981; ASKONAS 1984). The balance of these various mediators is therefore likely to have profound consequences for the direction and control of immune responsiveness. In general, the quantity and quality of immune responses is determined by a complex and poorly understood set of interactions, and thus suppression of immune responses must also surely reflect a similarly complex balance of factors.

Comparable studies of the immune responsiveness of experimental animals during infection with filarial nematodes have also demonstrated depression of host responses to both parasite-derived and heterologous antigens. Reduced anti-parasite responses have been associated with patent microfilaraemia in hamsters infected with *Dipetalonema viteae* (WEISS 1978) and in jirds with *Brugia malayi* (KWA and MAK 1980). Non-specific suppression of mitogenic stimulation and responses to unrelated antigens have been observed during experimental infections with *D. viteae* (DE-LASANDRO and KLEI 1976) and *Brugia pahangi* (LAMMIE and KATZ 1983a). Once again both plastic adherent and non-adherent suppressor cells and serum factors have been implicated as mediating the suppression. However, the more thorough investigations possible with experimental animals have revealed that suppression is a dynamic phenomenon. Both WELLER (1978) and WEISS (1978) demonstrated that maximal depression of host responsiveness occurs with the onset of microfilaeremia. Apart from this general trend other smaller vacillations in responsiveness may occur. Crucial in these studies are the source of the cells and the type of assay performed. The changes in responsiveness of splenocytes are not the same as the changes observed when cells from peripheral lymph nodes are stimulated with mitogens (WELLER 1978). Finally the results obtained from in vitro tests alone may be misleading. LAMMIE and KATZ (1983a) found that the responses of spleen cells from jirds to stimulation with mitogens or filarial antigens in vitro were chronically suppressed during the course of infection with *Brugia pahangi*. The suppression was analysed and found to be mediated by an adherent suppressor cell (LAMMIE and KATZ 1983a). However, when these workers tried to demonstrate the phenomenon in vivo only a transient non-specific suppression was found to occur at the onset of patency in the animals (LAMMIE and KATZ 1983b).

In conclusion, immunosuppression has been demonstrated to occur in a wide range of hosts during infection with nematodes. However, as shown by LAMMIE and KATZ (1983a, b) the interactions of cells within the lymphoid system need not mirror results obtained from in vitro assays with lymphocytes takes solely from the spleen or circulation. Furthermore suppression is a dynamic phenomenon and the observed reduced response may be a result of a number of factors, not all of which need affect the balance of host and parasite. For example, the depressed mitogen responses of lymphocytes from jirds infected with *B. pahangi* in vitro were reversed when animals were treated with cyclophosphamide. However, treatment did not alter specific host responses to parasite antigens (KATZ and LAMMIE 1984). This would imply that there are at least two independently operating suppressor mechanisms. Bearing these potential pitfalls in mind further work is required analysing immunodepression during parasitosis. First, it needs to be demonstrated that the various phenomena of reduced host responses are beneficial to parasite survival, as is the case for primary infections with *Trichinella* (FAUBERT 1982). Second, it is necessary to isolate the putative parasite-derived suppressive factors to demonstrate that the observed suppression is an active attempt by the parasite to overcome resistance to infection. The identification of molecules able to subvert the immune system would confirm that the parasite was not affecting host responses solely by overloading the immune system. Analysis of these immunomodulatory molecules would also be of interest in its own right.

6 Host Responses and the Control of Parasitosis

The outcome of infection is not only determined by the parasite. The nature of the host response is of equal importance. Infection with a specific parasite will induce a range of responses and possibly a spectrum of clinical outcomes in an outbred population. This variation has been reviewed by WAKELIN (1978). The ultimate aim of immunoparasitology is to try and identify the protective responses of the host which lead to the rejection of the parasite. Furthermore, in the clinical situation attempts must be made to ameliorate immunopathological reactions. One approach to this study has been to compare the response of different inbred and immuno-deficient strains of animals to infections. Differences between strains will be equi-valent to different points on the spectrum of the possible outcomes in an outbred population. Analysis of the host responses in particular cases will lead, perhaps, to an understanding of the pathogenesis of disease and the induction of resistance to para-sites.

The effectiveness of host humoral responses to parasites not only depends on the epitope to which the antibody binds, but also on the isotype of that antibody. Each isotype mediates a unique range of effector functions. For example, IgE and IgG1 are the only murine antibodies which are bound by mast cell Fc receptors (BLOCH 1967). Therefore only these classes will mediate anaphylactic responses. Similarly the affinity by which immunoglobulins are bound by neutrophils, eosinophils and macrophages varies between the different isotypes. The result is that the ability of the different classes of antibodies to mediate cellular cytotoxicity varies. The activation of complement is also a class-dependent phenomenon. Thus, humoral resistance to parasites is, as one would expect, an isotype-dependent phenomenon. This has been most clearly demonstrated in killing assays performed in vitro. Of five murine mono-clonal antibodies raised by HOROWITZ et al. (1983) to schistosomula of *Schistosoma mansoni,* only the antibody of IgE isotype could kill schistosomula in vitro by me-diating the adherence of macrophages. Three of the remaining monoclonals were of the IgG1 isotype, and these were able to kill schistosomula in vitro through the activa-tion of the complement system. By way of contrast, a protective monoclonal antibody of the IgG2^a isotype raised in rats against schistosomula by GRZYCH et al. (1982) was able to mediate cellular cytotoxicity in vitro via eosinophil-dependent killing. There have also been two examples of fortuitous isolation of monoclonal antibodies which have been able to mediate cell adherence in vitro and also protect in vivo against nematode infections (CANLAS et al. 1984; ORTEGA-PIERRES et al. 1984b). Both were of the IgG1 isotype and both killed the blood-borne stages of infection. Apart from the study of monoclonal antibodies which mediate protection, there have been a number of other approaches utilised to study the role of antiparasite antibodies of specific isotypes in host protection.

Nematospiroides dubius is a nematode which causes a chronic infection of the gastrointestinal tract in mice. Protective immunity is difficult to induce, but once established it may be transferred to other animals by either serum or mesenteric lymph node cells (BEHNKE and PARISH 1981). The most effective class of antibody to transfer resistance was concluded to be IgG1 (WILLIAMS and BEHNKE 1983). Here the evidence was indirect and based on the observation that IgG1 was the only isotype to be significantly elevated at the time when immunity developed. Subsequent reports from the same group confirmed this prediction, since an IgG1 fraction of immune serum (separated on an immobilised protein A column) was the most detrimental to adult worms in transfer experiments. Furthermore, absorption experiments proved that most of the IgG1 fraction comprised antiparasite antibodies (PRITCHARD et al. 1983).

Another method of determining the important isotypes of antiparasite antibodies is to infect mice which are unable to produce antibodies of certain isotypes due to congenital deficiences. For example, the CBA/N strain of mouse carries a sex-linked genetic defect called *xid*. Male mice carrying this defective allele fail to exhibit the full range of B-cell functions (for review see SCHER 1982). T-cell-dependent responses are for the most part intact, but some T-cell-independent responses are ablated. As a result when mice exhibiting the *xid* defect are infected with *Dipetalomena viteae* they fail to produce IgM antibodies to surface components of microfilariae. On the other hand, IgG antibodies to these components are produced by these mice in equivalent amounts to immunocompetent controls (PHILIPP et al. 1984c). Associated with this absence of specific IgM antibodies in *xid* mice is a greatly extended period of patent infection and higher peak density of microfilarae in the blood (PHILIPP et al. 1984c); THOMPSON et al. 1979). This result complemented the suggestion by WEISS and TANNER (1979) that IgM antibody directed against cuticular components was important in mediating cell adherence and killing microfilariae of *D. viteae* in hamsters. A similar experimental protocol using the CBA/N mouse has indicated that IgM has a similar role to play in clearing microfilariae of *Brugia malayi* in mice (THOMPSON et al. 1981).

Although the clearance of microfilariae of *D. viteae* in mice and hamsters is thought to be dependent on IgM antibodies, evidence suggests that inbred rat use a different method. Like mice, adult rats will not permit the development of infective larvae to mature adults. However, they will sustain implantation of adults subcutaneously (HAQUE et al. 1980). After the ensuing microfilaraemia, the rats are immune to further transplanted adults. Sera from these latent rats, but not those with a patent infection, will mediate adherence of normal peritoneal cells to microfilariae in vitro. The adherence and subsequent killing is IgE dependent because the destruction of IgE immunoglobulin in the sera by heat treatment will ablate the adherence reactions (HAQUE et al. 1980). However, as yet one cannot exclude the possibility that IgM antibodies may play some role in the development of latency in vitro. The different mechanisms implicated in the killing of microfilariae of *D. viteae* in the two different hosts exemplify the pitfalls of trying to extrapolate the results obtained with one host to another.

An alternative method to assess the importance of particular isotypes of antibody responses in vivo is actively to render the host immunodeficient. One such model was used by DESSEIN et al. (1981). Rats were treated repeatedly with an anti-rat IgE antibody. The animals were unable to produce any reaginic antibodies even to normally potent allergens, whilst other humoral responses seemed to be intact. When these animals were challenged with infective larvae of *Trichinella spiralis* they did not produce any detectable IgE antibodies; eosinophilia in the blood and tissue around developing muscle larvae was markedly reduced, and they were found to harbour significantly more muscle larvae after 1 month than intact, control animals. A similar role for IgE in controlling *Trichinella* infections in mice has been proposed, though the evidence is indirect. RIVERA-ORTIZ and NUSSENZWEIG (1976) looked at different strains of mice infected with *Trichinella* and observed an inverse relationship between the number of muscle larvae harboured after the infection cycle and the peak IgE anti-*Trichinella* antibody response, detected by passive cutaneous anaphylaxis using a crude extract of muscle larvae as the allergen. This difference was not due to anti-adult immunity since all strains of mice expelled the adult worms over a similar time course. However, the authors failed to account for the role of other genetic differences in the host response to infection. For example, the H-2 haplotype of host mice was shown to influence the fecundity of adult female worms (WASSOM et al. 1984). Similarly, the time taken for antiparasite responses to develop is also strain dependent (see below).

When considering the isotype of humoral responses to parasite antigens one must not forget that antibody production is a dynamic event; the balance of serum antibody classes changes during the course of an infection. This has been exemplified in the case of *T. spiralis* infection. Following infection of rats with *T. spiralis*, PHILIPP et al. (1981) took sequential bleeds and measured the amount of antiparasite antibody present in the serum in two ways. First, antibodies binding solubilised surface antigens of each stage were estimated by an immunocoprecipitation assay. This information was compared with the ability of the same sera to mediate binding of eosinophils to the surface of viable worms. Interestingly, changes in the relative amounts of antibody detected in one assay were not necessarily mirrored in the second. Most striking was the comparison of the two assays analysing anti-infective larvae responses. The titre of immunocoprecipitating antibody rose throughout the course of the experiment, whereas the titre of antibody-dependent eosinophil adherence rose for the first 20 days after infection and then decreased. As yet the class(es) of antibody detected in the eosinophil adherence assay in this model is not known, but it does demonstrate that different antibody assays detect antibodies with different characteristics.

Not only may the isotype profile of antibody responses determine the course of infection, but the recognition of potentially protective parasite antigens is equally important. It has been shown that the time taken to recognise surface antigens of *Trichinella* after infection differs between different strains of mice. The time taken by NIH and C3H mice to expel adult worms from the gastrointestinal tract is different. NIH strain mice expel all their adults by 16 days postinfection (KENNEDY 1976) whereas worm expulsion in C3H mice is not complete even after 30 days (JUNGERY and OGILVIE 1982). When the humoral response was followed, differences were noted in the time taken for antibodies to surface antigens of the parasite to appear, as determined by immunocoprecipitation of solubilised surface antigens and eosinophil adherence. Furthermore, when immunoprecipitated antigens were examined qualitatively by SDS-PAGE, not all the surface-labelled antigens of a particular stage were recognised concurrently by the two strains. For example, the protein bands of muscle larvae (55 and 105 kd) were recognised by antibodies 4 days postinfection in NIH mice. In C3H mice, however, the same proteins did not stimulate detectable antibodies until the 30th day of infection (JUNGERY and OGILVIE 1982). This difference, or similar differences in the appearance of antibodies to specific antigens of the adults or newborn larvae, could be crucial to parasite survival. If variation occurs in the time taken by different hosts to recognise antigens which are potentially protective or mediate parasite rejection, then differences in the course of parasitosis will result. However, in this particular model of mice infected with *Trichinella* the protective antigens of significance in controlling a primary infection have not yet been identified.

Although the evidence above would implicate variable recognition of surface antigens as the cause for differential survival times of *Trichinella* in inbred mice, other factors must not be excluded. To date there is no evidence that antibodies play any role in the expulsion of adult worms of *Trichinella* from the gastrointestinal tract. Most evidence implicates an immunologically initiated T-cell-mediated inflammatory response in causing expulsion (LARSH and WEATHERLY 1975). However the appearance of antibodies to specific antigens probably reflects recognition by the immune system in general. With regard to the shorter time taken by NIH mice to expel all adults of *T. spiralis* from the gastrointestinal tract, one must not forget differences in background genetic factors, not related to immune recognition, between mouse strains, in particular the ability of the different mice to generate inflammatory responses. WAKELIN and DONOACHIE (1981) concluded that the primary reason for the rapid expulsion by NIH mice was the very quick inflammatory response generated by bone-marrow-derived myeloid cells recruited in the immune response. The secondary role of immune factors was demonstrated in a series of adoptive transfer experiments. The responder phenotype was found to be influenced more by the transfer of bone-marrow-derived cells than by immune mesenteric lymph node cells. This supports the notion that non-lymphoid cells can have an equally important role in the expulsion of gastrointestinal worms.

7 Nematode Antigens for Diagnosis

The prime economic drive to characterise nematode antigens is from their application in diagnosis of nematode infections of man. The most important of these are the filariases.

Filariasis is a complex of diseases caused by nematode worms which require two hosts, one mammalian, one arthropod, to complete the life cycle. The mammalian host is infected when infective larvae (L3s) are injected into the blood by a biting arthropod. The larvae undergo two moults to develop into the mature adults which reside in the afferent lymphatics or the skin of the hosts. The adults reproduce sexually and the females release live embryos called microfilariae. These microfilariae either circulate in the blood or reside in the skin until they are taken up by another insect. The cycle of development is then completed by two moults in the intermediate host. Although filariasis is seldom a lethal disease, the debilitating nature of its clinical symptoms (blindness, elephantiasis, fever) make it a major cause of suffering in the world. The primary filarial worms causing disease of man are the skin-dwelling *Onchocerca volvulus* and the lymphatic-dwelling *Wuchereria bancrofti* and *Brugia malayi*. Along with others of lesser importance such as *Loa loa, Mansonella perstans* and *Mansonella ozzardi*, the WHO estimates that some 200 million people are affected in the world (DAVIS 1983).

Within an endemic area an outbred population presents a complete clinical spectrum of disease depending on each person's exposure to infection and their immune responses to each stage. Each symptomatic stage has its definitive parasito-logical and immunological criteria. Immunodiagnostic tests are needed desperately because of the inadequacy of present parasitological methods of diagnosis. Present methods rely solely on the detection of microfilariae in venous blood or skin snips, which are laborious and cannot be mechanised. These tests may not only fail to detect worms in patients with low or periodic microfilarial loads, but are totally useless when trying to detect prepatent or latent infections when no microfilariae are present. What is required for a successful diagnostic test for filariasis has been defined by the WHO (World Health Organisation Memorandum 1975) as a sensitive, specific, quantitative test which can define patients at risk to developing disease symptoms, and yet is applicable to the field situation. As one would expect, studies have concentrated on developing some form of serological test which provides the specificity and sensitivity required. Early serological tests involved complement fixation, haemag-glutination and fluorescent antibody tests. More recently, enzyme-linked immuno-sorbent assay (ELISA) techniques have been applied. In a very simple straightfor-ward comparison, the ELISA has been demonstrated to be up to 100 times more sensitive than other serological techniques in assaying for trichinellosis in a variety of hosts (RUITENBERG et al. 1975; VAN KNAPEN et al. 1980). However in the case of sero-logical assays for diagnosis of filarial infections the improved sensitivity of the tests has been overshadowed by other problems. As the reviews by KAGAN (1963) and AMBROISE-THOMAS (1974, 1980) have highlighted, the primary problem encountered is cross-reactivity in the tests. This is due to the use of crude antigenic extracts and sera from patients with incomplete medical records. Hopefully diagnostic studies using defined parasite antigen-antibody reactions will overcome these problems.

Surface or metabolic labelling provides us with the means to select one or more parasite antigens from specific compartments of the worm. Earlier work demon-strating that there were a restricted number of stage- and species-specific antigens on the surface or in the secretions of *T. spiralis* (PHILIPP et al. 1981; PARKHOUSE and CLARK 1983), *N. brasiliensis* (MAIZELS et al. 1983a) and *T. canis* (MAIZELS et al. 1984a) were very encouraging. In the case of *T. spiralis* the results obtained from the analysis of radiolabelled proteins helps to explain the remarkable specificity of the ELISA assay for trichinellosis, when using the crude saline soluble fraction from homoge-nised muscle larvae (RUITENBERG et al. 1975). As mentioned above, the surface-labelled proteins of muscle larvae of *T. spiralis* are both stage and species specific (PHILIPP et al. 1980; PARKHOUSE et al. 1981). Furthermore studies of metabolically labelled proteins demonstrated that secreted proteins displayed similar properties (PARKHOUSE and CLARK 1983). Finally, when muscle larvae were metabolically label-led with [^{35}S]methionine, none of the labelled solubilised somatic proteins were pre-cipitated by sera from infected mice. This suggests that most of the antigenic proteins analysed so far are stage and species specific, accounting for the specificity of this ELISA test.

In contrast to the earlier nematodes studied, filarial worms present a more complicated picture. The surfaces of *Litomosoides carinii* do not have totally exclusive stage-specific determinants (PHILIPP et al. 1984b). Similarly, the major surface-labelled proteins of the Brugian filariases are cross-reactive not only between stages but also between species (MAIZELS et al. 1983b). In fact, the cross-reactivity between stages is so strong that mice primed with microfilariae and then challenged with infective larvae produce a secondary-like antibody response to surface determinants of the L3s (MAIZELS et al. 1983b). The major surface proteins of *Onchocerca gibsoni* displayed even greater cross-reactivity when tested against a panel of sera taken from patients with a variety of filarial and trematodal infections (FORSYTH et al. 1981b). However discouraging these results may seem, the possibility should not be overlooked that one or more of the minor proteins in a specific compartment may be totally specific for either the species or the stage of the parasite. The ability selectively to label proteins from each compartment enables us to look systematically for potentially immunodiagnostic proteins in each compartment. One may imagine that the surface and secretions of the parasite are more likely to contain the species- and stage-specific components needed for the parasite's survival in the different environments encountered during its life cycle. In contrast, the basic functional, or "housekeeping", molecules found in the somatic fraction will be more conserved not only through the life cycle, but also across species. It is important to note that radio-labelling procedures do not necessarily tag all of the antigenic components. Glycolipids and carbohydrate moieties are not labelled with radioactive amino acids, nor perhaps are all the proteins. For example KAUSHAL et al. (1982) found that iodination of the E/S components of microfilariae of *Brugia malayi* produced an unexpectedly specific diagnostic "antigen". Here, although the E/S fraction was complex as judged by Coomassie blue staining of proteins separated by SDS-PAGE, similar analysis of the iodinated protein revealed that only a restricted number of the smaller molecular weight components were labelled, and these were, presumably, species specific.

By the application of these techniques PHILIPP et al. (1984a) identified a potentially useful protein for diagnosis on the surface of adult stages of *Onchocerca volvulus*. It has a mw of 20 000 and promising results were obtained when it was used in a serological survey of people living in an area endemic for onchocerciasis in southern Mexico. The 20-kd-labelled component from the surface was separated from the other labelled surface proteins by gel filtration and then used in an immunocoprecipitation test with sera from people living in villages either endemic or non-endemic for the disease. The specificity of the test was 98% and the sensitivity 92%. Of interest was the detection of antibodies in sera of patients living in the endemic area but without signs of clinical disease. These probably represent examples of people with prepatent infections, or who have been exposed to the worms without developing disease.

Specified antigens are required for an assay which is of high specificity. They need not however be derived from the parasite detected in the assay. There are examples of specific cross-reactions between nematodes which have been exploited for diagnosis. For example, DISSANAYAKE and his co-workers (DISSANAYAKE and ISMAIL 1980; DISSANAYAKE et al. 1982) reported that fractions of a detergent-soluble extract of adult worms of *Setaria digitata* (a nematode infection of cattle) separated on DEAE-Sephadex columns were cross-reactive with *Wuchereria bancrofti*. Some of these components cross-reacted with major antigens of the microfilariae since one of the *S. digitata* extracts blocked the binding of immune sera to microfilariae of *W. bancrofti*. Another fraction designated SD 2.4 was used as an antigen in an assay for bancroftian filariasis. The antigen cross-reacted with a component of adults of *W. bancrofti* (DISSANAYAKE et al. 1980). Subsequently DISSANAYAKE and ISMAIL (1982) isolated immune complexes from the sera of microfilaraemic patients and found that these complexes contained the component which cross-reacted with SD 2.4. Another potentially useful cross-reactivity identified was a fraction of adults of *Litomosoides carinii* detected by sera from patients infected with *O. volvulus* (KLENK et al. 1984). When this fraction was used in a series of three tests, sera from infected people always registered as positive in at least two of the three tests (i.e. 100% sensitivity) and only one non-specific reaction was found. Cross-reactions between parasites are generally considered a problem in diagnostic tests. However, the availability of material from the filarial parasites of man is highly restricted due to the limited host range of these nematodes. Isolating components from abundantly available worms which are usefully cross-reactive circumvents one of the problems in scaling up the diagnostic test for widespread use.

The detection of immune complexes in the blood of patients with filariasis by DISSANAYAKE et al. (1982) is not unique. A number of other workers have detected circulating immune complexes in patients with bancroftian filariasis (GAJANA et al. 1982; PRASAD et al. 1983a) and onchocerciasis (PAGANELLI et al. 1980; OUASSI et al. 1981). Analyses of these complexes from patients with *W. bancrofti* have shown that they contain both adult worm (DISSANAYAKE et al. 1982) and microfilarial E/S products as well as antimicrofilarial antibodies (PRASAD et al. 1983b). Similarly filarial worm components have been detected in complexes from patients with onchocerciasis. This time the components were derived either from the female worm's uterine secretions or the egg (DES MOUTIS et al. 1983). These demonstrations of circulating antigens in patients with filariasis suggest an alternative approach to immunodiagnosis. Instead of developing methods to detect specific antibodies in the blood, detection of specific antigens should be possible.

The detection of circulating antigen either free or complexed with host antibodies may be advantageous. The disappearance of antibodies from the circulation following chemotherapy and the clearance of parasites is very slow (MALHOTRA and HARINATH 1984). The changes in levels of circulating antigen may be far more dependent on the presence and condition of the parasites and so may possibly provide a closer reflection of the clinical situation. One example of a diagnostic test depending on the detection of circulating antigen has been developed by Capron's group in France. OUASSI et al. (1981) detected circulating antigens in the sera of patients with onchocerciasis. The so-called circulating onchocercal antigen (COA) was found in 75% of patients with clinical signs of disease. Although the antigen was specific to onchocerciasis, polyclonal sera raised in rabbits to this antigen also bound other filarial worm antigens. This problem was reduced by the production of a monoclonal antibody to the antigen (DES MOUTIS et al. 1983).

Accurate diagnosis depends not only on a standardised and characterised antigen, but also on a suitable assay system. As stated by the WHO the assay should be sensitive, specific, quantitative, economical and suitable for use in endemic areas. The most sensitive serological assay systems, radioimmunoassay and ELISA, have been applied in experimental diagnostic tests for nematode infections (OUASSI et al. 1981; DES MOUTIS et al. 1983; DE SAVIGNY et al. 1979; MAIZELS et al. 1984a; VAN KNAPEN et al. 1980). However, sensitivity and specificity depends on the antigen-antibody system under scrutiny. It must be noted that not all assay systems are totally interchangeable. For example, MAIZELS et al. (1984a) examined sera from positively identified cases of *Toxocara* and control sera from patients with other parasitoses using E/S antigens in both in ELISA assay and an immunocoprecipitation test. Although there was a broad correlation between the results of the two assays, exceptions were found because the two assays measure slightly different parameters. There are even examples where detection of a particular antigen is limited to one of the two assays. For example, two monoclonal antibodies raised to adult surface antigens of *Brugia pahangi* were found to perform very differently in ELISA and radioimmunocoprecipitation assays (SU-TANTO 1983). One antibody could only be used in ELISAs because the antigen it bound was not a protein and so could not be radioiodinated; the other only worked well in radioimmunoassay, possibly because it was bound poorly by the ELISA plate.

Whether a potential test is economic and suitable for use "in the field" can only be decided individually. However, a few observations must be made. The requirement of highly purified antigen, especially from the major filarial nematodes of man which have a highly restricted host range, would make a test prohibitively expensive. This problem may be circumvented in the future through the use of the techniques of genetic engineering. The ability to clone the genes which produce the diagnostic antigen and then introduce it into suitable expression systems would allow almost limitless amounts of antigen to be prepared relatively easily and cheaply. An excellent exemple of the potential of genetic engineering techniques, albeit in the study of malaria, is from work by KEMP and his collegues (KEMP et al. 1983; STAHL et al. 1984). They have been looking for specific protective antigens to malaria by cloning cDNA in microbial expression systems and then screening the clones for the presence of specific parasite antigens. This approach has removed the need for mono-specific probing antibodies.

In the absence of such methods of producing purified antigen, then the specificity of an assay would be most economically provided for by the use of monoclonal anti-bodies either to absorb a specific antigen from a crude antigenic extract onto a solid surface in some form of "sandwich" assay, the serum being subsequently tested against the "captured" antigen, or alternatively a crude antigenic extract may be used and antibodies in the test serum would be assayed by its ability to compete with the binding of a monoclonal antibody to a specific component of the crude extract.

Finally, the application of diagnostic tests to field situations needs to employ a method which does not require expensive laboratory equipment. Centrifuges and scintillation counters are costly and not mobile pieces of equipment. In comparison ELISA-based assays which can be "read" by eye and yet still resolve sera accurately seem far more suitable. The robust nature of the plates and the exclusion of radioiso-topes which lose activity relatively rapidly are further advantages of this system. However, it must be reiterated that each system of antigen-antibody interaction must be scrutinised for its individual merits.

8 Concluding Remarks

The study, isolation and utilisation of defined nematode antigens have involved the application of techniques common to those used in studying any macromolecule. The application to the study of parasitic nematode antigens has illuminated the subtle nature of the interaction between parasite and host. Although mechanisms of evasion by the parasite or killing by the host have only, as yet, been described in one or two systems, they will, no doubt, be found to play a role to some degree in all nematode-host interactions. Second-stage larvae of *Toxocara canis*, for example, make an ideal model for studying the role of ES components. Work on this model system has demonstrated the role of secretions in blocking immune responses to the parasite surface (SMITH et al. 1981) and the potential use of ES components in diagnostic assays (DE SAVIGNY et al. 1979; MAIZELS et al. 1984a). These results have now been shown also to be applicable to other nematodes. Antigenic components in the ES of *T. spiralis* are highly restricted (PARKHOUSE and CLARK 1983) and the ES of microfilariae of *Brugia malayi* have also been found to contain fewer cross-reactive components than other antigenic compartments of the parasite (KAUSHAL et al. 1982).

At the moment our knowledge is restricted to laboratory models. The next step is to apply this knowledge to understand and control parasites affecting man. The lessons learned from *T. spiralis, N. brasiliensis* and *L. carinii* will enable one to ask more pertinent questions about parasites where material is in restricted supply. There are still many gaps in our knowledge concerning the biology of nematodes and their survival in the host. Many of the questions can only be answered with experiments in model systems.

One aspect of host control which can be studied most effectively in model systems is the control of the isotype of host humoral responses. The role played by antibodies of distinct isotypes in relation to parasite rejection has already been mentioned. Although information on the important role of the various immunoglobulin classes is now forthcoming, much less is known about the factors which influence class switching. Of particular interest are studies on immunoglobulin E. The nematode *N. brasiliensis* has already played a significant role in the study of IgE isotype expression (ISHIZAKA 1983). Yet the isolation of binding factors which enhance or suppress IgE production is only one step to the complete understanding of this complex phenomenon. Not only is there interest in allergenic epitopes and factors controlling the IgE isotype response, there may also be a role for IgE detection in diagnosis. There have been a number of reports in which the IgE response was found to display greater antigenic specificity than IgG or IgM antibodies for nematodes (WEISS et al. 1982b; OTTESEN et al. 1979). Furthermore apart from providing the basis for more specific tests for diagnosis, the IgE antibody response may provide information on the clinical prognosis of disease. It has been found that the pattern of binding of proteins from adult worms of *Brugia malayi* by IgE antibodies in patients' sera depended on the clinical nature of the disease (HUSSAIN and OTTESEN 1983; WEISS et al. 1982a), emphasising the dynamic nature of IgE responses, like all other isotypes.

Turing our attention back to parasite antigens again, a very pertinent question is the role of parasite antigens to the parasite. The question that arises is "Why does a parasite not try to delete or disguise molecules which can elicit host protective responses?" The reason seems likely that the continued exposure of an antigen is vital to the parasite. This begs the question concerning the functions of parasite antigens. So far, functions have been ascribed to only two nematode parasite antigens; both components were found to be enzymes. The first demonstrated was the acetyl cholinesterase of *N. brasiliensis* (JONES and OGILVIE 1972; SANDERSON et al. 1972). This has been postulated to act as a biochemical holdfast for the worm in the gastro-intestinal tract (PHILIPP 1984). Furthermore it has been shown that worms adapted to survive in an immune environment produce enzymes of different electrophoretic mobility (EDWARDS et al. 1971). One may envisage that acetyl-cholinesterase activity is essential for parasite survival and that the enyzmes released by worms adapted to the immune environment are not inactivated by host antibodies. The second example of an antigen whose function is known is the superoxide dismutase of *Trichinella spiralis* (RHOADS 1983). This enzyme, secreted by infective larvae, is of identical size to bovine superoxide dismutase (a dimer of 36 kd), yet displays anti-genic determinants recognised specifically by sera from animals infected with *Trichinella*. Once again a biological role has been proposed for this enzyme, protec-ting the parasite from the killing products of host defence cells and thus specific anti-bodies could protect the host (RHOADS 1983). Apart from these antigens of defined function, there are other exposed structures of parasites which could also elicit pro-tective responses. CHEN and HOWELLS (1979) demonstrated that the cuticle of adults of *Brugia pahangi* transports specific nutrients into the worm for metabolic use. To be able actively to transport components across the cuticle, the worm requires transport proteins inserted into the surface and receptors freely available to the external medium. These molecular structures vital to worm survival could also be targets for host responses.

Not all antigens of nematodes need mediate essential metabolic functions for the worm. An alternative reason for their continued exposure is that they actually allow the parasite to survive by stimulating a host response. As mentioned before the second-stage larvae of *Toxocara canis* release vast amounts of secretions derived from the surface which can prevent the binding of antibodies to the parasite's surface (SMITH et al. 1981). In addition, the constant exposure and shedding of these antigenic components may divert the attention of the immune system away from other mole-cules vital to host survival. Such a mechanism has also been proposed for the produc-tion of immunogenic surface proteins of other parasites; for example, the sporozoite stage of *Plasmodium knowlesi* (GODSON et al. 1984). Finally, the stimulation of responses to stage-specific antigens may prevent secondary, life-threatening, infec-tions of the host. For example, the infective larvae of *Trichinella spiralis* in the muscle secrete a very immunogenic glycoprotein which has been shown to be host protective (SILBERSTEIN and DESPOMMIER 1984). This same secretion is released during a normal infection (CAPO et al. 1985). Thus the resident larvae could prevent the establishment of a second potentially lethal infection in the gastrointestinal tract, thus ensuring the survival of established parasites. Thus, infection with *Trichinella* is an example of "concomitant immunity".

One doubt concerning the application of experimental results to the real world is whether experimental infections and situations are realistic. For example, if rats are repeatedly infected with small numbers of *N. brasiliensis,* "adapted" worms, which contain some enzymes of altered electrophoretic mobility (EDWARDS et al. 1971), survive for prolonged periods in the gastrointestinal tract of otherwise immune animals (JENKINS and PHILIPSON 1972). Their survival in an immune host is not due to a different antigenic structure of the surface of the worms (MAIZELS et al. 1983a) and therefore may be due to a change in the responsiveness of the host rather than a change in the worm.

Despite these doubts, the successes of recent work in identifying potentially protective (ORTEGA-PIERRES et al. 1984a) or diagnostic antigens (PHILIPP et al. 1984a) are obvious. The rational approach that the knowledge of defined parasite antigen antibody system allows is a great advance and will undoubtedly pay dividends in the near future for clinical medicine.

References

Ali NMH, Behnke JM (1983) *Nematospiroides dubius* factors affecting the primary response to SRBC in infected mice. J Helminthol 57:343-353
Ali NMH, Behnke JM (1984) Non-specific immunodepression by larval and adult *Nematospiroides dubius*. Parasitology 88:153-162
Ambler J, Miller JN, Orr TSC (1974) Some properties of *Ascaris suum* allergen A. Int Arch Allergy Appl Immunol 46:427-437
Ambroise-Thomas P (1974) Immunological diagnosis of human filariases: present possibilities, difficulties and limitations. Acta Trop (Basel) 31:108-128
Ambroise-Thomas P (1980) In: Houba V (ed) Immunological investigation of tropical parasitic diseases. Churchill Livingstone, Edinburgh
Anders RF, Howard RJ, Mitchell GF (1982) Parasite antigens and methods of analysis. In: Cohen S, Warren KS (eds) Immunology of parasitic diseases, 2nd ed Blackwell Scientific, Oxford
Anya AO (1966) The structure and chemical composition of the nematode cuticle. Observations on some axyurids and *Ascaris*. Parasitology 56:179-198
Askonas BA (1984) Interference in general immune function by parasite infections. African trypanosomiasis as a model system. Parasitology 88:633-638
Au ACS, Denham DA, Draper CC (1982) Detection of antibodies in *Brugia pahangi*-infected cats by counter immunoelectrophoresis, indirect fluorescent antibody test and enzyme-linked immunosorbent assay. Z Parasitenkd 68:313-320
Behnke JM, Parish HA (1980) Expulsion of *Nematospiroides dubius* from the intestine of mice treated with immune serum. Parasite Immunol 1:13-26
Behnke JM, Parish HA (1981) Transfer of immunity to *Nematospiroides dubius:* co-operation between lymphoid cells and antibodies in mediating worm expulsion. Parasite Immunol 3.249-259
Behnke JM, Hannah J, Pritchard DI (1983) *Nematospiroides dubius* in the mouse: evidence that adult worms depress the expression of homologous immunity. Parasite Immunol 5:397-408
Berrens L (1971) In: The chemistry of atopic allergens. Monographs in allergy, vol 7. Karger, Basel
Bird AF (1980) The nematode cuticle and its surface. In: Zuckerman BM (ed) Nematodes as biological models, vol 2. Academic, New York pp 213-236
Bloch KJ (1967) The anaphylactic antibodies of mammals including man. Progr Allergy 10.84-150
Bradbury SM, Percy DH, Strejan GH (1974) Immunology of *Ascaris suum* infection. I. Production of reaginic antibodies to worm components in rats. Int Arch Allergy Appl Immunol 46:498-511
Buys J, Wever R, van Stigt R, Ruitenberg EJ (1981) The killing of newborn larvae of *Trichinella spiralis* by eosinophil peroxidase in vitro. Eur J Immunol 11:843-845
Canlas M, Wadee A, Lamontagne L, Piessens WF (1984) A monoclonal antibody to surface antigens on microfilariae of *Brugia malayi* reduces microfilaraemia in infected jirds. Am J Trop Med Hyg 33:420-424
Capo V, Despommier DD, Silberstein DS (1985) The site of exclysis of the L₁ larva of *Trichinella spiralis*. J Parasitol 70:992-994
Capron A, Dessaint JP (1981) IgE: a molecule in search of a function. Ann Immunol 132C:3-8
Chen SN, Howells RE (1979) The uptake of dyes, monosaccharides and amino acids by the filarial worm *Brugia pahangi*. Parasitology 78:343-354

Chitwood BG, Chitwood MB (1974) Introduction to nematology. University Press, Baltimore, pp 1–7

Clark NWT, Philipp M, Parkhouse RME (1982) Non-covalent interactions result in aggregation of surface antigens of the parasitic nematode *Trichinella spiralis*. Biochem J 206:27–32

Clegg JA, Smith MA (1978) Prospects for the development of dead vaccines against helminths. Adv Parasitol 16:165–218

Dalesandro DA, Klei TR (1976) Evidence for immunodepression of Syrian hamsters and Mongolian jirds by *Dipetalonema viteae* infections. Trans R Soc Trop Med Hyg 70:534–535

Dandeu JP, Lux M (1978) Purification and characterisation of two proteins from *Ascaris suum* extract, antigenically different but bearing common allergenic epitopes. Immunol Commun 7:393–415

Davis A (1983) The importance of Parasitic Disease. In: Warren K, Bowers JZ (eds) Parasitology, a global perspective. Springer, Berlin Heidelberg New York

de Savigny D, Voller A, Woodruff AW (1979) Toxocariasis serological diagnosis by enzyme immuno-assay. J Clin Pathol 32:284–288

des Moutis I, Ouassi A, Grzych JM, Yarzabal L, Haque A, Capron A (1983) *Onchocerca volvulus:* detection of circulating antigen by monoclonal antibodies in human onchocerciasis. Am J Trop Med Hyg 32:533–542

Despommier DD (1975) Adaptive changes in muscle fibres infected with *Trichinella spiralis*. Am J Pathol 78:477–496

Despommier DD, Campbell WC, Blair LS (1977) The in vivo and in vitro analysis of immunity to *Trichinella spiralis* in mice and rats. Parasitology 74:109–119

Dessein AJ, Parker WL, James SL, David JR (1981) IgE antibody and resistance to infection. I. Selective suppression of the IgE antibody response in rats diminishes the resistance and the eosinophil response to *Trichinella spiralis* infection. J Exp Med 153:423–436

Dissanayake S, Ismail MM (1980) Antigens of *Setaria digitata:* cross reactions with surface antigens of *Wuchereria bancrofti* microfilariae and serum antibodies of *W. bancrofti* infected subjects. Bull WHO 58:644–654

Dissanayake S, Galahitiyawa SC, Ismail MM (1982) Immune complexes in *Wuchereria bancrofti* infection in man. Bull WHO 60:919–927

Edwards AJ, Burt JS, Ogilvie BM (1971) The effect of immunity upon some enzymes of the parasitic nematode, *Nippostrongylus brasiliensis*. Parasitology 62:339–347

Faubert GM (1976) Depression of the plaque forming cells to sheep red blood cells by the newborn larvae of *Trichinella spiralis*. Immunology 30:485–490

Faubert GM (1982) The reversal of the immuno-depression phenomenon in trichinellosis and its effect on the life cycle of the parasite. Parasite Immunol 4:13–20

Forsyth KP, Copeman DB, Abbot AP, Anders RF, Mitchell GF (1981a) Identification of radioiodinated cuticular proteins and antigens of *Onchocerca gibsoni* microfilariae. Acta Trop (Basel) 38:329–342

Forsyth KP, Copeman DB, Anders RF, Mitchell GF (1981b) The major radioiodinated cuticular antigens of *Onchocerca gibsoni* microfilariae are neither species nor onchocera specific. Acta Trop (Basel) 38:343–352

Forsyth KP, Copeman DB, Mitchell GF (1984) Differences in the surface radioiodinated proteins of skin and uteric microfilariae of *Onchocerca gibsoni*. Mol Biochem Parasitol 10:217–229

Fujita K, Tsukidate S (1981) Preparation of a highly purified allergen from *Dirofilaria immitis*. Reaginic antibody formation in mice. Immunology 42:363–370

Fujita K, Ikeda T, Tsukidate S (1979) Immunological and physico-chemical properties of a highly purified allergen from *Dirofilaria immitis*. Int Arch Allergy Appl Immunol 60:121–131

Furman S, Ash LR (1983) Analysis of *Brugia pahangi* microfilariae surface carbohydrates: comparison of the binding of a panel of fluoresceinated lectins to mature in vivo derived and immature in utero derived microfilariae. Acta Trop (Basel) 40:45–51

Gajana A, Bheema Rao US, Manomani LM (1982) Preliminary study on circulating immune complexes in bancroftian filariasis. Indian J Med Res 76:146–149

Godson GN, Ellis J, Lupski JR, Ozaki LS, Svec P (1984) Structure and organisation of genes for sporozoite surface antigens. Philos Trans R Soc Lond 307:129–139

Greene BM, Taylor HR, Aikawa M (1981) Cellular killing of microfilariae of *Onchocerca volvulus:* eosino-phil and neutrophil mediated immune serum-dependent destruction. J Immunol 127:1611–1618

Grove DI, Davis RS (1978) Serological diagnosis of bancroftian and malayan filariasis. Am J Trop Med Hyg 27:508–513

Grzych JM, Capron M, Bazin H, Capron A (1982) In vitro and in vivo effector function of rat IgG2a monoclonal anti-*S. mansoni* antibodies. J Immunol 129:2739-2743

Haque A, Joseph M, Ouassi MA, Capron M, Capron A (1980) IgE antibody-mediated cytotoxicity of rat macrophages against microfilariae of *Dipetalonema viteae* in vitro. Clin Exp Immunol 40:487-495

Hogarth-Scott RS (1967) The molecular weight range of nematode allergens. Immunology 13:535-537

Horowitz S, Tarrab-Hazdai R, Eshhar Z, Arnon R (1983) Anti-schistosome monoclonal antibodies of different isotypes: correlation with cytotoxicity. EMBO J 2:193-198

Howells RE, Chen SN (1981) *Brugia pahangi:* feeding and nutrient uptake in vitro and in vivo. Exp Parasitol 51:42-58

Hussain R, Ottesen EA (1983) IgE responses in human filariasis. II. Quantitative characterisation of filaria specific IgE. J Immunol 131:1516-1521

Hussain R, Bradbury SM, Strejan G (1973) Hypersensitivity to *Ascaris* antigens. VIII. Characterisation of a highly purified allergen. J Immunol 111:260-268

Ishizaka K (1983) IgE-Binding factors from rat T lymphocytes. In: Pick E (ed) Lymphokines, vol 8. Academic, New York pp 41-80

Ishizaka T, Ishizaka K (1975) Biology of immunoglobulin E. Prog Allergy 19:101-105

Jarrett EE (1978) Stimuli for the production and control of IgE in rats. Immunol Rev 41:52-76

Jarrett EE, Bazin H (1974) Elevation of total serum IgE following helminth parasite infection. Nature 251:613-614

Jenkins SN, Behnke JM (1977) Impairment of primary expulsion of *Trichuris muris* in mice concurrently infected with *Nematospiroides dubius*. Parasitology 75:71-78

Jenkins DC, Phillipson RF (1972) Increased establishment and longevity of *Nippostrongylus brasiliensis* in immune rats given repeated small infections. Int J Parasitol 2:105-111

Johnson P, Mackenzie CD, Suswillo RR, Denham DA (1981) Serum mediated adherence of feline granulocytes to microfilariae of *Brugia pahangi* in vitro: variations with parasite maturation. Parasite Immunol 3:69-80

Jones VE, Ogilvie BM (1972) Protective immunity to *Nippostrongylus brasiliensis* in the ratt. (iii) Modulation of worm's acetyl cholinesterase by antibodies. Immunology 22:119-129

Jungery M, Ogilvie BM (1982) Antibody response to stage specific *Trichinella spiralis* surface antigens in strong and weak responder mouse strains. J Immunol 129:839-843

Jungery M, Clark NWT, Parkhouse RME (1983) A major change in surface antigens during the maturation of newborn larvae of *Trichinella spiralis*. Mol Biochem Parasitol 7:101-109

Kagan IG (1963) A review of immunologic methods for the diagnosis of filariasis. J Parasitol 49:773-779

Katz SP, Lammie PJ (1984) Effect of cyclophosphamide on the immune responsiveness of jirds infected with *Brugia pahangi*. Infect Immun 43:753-755

Kaushal NA, Hussain R, Ottesen EA (1984) Excretory-secretory and somatic antigens in the diagnosis of human filariasis. Clin Exp Immunol 56:567-576

Kawanishi H, Saltzman LE, Strober W (1982) Characteristics and regulatory function of murine Con-A-induced cloned T cells obtained from Peyer's patches and spleen: mechanical regulating isotype-specific immunoglobulin production by Peyer's Patch B cells. J Immunol 129:475-483

Kazura JW, Meshnick SR (1984) Scavenger enzymes and resistance to oxygen mediated damage in *Trichinella spiralis*. Mol Biochem Parasitol 10:1-10

Kemp DJ, Coppel RL, Cowman AF, Saint RB, Brown GV, Anders RF (1983) Expression of *Plasmodium falciparum* blood stage antigens in *Escherichia coli:* detection with antibodies from immune humans. Proc Natl Acad Sci USA 80:3787-3791

Kennedy MW (1976) Kinetics of establishment and rejection of the enteral phase of a primary infection of *Trichinella spiralis* in the NIH strain mouse. Trans R Soc Trop Med Hyg 70:285

Kiyono H, McGhee JR, Mosteller LM, Elridge JH, Koopman WJ, Kearny JF, Michalek SM (1982) Murine Peyer's Patch T cell clones. Characterisation of antigen-specific helper T cells for immunoglobulin A responses. J Exp Med 156:1115-1130

Klenk A, Geyer E, Zahner M (1984) Serodiagnosis of human onchocerciasis. Evalution of sensitivity and specificity of a purified *Litomosoides carinii* adult worm antigen. Tropenmed Parasitol 35:81-84

Kwa BK, Mak JK (1980) Specific depression of cell-mediated immunity in Malayan filariasis. Trans R Soc Trop Med Hyg 74:522-527

Lammie PJ, Katz SP (1983a) Immunoregulation in experimental filariasis. I. In vitro suppression of mitogen-induced blastogenesis by adherent cells from jirds chronically infected with *Brugia pahangi*. J Immunol 130:1381-1385

Lammie PJ, Katz SP (1983b) Immunoregulation in experimental filariasis. II. Responses to parasite and non-parasite antigens in jirds with *Brugia pahangi*. J Immunol 130:1386–1389

Larsh JE, Weatherly NF (1975) Cell mediated immunity against certain parasitic worms. Adv Parasitol 13:183–222

Lee DL (1972) Structure of the helminth cuticle. Adv Parasitol 10:347

Lee DL, Wright KA, Shivers RR (1984) A freeze-fracture study of the surface of the infective stage larva of the nematode *Trichinella*. Tissue and Cell 16:819–828

Lichtenfels JR, Murrell KD, Pilitt PA (1984) Comparison of three subspecies of *Trichinella spiralis* by scanning electron microscopy. J Parasitol 69:1131–1140

Lumsden DL (1975) Surface ultrastructure and cytochemistry of parasitic helminths. Exp Parasitol 37:267–339

Mackenzie CD, Preston PM, Ogilvie BM (1978) Immunological properties of the surface of parasitic nematodes. Nature 276:826–828

Mackenzie CD, Jungery M, Taylor PM, Ogilvie BM (1980) Activation of complement, the induction of antibodies to the surface of nematodes and the effect of these factors and cells on worm survival in vitro. Eur J Immunol 10:594–601

Maizels RM, Meghji M, Ogilvie BM (1983a) Restricted sets of parasite antigens for the surface of different stages and sexes of the nematode parasite *Nippostrongylus brasiliensis*. Immunology 48:107–121

Maizels RM, Partono F, Oemijati S, Denham DA, Ogilvie BM (1983b) Cross-reactive surface antigens on three stages of *Brugia malayi, B. pahangi* and *B. timori*. Parasitology 87:249–263

Maizels RM, Partono F, Oemijati S, Ogilvie BM (1983c) Antigenic analysis of *Brugia timori*, a filarial nematode of man: initial characterisation by surface radioiodination and evaluation of diagnostic potential. Clin Exp Immunol 51:269–277

Maizels RM, de Savigny D, Ogilvie BM (1984a) Characterisation of surface and exretory-secretory antigens of *Toxocara canis* infective larvae. Parasite Immunol 6:23–37

Maizels RM, Philipp M, Dasgupta A, Partono F (1984b) Human serum albumin is a major component on the surface of microfilariae of *Wuchereria bancrofti*. Parasite Immunol 6:185–190

Malhotra A, Harinath BC (1984) Detection and monitoring of microfilarial ES antigen levels by inhibition of ELISA during DEC therapy. Indian J Med Res 79:194–198

Manning DD, Manning JK, Reed ND (1976) Suppression of reaginic antibody (IgE) formation in mice by treatment with anti-u antiserum. J Exp Med 144:288–292

Martinz-Palomo A (1978) Ultrastructural characterisation of the cuticle of *Onchocerca volvulus* microfilaria. J Parasitol 64:127–136

McLaren DJ (1984) Disguise as an evasive strategem of parasitic organism. Parasitology 88:597–611

Moloney A, Denham DA (1979) Effects of immune serum and cells on newborn larvae of *Trichinella spiralis*. Parasite Immunol 1:3–11

O'Donnell IJ, Mitchell GF (1978) An investigation of the allergens of *Ascaris* using a radioallergosorbent test (RAST) and sera of naturally infected humans: comparison with an allergen for mice identified by passive cutaneous anaphylaxis test. Aust J Biol Sci 31:459–487

Ogilvie BM, Philipp M, Jungery M, Maizels RM, Worms MJ, Parkhouse RME (1980) The surface of nematodes and the immune response of the host. In: van der Bossche H (ed) The host invader interplay. Elsevier Amsterdam, pp 99–104

Ortega-Pierres G, Chayen A, Clark NWT, Parkhouse RME (1984a) The occurrence of antibodies to hidden and exposed determinants of surface antigens of *Trichinella spiralis*. Parasitology 88:359–369

Ortega-Pierres G, Mackenzie CD, Parkhouse RME (1984b) Protection against *Trichinella spiralis* induced by a monoclonal antibody that promotes killing of newborn larvae by granulocytes. Parasite Immunol 6:275–284

Ortega-Pierres G, Clark NWT, Parkhouse RME (1985) Regional specialisation of the surface of a parasitic nematode. Parasite Immunol (in press)

Ottesen EA, Weller PF, Heck L (1977) Specific cellular immune unresponsiveness in human filariasis. Immunology 33:413–421

Ottesen EA, Neva FA, Paranjape RS, Tripathy SP, Thiruvengadam KV, Beaver MA (1979) Specific allergic sensitisation to filarial antigens in tropical eosinophilia syndrome. Lancet ii:1158–1160

Ouassi A, Kouemeni LE, Haque A, Ridd PR, Andre PS, Capron A (1981) Detection of circulating antigens in onchocerciasis. Am J Trop Med Hyg 30:1211–1218

Paganelli R, Ngu NL, Levinsky RJ (1980) Circulating immune complexes in onchocerciasis. Clin Exp Immunol 39:570–575

Parkhouse RME, Clark NWT (1983) Stage specific secreted and somatic antigens of *Trichinella spiralis*. Mol Biochem Parasitol 9:319–327

Parkhouse RME, Cooper MD (1977) A model for the differentiation of B lymphocytes with implications for the biological role of IgD. Immunol Rev 37:105–126

Parkhouse RME, Philipp M, Ogilvie BM (1981) Characterisation of surface antigens of *Trichinella spiralis* infective larvae. Parasite Immunol 3:339–352

Parkhouse RME, Bofill M, Gomez-Priego A, Janossy G (1985) Human macrophages and T-lymphocyte subsets infiltrating nodules of *Onchocerca volvulus*. Parasite Immunol (in press)

Pery P, Luffau G (1979) Antigens of helminths. In: Sela M (ed) The Antigens, vol 5. Academic, New York, pp 83–172

Petit A, Pery P, Luffau G (1980) Purification of an allergen from culture fluids of *Nippostrongylus brasiliensis*. Mol Immunol 17:1341–1349

Philipp M (1984) Acetylcholinesterone secreted by intestinal nematodes: a reinterpretation of its putative role of a "biochemical holdfast". Trans R Soc Trop Med Hyg 78:138–139

Philipp M, Parkhouse RME, Ogilvie BM (1980) Changing proteins on the surface of a parasitic nematode. Nature 287:538–540

Philipp M, Taylor PM, Parkhouse RME, Ogilvie BM (1981) Immune response to stage specific surface antigens of the parasitic nematode *Trichinella spiralis*. J Exp Med 154:210–215

Philipp M, Gomez-Priego A, Parkhouse RME, Davies ML, Clark NWT, Ogilvie BM, Beltran-Hernandez F (1984a) Identification of an antigen of *Onchocerca volvulus* of possible diagnostic use. Parasitology 84:295–310

Philipp M, Worms MJ, McLaren DJ, Ogilvie BM, Parkhouse RME, Taylor PM (1984b) Surface proteins of a filarial nematode: a major soluble antigen and a host component on the cuticle of *Litomosoides carinii*. Parasite Immunol 6:63–82

Philipp M, Worms MJ, Maizels RM, Ogilvie BM (1984c) Rodent Models of Filariasis. In: Marchalonis JJ (ed) Contemporary Topics in Immunobiology vol 12. Plenum New York, pp 275–321

Piessens WF, McGreevy PB, Piessens PW, McGreevy M, Koiman I, Saroso JS, Dennis DT (1980a) Immune responses in human infections with *Brugia malayi*. Specific cellular unresponsiveness to filarial antigens. J Clin Invest 65:172–179

Piessens WF, McGreevy PB, Ratiwayanto S, McGreevy M, Piessens PW, Koiman I, Saroso JS, Dennis DT (1980b) Immune responses in human infections with *Brugia malayi:* correlation of cellular and humoral reactions to microfilarial antigens with clinical status. Am J Trop Med Hyg 29:563–570

Piessens WF, Ratiwayanto S, Tuti S, Palmieri JH, Piessens P, Koiman I, Dennis DT (1980c) Antigen-specific suppressor cells and suppressor factors in human filariasis with *Brugia malayi*. N Engl J Med 302:833–837

Piessens WF, Partono F, Hoffman SL, Ratiwayanto S, Piessens PW, Palmieri JR, Koiman I, Dennis DT, Carney WP (1982) Antigen-specific suppressor-T-lymphocytes in human lymphatic filariasis. N Engl J Med 307:144–148

Ponundurai T, Denham DA, Nelson GS, Rogers R (1974) Studies with *Brugia pahangi*. Antibodies against adult and microfilarial stages. J Helminthol 48:107–111

Prasad GBK, Kharat I, Harinath BC (1983a) Detection of anti-filarial ES antigen-antibody in immune complex in Bancroftian filariasis by enzyme immunoassay. Trans R Soc Trop Med Hyg 77:771–772

Prasad GBK, Reddy MVR, Hariath BC (1983b) Detection of filarial antigen in immune complexes in Bancroftian filariasis by ELISA. Indian J Med Res 78:780–783

Premaratne VN, Parkhouse RME, Denham DA (1984) The use of anti-cat immunoglobulin monoclonals in detecting immune responses of cats to *Brugia pahangi*. Parasitology 89: lxvii

Pritchard DI, Williams DJL, Behnke JM, Lee TDG (1983) The role of IgG1 hypergammaglobulinaemia in immunity to the gastrointestinal nematode *Nematospiroides dubius*. The immunochemical purification, antigen specificity and in vivo anti-parasite effect of IgG1 from immune serum. Immunology 49:353–363

Pritchard DI, Ali NMH, Behnke JM (1984) Analysis of the mechanism of immunodepression following heterologous antigenic stimulation during concurrent infection with *Nematospiroides dubius*. Immunology 51:633–642

Purkerson M, Despommier D (1974) Fine structure of the muscle phase of *Trichinella spiralis* in the mouse. In: Kim C (ed) Trichinellosis. Intext, New York

Rhoads ML (1983) *Trichinella spiralis*: identification and purification of superoxide dismutase. Exp Parasitol 56:41–54

Ridley DS, Hedge EC (1977) Immunofluorescent reactions with microfilariae 2. Bearing on host-parasite relations. Trans R Soc Trop Med Hyg 71:522–525

Rivera-Ortiz C, Nessenzweig R (1976) *Trichinella spiralis*: anaphylactic antibody formation and suscepti-bility in strains of inbred mice. Exp Parasitol 39:7–17

Rothwell TLW, Prichard RK, Love RJ (1974) Studies on the role of histamine and 5 hydroxytryptamine in immunity against the nematodes *Trichostrongylus colubriforms*. I. In vivo and in vitro effects of the amines. Int Arch Allergy Appl Immunol 46:1–13

Ruitenberg EJ, Ljungstrom I, Steerenberg PA, Buys J (1975) Application of immunofluorescence and immunoenzyme methods in the serodiagnosis of *Trichinella spiralis* infection. Ann NY Acad Sci 254:296–303

Sanderson BM, Jenkins DC, Phillipson RF (1972) *Nippostrongylus brasiliensis*: relation between immune damage and acetyl cholinesterase levels. Int J Parasitol 2:227–232

Scher I (1982) The CBA/N mouse strain: An experimental model illustrating the influence of the X-chromosome on immunity. Adv Immunol 33:1–71

Silberstein DS, Despommier DD (1984) Antigens from *Trichinella spiralis* that induce a protective response in the mouse. J Immunol 132:898–904

Smith HV, Quinn R, Kusel JR, Girdwood RWA (1981) The effect of temperature and anti-metabolites on antibody binding to the outer surface of second stage *Toxocara canis* larvae. Mol Biochem Parasitol 4:183–193

Stahl HD, Coppel RL, Brown GV, Saint R, Lingelbach K, Cowman AF, Anders RF, Kemp DJ (1984) Differential antibody screening of cloned *Plasmodium falciparam* sequences expressed in *Escherichia coli*: procedure for isolation of defined antigens and analysis of human antisera. Proc Natl Acad Sci USA 81:2456–2460

Suemara M, Ishizaka J (1979) Potentiation of IgE response in vitro by T cells from rats infected with *Nippostrongylus brasiliensis*. J Immunol 123:918–924

Suemara M, Yodoi J, Hirashima M, Ishizaka K (1980) Regulatory role of IgE-binding factors from rat T-lymphocytes. I. Mechanism of enhancement of IgE response by IgE-potentiating factor. J Immunol 125:148–154

Sutanto I (1983) A study on the Antigenicity of surface components of adult *Brugia pahangi* and their stage specificity. M Phil Thesis. University of Brunel

Svet-Moldavsky GJ, Shadhijan GS, Chernyakhoiskaya IY; Mkheidze DM, Litovchenko TA, Ozeretskovs-kaya NN, Kadaghidze ZG (1970) Inhibition of skin allograft rejection in *Trichinella* infected mice. Transplantation 9:69–71

Thompson JR, Crandall RB, Crandall CA, Neilson JJ (1979) Clearance of microfilariae of *Dipetalonema viteae* in CBA/N and CBA/H mice. J Parasitol 65:966–968

Thompson JR, Crandall RB, Crandall CA, Neilson JJ (1981) Microfilaraemia and antibody responses in CBA/H and CBA/N mice following injection of microfilariae of *Brugia malayi*. J Parasitol 67:728–730

Unanue ER (1981) The regulatory role of the macrophage in antigenic stimulation. II Symbiotic relationship between the lymphocyte and macrophage. Adv Immunol 31:1–136

van Knapen F, Franchimont JH, Ruitenberg EJ, Baldelli B, Bradley J, Gibson TE, Gottal C, Henrikson SA, Kohler G, Skorgaard N, Soula N, Taylor SM (1980) Comparison of the enzyme-linked immunosorbent assay (ELISA) with three other methods for the detection of *Trichinella spiralis* infections in pigs. Vet Parasitol 7:109–121

Wakelin D (1978) Genetic control of susceptibility and resistance to parasitic infection. Adv Parasitol 16:219–308

Wakelin D, Donachie A (1981) Genetic control of immunity to *Trichinella spiralis*. Donor bone marrow cells determine responses to infection in mouse radiation chimaeras. Immunology 43:787

Wassom DL, Wakelin D, Brooks BO, Krco KJ, David CS (1984) Genetic control of immunity to *Trichinella spiralis* infections in mice: hypothesis to explain the role of H-2 genes in primary and challenge in-fections. Immunology 51:625–631

Wegerhof PM, Wenk P (1979) Studies of acquired resistance of the cotton rat against microfilariae of *Litomosoides carinii*. I. Effects of single and repeated injections of microfilariae. Z Parasitenkd 60:50–64

Weiss N (1978) *Dipetalonema viteae*: in vitro blastogenesis of hamster spleen and lymph node cells to phytohaemaglutinin and filarial antigens. Exp Parasitol 46:283–299

Weiss N, Tanner M (1979) Studies on *Dipetalonema vitaea* (Filaroidea) 3. Antibody dependent cell mediated destruction of microfilariae in vivo. Tropenmed Parasitol 30:73–80

Weiss N, Gualzat M, Wyss T, Betschart B (1982a) Detection of IgE-binding *Onchocerca volvulus* antigens after electrophoretic and immuno-enzyme reaction. Acta Trop (Basel) 39:373–377

Weiss N, Huassain R, Ottesen EA (1982b) IgE antibodies are more species specific than IgE antibodies in human onchocerciasis and lymphatic filariasis. Immunology 45:129–137

Weller PF (1978) Cell-mediated immunity in experimental filariasis: lymphocyte reactivity of filarial stage-specific antigens and to bound T cell mitogens during acute and chronic infection. Cell Immunol 37:369–382

Williams DJ, Behnke JH (1983) Host protective antibodies and serum immunoglobulin isotypes in mice chronically infected or repeatedly immunised with the nematode parasite *Nematospiroides dubius*. Immunology 48:37–47

Wong MM, Guest MF (1969) Filarial antibodies and eosinophilia in human subjects in an endemic area. Trans R Soc Trop Med Hyg 63:796–800

Wong MM, Suter PF (1979) Indirect fluorescent antibody test in occult dirofilariasis. Am J Vet Res 40:414–420

World Health Organisation Memorandum (1975) Parasite antigens. Bull WHO 52:237–249

Wright KA (1979) *Trichinella spiralis:* an intracellular parasite in the intestinal phase. J Parasitol 65:441–445

Schistosomes: Surface, Egg and Circulating Antigens

A.J.G. Simpson and S.R. Smithers

1 Introduction

Schistosomes or blood flukes are macroscopic multicellular organisms which live in the vascular system of vertebrates. Several species are medically important parasites of man in tropical, developing countries where infection is transmitted by freshwater snails. The free-swimming cercarial stage of the parasite shed by the snails is able to penetrate host skin directly. After 3–4 days in the dermis, the young schistosomes or schistosomula begin their intravascular migration through the lungs to the hepatic portal system, where they eventually mature and come to live in the small mesenteric or pelvic veins. A mature female worm will deposit upwards of 100 eggs/day into the blood stream and it is the host's immune granulomatous reaction to eggs lodged in the tissues which is largely responsible for the chronic, debilitating, and often fatal disease.

Division of Parasitology, National Institute for Medical Research, Mill Hill, London NW7 1AA, United Kingdom

The extension of irrigation schemes, the construction of dams and the concentration of human populations are today contributing to the increase in the distribution and intensity of schistosome infection. Although safe and effective drugs are available to treat patients, there is at present no reliable serodiagnostic method of detecting an active infection or a cure. Furthermore, the logistical problems of treating populations continually exposed to reinfection emphasizes the need for a vaccine as an aid in irradicating the disease. Thus the identification of antigens which may facilitate accurate diagnosis of schistosome infection as well as antigens capable of mediating protective immunity are goals of the highest priority in the field of schistosome research.

As the schistosome develops and matures within its host there are no abrupt morphological or metabolic changes corresponding to the larval moults seen in nematodes; growth from schistosomula to adult worm is a gradual and continuous process and these two stages are essentially antigenically identical. There are, however, additional adult antigens which are associated with egg production and with a fully developed gut, which are not present in schistosomula. The production of relatively large amounts of these egg- and gut-associated antigens suggests that they may prove to be useful diagnostic reagents and have been characterized with this aim in mind. A further reason for studying egg- and gut-associated antigens is that they are responsible for inducing immunopathological changes in host tissue and a knowledge of the nature of these antigens will be essential to understand the aetiology of the disease.

Part of the intense immune response provoked by infection with adult worms leads to some degree of resistance to reinfection whereby immune effector mechanisms are directed at the young schistosomula (see SMITHERS and DOENHOFF 1982). The antigens responsible for this protection are shared between adult and juvenile parasites and are assumed to be associated with the surface of the organism. Thus the search for antigens responsible for immune protection has concentrated on the characterization of schistosome surface antigens. Although young schistosomula are vulnerable to the host's response, older schistosomes generally live in their host for considerable periods because they have developed mechanisms which enable them to survive in the immunologically hostile environment of the blood. The schistosome surface therefore is not only a target for the host's immune response, but it also serves to protect the parasite from the host. Thus the identification and characterization of schistosome surface antigens as well as understanding the modulation of their expression are necessary not only for understanding parasite immunity but also how the schistosome avoids the immune response. It is possible therefore that study of schistosome surface antigens involved in protective immunity will eventually provide insight into fundamental questions of parasitism as well as being a rational route to the development of a vaccine, and thus the initial section of this chapter is devoted to this topic.

2 Schistosome Surface Antigens

This section, which will discuss the identification of surface antigens and their role in immunity, will concentrate on *Schistosoma mansoni* although there are two other important schistosome species that infect man, *S. haematobium* and *S. japonicum*. To date, less experimental immunology has been undertaken with these latter species although recently the surface antigens of *S. haematobium* have been identified (see below).

2.1 The Biochemical Nature of the Schistosome Surface

During the vertebrate stage of its life cycle the schistosome surface is formed by a continuous surface membrane composed of a double-lipid bilayer as revealed by ultrastructural examination (HOCKLEY and MCLAREN 1973) and freeze-fracture studies (MCLAREN and HOCKLEY 1976). Immediately beneath the surface membrane lies an acellular syncytium again covering the entire worm (for a detailed description of the structure of the schistosome surface see HOCKLEY 1973; MCLAREN 1980). The surface membrane is physiologically active and mediates several absorptive functions (ASCH and READ 1975; ROGERS and BUEDING 1975). Indeed, several enzyme activities have been shown to be associated with the surface membrane including phosphatases, ATPase and glycosyl transferases (SIMPSON et al. 1981a, b; CESARI et al. 1981; PODESTA and MCDIARMID 1982). The biochemical composition of the surface membrane is similar to that of the plasma membrane of other eucaryotic organisms. Thus, a large amount of lipid is readily available to surface-directed iodination both on schistosomula and adult worms (RUMJANEK and MCLAREN 1981; ROBERTS et al. 1983). In addition, complex surface oligosaccharides are expressed on parasitic forms of the schistosome as shown by the diversity of lectins that bind to the surface of the organism (MURRELL et al. 1978; SIMPSON and SMITHERS 1980; SIMPSON et al. 1983b). A variety of radiolabelling techniques have also demonstrated the presence of a complex set of proteins and glycoproteins on the parasite surface (KUSEL et al. 1972; HAYUNGA et al. 1979a; RAMASAMY 1979; SNARY et al. 1980; TAYLOR et al. 1981; SAMUELSON and CAULFIELD 1982; ROBERTS et al. 1983). As in all cell surfaces each of these chemically defined sets of membrane components contribute to the maintenance of the integrity of the membrane and metabolic and physiological events associated with it. In this discussion our interest is with those constituents which are responsible for the antigenicity of the schistosome surface. It is quite possible and indeed likely that both the lipidic and glycolipid moieties of the schistosome surface play a role in the determination of antigenicity particularly since profound changes in the nature of the exposed lipids and oligosaccharides occur during the period when freshly transformed schistosomula are becoming less antigenic (RUMJANEK and MCLAREN 1981; SIMPSON et al. 1983b). Furthermore, many host-derived antigens which are acquired by the schistosome and which are thought to contribute to the evasion of the immune response are glycolipid in nature (GOLDRING et al. 1976, 1977a). As yet, however, no defined lipids or glycolipids have been directly demonstrated to be schistosome surface antigens or epitopes. On the other hand, some of the polypeptides exposed on the surface of the organism have been demonstrated to be antigens by immunoprecipitation with antibodies from immune animals and patients with schistosomiais. Thus, while it may eventually be demonstrated that there are lipidic antigens on the schistosome surface, this section will concentrate on the polypeptide surface antigens of the schistosome and their role in immunity.

2.2 The Detection of Polypeptide Antigens on the Surface of Schistosomula Using Surface-Labelling Techniques and Immunoprecipitation

The utilization of fluorescent antibody-binding techniques and in vitro cytoxicity assays have demonstrated that schistosomula, immediately after their transformation from cercariae, express antigens and are vulnerable to a variety of killing mechanisms whereas older parasites are generally less antigenic and more resistant to immune killing (SMITHERS and DOENHOFF 1982). It is the freshly transformed schistosomulum, therefore, at which most efforts at identifying surface antigens have been directed. The most direct approach used to identify surface antigens has been the use of protein-labelling techniques which in other biological systems have been demonstrated not to penetrate membranes and thus only label exposed polypeptides.

The technique most frequently used to label the schistosomulum surface has been lactoperoxidase-catalysed iodination (RAMASAMY 1979; SNARY et al. 1980; DISSOUS et al. 1981; SHAH and RAMASAMAY 1982; SIMPSON et al. 1983a). The early studies of RAMASAMY (1979) and SNARY et al. (1980) were confined to the surface labelling of parasite proteins followed by electrophoresis. Both groups reported that a dominant ^{125}I-labelled protein of between 70 and 90 kd was present on the surface of schistosomula and cercariae. This probably represents the adsorption of ^{125}I-labelled lactoperoxidase to the surface of the organism rather than the identification of a dominant schistosolumum surface protein. A molecule of this size was not precipitated in later studies with antischistosome antibody but SHAH and RAMASAMY (1982) demonstrated that this protein could be precipitated by anti-lactoperoxidase antibody. In addition to the identification of this extraneous molecule the analysis of ^{125}I-labelled freshly transformed schistosomula revealed a complex pattern of labelled proteins with molecular weights, as described by both groups, ranging up to approximately 100 kd with no one dominant species.

DISSOUS et al. (1981) utilized the same labelling technique but took their analysis further by solubilizing the ^{125}I-labelled molecules in Nonidet P40 and immunoprecipitating with antibodies from immune rats in order to identify antigenic polypeptides. This procedure demonstrated that the only ^{125}I-labelled schistosomulum antigens recognized by sera from immune rats are a group of molecules with molecular weights ranging between 32 and 40 kd. These molecules ran very closely on a one-dimensional electrophoretic gel and appeared as a broad smear. The exact number of antigenic components could not be ascertained by this procedure, but DISSOUS et al. (1981) demonstrated the presence of at least three molecules. These antigens were confirmed as being exposed on the surface by their sensitivity to trypsin digestion. Thus it was demonstrated that only a small subset of the molecules available for ^{125}I-labelling on the surface of schistosomula are recognized as antigens by infected rats. This non-antigenicity of many surface compounds has now been reported by several investigators and is an observation of great potential importance when considering the ability of the parasite to survive within an immune host. In the same report DISSOUS et al. (1981) demonstrated that sera collected from two human patients also immunoprecipitated the 32 000–40 000 mw antigens, which provided further evidence that these molecules represent important schistosome antigens. Later, DISSOUS and CAPRON (1981) reported that both infected mice and monkeys also produce antibodies against this antigen complex. In addition to this group of antigens, however, the human sera also precipitated two ^{125}I-labelled polypeptides of molecular weights greater than 95 k as well as several others of molecular weights approximately 20 k. This suggests that although the 32- to 40-kd antigen group represents dominant schistosomulum antigens recognized by several host species, there are additional antigens expressed on the surface of schistosomula. Indeed analysis of the ^{125}I-labelled schistosomulum surface antigens recognized by chronically infected mice is consistent with this view. SIMPSON et al. (1983a) also found that a 32- to 40-kd complex comprised major antigens precipitated by immune mouse sera but other antigens were also identified. Several high molecular weight molecules were labelled in some experiments and found to be antigenic by immunoprecipitation. Adsorption experiments confirmed that some of these were indeed surface components but were for some reason inconsistently labelled. A similar phenomenon was also observed when schistosomula were labelled with ^{125}I utilizing Iodogen (SIMPSON et al. 1984a). Nevertheless of these high molecular weight antigens, one of 92–94 kd was labelled in most experiments and should be considered as an antigen of potential interest (SIMPSON et al. 1983a). On the other hand, in addition to the major 32- to 40-kd complex several low molecular weight antigens were consistently identified when lactoperoxidase-catalysed iodination was utilized. Interestingly, antibodies from mice vaccinated with highly irradiated cercariae produced a different pattern of precipitation than those from chronically infected mice. This is discussed in more detail below but collectively both types of sera precipitated major low molecular weight antigens of 22, 19, 16 and 15 kd as well as the 32- to 40-kd complex.

With the exception of the 22-kd molecule similar antigens were identified on schistosomula using the Iodogen-labelling technique of lactoperoxidase-catalysed iodination (KNIGHT et al. 1984). These two techniques thus seem interchangeable and equally effective although Iodogen labelling is being increasingly utilized because of its simplicity and convenience (PHILLIP and RUMJANEK 1984). The 20-, 17- and 15-kd antigens reported in this latter study apparently correspond to the 19-, 16- and 15-kd antigens described by SIMPSON et al. (1983a). The former molecular weight designations will be used in the following discussion. These antigens, the source of their identification and their relationship to earlier molecular weight designations are listed in Table 1. A sodium dodecyl sulphate polyacrylamide gel (SDS-PAGE) of the precipitated antigens is shown in Fig. 1.

The 38- to 32-kd complex appears as a smear even when subjected to two-dimensional electrophoresis (KELLY et al. 1985). This complex has been divided into two antigens, however, since the monoclonal antibody directed against the 38-kd component (DISSOUS et al. 1982) and the specificity of serum from mice immunized with irradiated cercariae for the 32-kd component demonstrate that at least two distinct antigenic species exist. The five lower molecular weight antigens of Table 1 can be consistently labelled by either lactoperoxidase-catalyzed iodination or Iodogen; they have been identified in more than one laboratory and are universally precipitated with antibodies of all mouse strains so far employed as well as other experimental hosts. They have all been demonstrated to be exposed on the schistosomula surface by trypsin digestion, by antibody adsorption to labelled parasites, or by monoclonal antibodies (see below); and these antigens form the basis for further discussion. Indeed since these antigens are readily labelled they are the most thoroughly studied

Table 1. The major polypeptide surface antigens of *Schistosoma mansoni* revealed by iodination catalysed by lactoperoxidase or Iodogen[a]

Molecular[b] weight	Recognized during chronic infection of mice[c]	Recognized by mice and rats immunized with irradiated cercariae[d]	Recognized during chronic infection in rats[e]	Recognized during infection in monkeys[f]	Recognized during human infection[g]	Monoclonal antibody prepared against antigen
92 000 (94 000)[h]	+	+	−	−	+	−
38 000 (40 000)	+	+	+	+	+	+[i]
32 000 (32 000)	+	+	+	+	+	−
20 000 (19 000)	+	+	−	−	+	+[j]
17 000 (16 000)	+	−	−	−	+	−
15 000 (15 000)	−	+	−	−	−	−

[a] In addition to the antigens listed here labelling of ^{125}I-iodosulphonilic acid has identified surface antigens of 105,68 and 28 kd (TAYLOR et al. [1981])

[b] Molecular weights taken from KNIGHT et al. (1984)

[c] Taken from SIMPSON et al. (1983a), KNIGHT et al. (1984) and DISSOUS and CAPRON (1981)

[d] Taken from SIMPSON et al. (1983a, 1985a) and KNIGHT et al. (1984). The 38-kd antigen is only recognized following a boosting infection with highly irradiated cercariae

[e] Taken from DISSOUS et al. (1981)

[f] Taken from DISSOUS and CAPRON (1981)

[g] Taken from DISSOUS et al. (1981) and SIMPSON et al. (1985c)

[h] The molecular weights given in parentheses are those reported for the corresponding antigens, in SIMPSON et al. (1983a). DISSOUS et al. (1981) also report the 32- and 38-kd antigens as a 32- to 40-kd complex

[i] Reported by GRYZCH et al. (1982)

[j] Reported by TAVARES et al. (1984)

of the schistosomulum surface antigens and form the subject of much of the work described in this section.

In addition to the lactoperoxidase and Iodogen methods of labelling of the schistosomulum surface several other workers have radiolabelled the surface of schistosomula using the Bolton and Hunter reagent (BRINK et al. 1980), galactose oxidase and NaB^{3}H$_4$ (SAMUELSON and CAULFIELD 1982) and the diazonium salt of ^{125}I iodosulphanilic acid (TAYLOR et al. 1981; BUTTERWORTH et al. 1982). All report that many surface components can be labelled by ^{125}I, again indicating the complexity of the schistosomulum surface. TAYLOR et al. (1981) also reported the identification of antigens amongst the labelled polypeptides by immunoprecipitation using immune mouse and human infection sera. Curiously, although the labelling reagent used is of broad specificity and can label proteins via their tyrosine residues (as can the lactoperioxidase- and Iodogen-labelling methods) the major 38- to 32-kd antigen complex was not identified. Instead three principal antigens of 105, 68 and 25 kd were resolved. The high molecular weight antigen may correspond to one of the high molecular weight antigens sometimes labelled with lactoperoxidase. The 68-kd antigen, on the other hand, appears to be a major antigen not recognized by other techniques. This may be due to the fact that the diazonium salt of ^{125}I iodosulphanilic acid can label histidyl and lysyl residues in addition to tryrosyl residues which are labelled by the

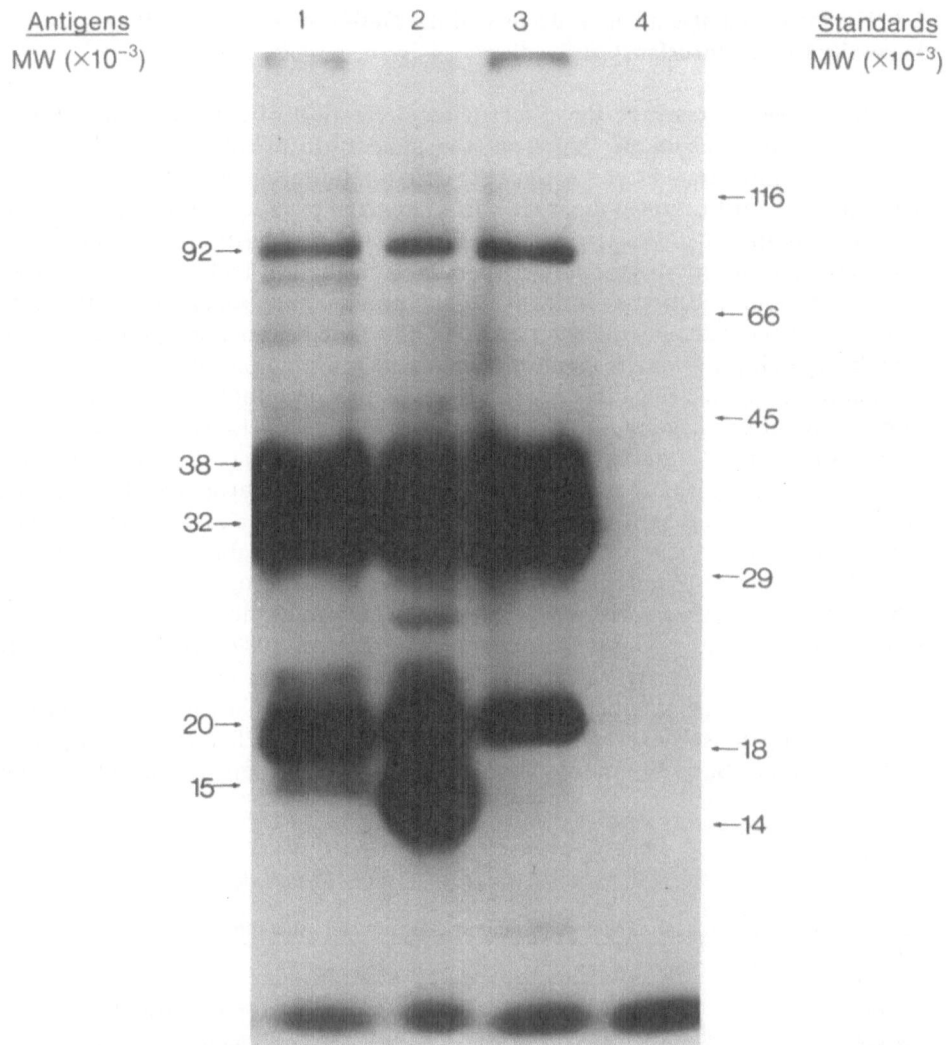

Fig. 1. The polypeptide surface antigens of *Schistosoma mansoni*. Mechanically transformed schistosomula were radiolabelled with ^{125}I using the Iodogen technique and surface components were solublized in saline containing 1% Triton X-100. Antigens were then identified by immunoprecipitating with (*1*) antibodies from mice immunized by chronic infection, (*2*) antibodies from mice immunized by repeated exposure to highly irradiated cercariae, (*3*) antibodies from rabbits immunized with purified adult worm surface membranes and (*4*) antibodies from normal mice. The immune complexes were collected by absorption to protein-A Sepharose and separated by electrophoresis through a 10% polyacrylamide gel. Antigenic polypeptides were revealed by autoradiography. (KNIGHT et al. 1984)

lactoperoxidase and Iodogen methods (HIGGINS and HARRINGTON 1959). Likewise, the 28-kd antigen appears to be a component only identified with this methodology. TAYLOR et al. (1981) showed that the 68-kd component is a glycoprotein and can be removed from the parasite by washing in saline and also that all three antigens were to some extent trypsin sensitive.

2.3 The Detection of Surface Antigens and the Definition of Their Role in Immunity Using Monoclonal Antibodies

An alternative approach to the direct characterization of surface antigens using various labelling techniques has been the generation of monoclonal antibodies which are demonstrated to be surface specific by fluorescence techniques and the subsequent identification of the molecule recognized by precipitation of radiolabelled worm prepration or Western blotting. This technique has three advantages: (a) it overcomes any question of the surface specificity of labelling techniques, (b) it makes possible the identification of antigens which although exposed are difficult to label externally since radiolabelled preparations of whole worm homogenates can be examined, (c) it provides a reagent with which the immunological role of each particular antigen can be evaluated and which can be used as a tool for antigen purification. Many groups have reported the production of monoclonal antibodies which bind to the schistosomulum surface (SMITH et al. 1982; TAYLOR and BUTTERWORTH 1982; GRYZCH et al. 1982; ZODDA and PHILLIPS 1982; NORDEN et al. 1982; HOROWITZ et al. 1983; ARONSTEIN et al. 1983; TAVARES et al. 1984; HARN et al. 1984). (The antigens recognized by these monoclonal antibodies are listed in Table 2). Some of these groups have demonstrated that particular monoclonal antibodies which bind to the schistosomula surface can either mediate killing in vitro or more impressively passively transfer immunity against schistosomes (SMITH et al. 1982; ZODDA and PHILLIPS 1982; GRYZCH et al. 1982; HOROWITZ et al. 1983; TAVARES et al. 1984; HARN et al. 1984). These data are of great significance since they demonstrate that one or more epitopes are sufficiently densely expressed on the schistosomulum surface to act as targets for immune killing. Secondly they demonstrate that the presence of a high enough

Table 2. Schistosome antigens recognized by monoclonal antibodies that bind to schistosomula

Molecular weight	Source of antigen	Reported by
24 000	Schistosomula	TAYLOR and BUTTERWORTH (1982)
38 000[a,]	Surface-labelled schistosomula	GRYZCH et al. (1982)
115 000	Adult worms	DISSOUS and CAPRON (1983)
160 000, 130 000	Cercariae	ZODDA and PHILLIPS (1982)
120 000	Cercariae	NORDEN et al. (1982)
170 000	Adult worms	
60 000, 5 000, 19 000	Cercariae	HOROWITZ et al. (1983)
180 000	Cercariae	ARNONSTEIN et al. (1983)
20 000[a]	Surface-labelled schistosomula	TAVARES et al. (1984)
160 000, 130 000	Schistosomula	HARN et al. (1984)

[a] These antigens are the same as those with similar molecular weights in Table 1

concentration of antibody against certain antigens can protect against subsequent infection. This demonstrates unequivocally that protection against schistosome infection can be immunologically based and that vaccination against the disease is possible. This is more important than is initially apparent since the immune basis of protection by chronic infection with *S. mansoni* has now been questioned and may in fact be due to the egg-induced pathology within the host (DEAN et al. 1981; WILSON et al. 1983).

Clearly those antigens which are recognized by monoclonal antibodies that can mediate immune killing are of particular interest. SMITH et al. (1982) did not report the identification of the target antigen of their protective monoclonal but both ZODDA and PHILLIPS (1982) and HOROWITZ et al. (1983) used ^{125}I-labelled cercarial antigen preparations to identify target antigens. ZODDA and PHILLIPS (1982) precipitated two antigens of molecular weight 160 000 and 130 000 from radiolabelled cercarial extract with their protective monoclonal whereas HOROWITZ et al. (1983) precipitated three major bands of 60 000, 50 000 and 19 000 daltons from cercarial sonicate with their antibody which mediated in vivo killing. HARN et al. (1984) demonstrated that target antigens of 160 and 130 kd were recognized by their protective monoclonal by Western blotting of extracted schistosomula surfaces but thought that they were different from the antigens detected by ZODDA and PHILLIPS (1982). All of these groups thus report antigens which do not seem to correspond to the antigens detected by surface labelling indicating that protective surface antigens not detected by this technique are expressed by schistosomula. It is possible, however, that if immunoprecipitation of surface-labelled components with these antibodies were to be attempted, that precipitation of one of the previously described antigens might result. Where surface-labelled polypeptides have been utilized in conjunction with monoclonal antibodies in both cases one of the set of defined molecules was precipitated (see below). It is of interest to note that one of these, a monoclonal antibody that recognizes the 38-kd antigen on the schistosomulum surface (DISSOUS et al. 1982) precipitated an antigen of 115-kd from solubilized adult worm membranes (DISSOUS and CAPRON 1983). Thus the molecule identified within a membrane preparation by this monoclonal antibody is different from that exposed on the schistosomulum surface and the same phenomenon cannot as yet be ruled out with the other monoclonal antibodies described.

TAYLOR and BUTTERWORTH (1982), NORDEN et al. (1982) and ARONSTEIN et al. (1983) also reported monoclonal antibodies that bind to the parasite surface but did not attempt to utilize the antibodies for immunological experiments. The antigens recognized in these studies have molecular weights of 24 000, 120 000 and 180 000 respectively amongst either schistosomular or cercarial proteins, the 180 000 mw protein described by ARONSTEIN et al. (1983) being associated with spines. The antibody described by NORDEN et al. (1982) also precipitated a 170 000 mw glycoprotein from adult worms again indicating distinct but cross-reacting antigens at different life cycle stages. Again, these results suggest a greater complexity of schistosomulum surface antigens than that revealed by surface labelling.

Two of the monoclonal antibodies reported to date which mediate immune killing identify surface antigens that had previously been identified by surface-labelling techniques. The rat monoclonal antibody produced by GRYZCH et al. (1982) which mediates eosinophil killing of schistosomula in vitro immunoprecipitated the

38-kd antigen (DISSOUS et al. 1982) and the mouse monoclonal antibody produced by TAVARES et al. (1984) that mediates eosinophil and complement killing of schistosomula in vitro immunoprecipitated the 20-kd antigen. In both cases the levels of killing were comparable to immune serum from the appropriate donor animals and both the anti-20-kd and the anti-38-kd monoclonal antibodies were demonstrated to react with antigens that are precipitated by sera from infected hosts (DISSOUS et al. 1982; SIMPSON et al. 1985a). Thus, it was demonstrated that at least two of the five reproducibly identified schistosomulum polypeptide surface antigens can act as targets of immune killing of schistosomula. Furthermore, antibodies against one of these, the anti-38-kd antigen, transferred immunity passively to recipient rats. Thus, although antigens not identified by ^{125}I surface labelling and immunoprecipitation are apparently expressed on schistosomula the evidence gained by the use of monoclonals shows that at least two of the antigens readily identified by surface labelling may be targets of protective immunity. The eventual production of monoclonal antibodies against the other defined antigens will demonstrate whether these are also targets of the immune response. It is noteworthy that the intensity of labelling of the 32-, 17- and 15-kd antigens is comparable with that of the 38- and 20-kd antigens already shown to be protective. Furthermore, antibodies from mice immunized with a single dose of highly irradiated cercariae are primarily directed against the 32-kd antigen with little or no antibody detectable against the 38- and 20-kd antigens (SIMPSON et al. 1983a, 1985a). This evidence suggests that protective activity may be directed not only against the 38- and 20-kd molecules but also against other polypeptide antigens. Thus the schistosomulum surface simultaneously presents a highly complex set of potentially protective antigens and is thus quite different from the majority of parasites discussed in this volume.

The anti-20-kd monoclonal antibody was found by TAVARES et al. (1984) to be species specific as were the monoclonal antibodies produced by SMITH et al. (1982). Indeed data using antibodies from patients indicate that the polypeptide surface antigens of both *S. mansoni* and *S. haematobium* are species specific (SIMPSON et al. 1984a, 1985b). Thus it may be possible to utilize monoclonal antibodies raised against schistosomulum surface antigens not only as tools in the investigation of protective immunity but also as immunodiagnostic reagents. In this context Mitchell and colleagues have identified an immunodiagnostic monoclonal antibody which recognizes a polypeptide antigen of 23-kd on the surface of *Schistosoma japonicum* (MITCHELL et al. 1981; CRUISE et al. 1983).

2.4 The Stage Specificity of Surface Antigen Expression and the Recognition of Surface Antigens During Human and Animal Infection

During penetration of the host's skin, cercariae of *S. mansoni*, which are adapted to a free-living existence, transform into the parasitic schistosomulum form, over a period of hours. Immediately after transformation the schistosomula, as described in the preceding section, are found to be highly antigenic. This period of antigenicity coincides with a vulnerability of the parasites to immune killing. As the parasites mature they become both decreasingly antigenic and increasingly resistant to killing. These events are coincident but experiments in which artificial antigens have been introduced onto the parasite's surface have shown that they are not necessarily inter-dependent (MOSER et al. 1980). Nevertheless, after 5 days of in vivo maturation, at which time they have migrated to the host's lungs, the schistosomula cannot be killed by immune mechanisms and do not bind antibodies from immune animals. Thus the antigens expressed on freshly transformed schistosomula appear to be no longer exposed at this time. Recent work has indicated that some parasite components are exposed, however, since antibodies raised by repeated immunization with parasite material can still bind to the surface of lung-stage schistosomula (BICKLE and FORD 1982) as can antibodies raised against purified adult worm membranes (PAYARES et al. 1985). These antigens have not yet been defined but appear not to be antigens that generate a strong antibody response in animals immunized by a normal chronic infection or with a single exposure to irradiated cercariae since antibodies from ani-mals immunized in this way do not bind to lung-stage organisms.

After leaving the lungs the parasites migrate to the liver and mature into adult worms after a period of 4–6 weeks. During maturation into adult worms the schisto-somes are not highly antigenic since only low amounts of antigen can be detected by immunofluorescent techniques using immune sera (GOLDRING et al. 1977b). Again, however, both antibodies raised against purified adult surface membranes (PAYARES et al. 1985) and some monoclonal antibodies (NORDEN et al. 1982; ARONSTEIN et al. 1983) indicate the existence of exposed parasite components not strongly recognized during infection.

The expression of polypeptide surface antigens generally reflects the overall antigen expression defined by antibody binding to the schistosomulum surface. When adult worms are labelled with ^{125}I using Iodogen under conditions in which schisto-somulum surface antigens are readily labelled, polypeptide surface antigens are barely detectable (SNARY et al. 1980; PAYARES et al. 1985). BUTTERWORTH et al. (1982) also re-ported that proteins became harder to label on maturing schistosomula. Experiments designed to explore the surface components of adult worms are extremely difficult to perform since their surface membrane, in contrast to that of schistosomula, is very fragile and is readily disrupted during in vitro culture (SIMPSON and MCLAREN 1982). Indeed simply incubating live adult schistosomes in phosphate-buffered saline for 5 min at 37 °C has been used as a simple method of releasing the surface membrane from the parasite (SIMPSON et al. 1981a). Thus it may be that the surface-labelled poly-peptide antigens on the adult worms described by HUYUNGA et al. (1979a, b) and SHAH and RAMASAMY (1982) may reflect the exposure of normally internal com-ponents as a result of disruption of the surface membrane. Indeed the autoradio-graphs shown by SHAH and RAMASAMY suggest the presence of ^{125}I-labelled mole-cules throughout the parasite tegument.

The mechanism by which the surface of the highly antigenic schistosomula deve-lops into the antigenically inert adult surface has been the subject of much research. One contributor to the changing antigenicity is the acquisition of host components, which is discussed further below. Nevertheless, there also appears to be spontaneous membrane changes since incubation of schistosomula in vitro in the absence of serum or other host components also results in reduced antigenicity (SAMUELSON et al. 1980; DESSEIN et al. 1981). The altered expression of polypeptide surface antigens on developing schistosomula has been studied either by labelling freshly trans-formed schistosomula and then following the fate of the labelled molecules or conversely by surface labelling schistosomula after various periods of maturation (DIS-SOUS et al. 1982; SIMPSON et al. 1984a; PAYARES et al. 1985). These investigations indi-cate that the modulation of expression of these antigens on the developing schistoso-mula is highly complex. Dissous and colleagues followed the expression of the major protective 38-kd antigen using a monoclonal antibody. They demonstrated that the 38-kd antigen could be radiolabelled on cercariae and that 50% of the originally ex-posed 38-kd antigen molecules were shed during transformation into schistosomula. Thereafter, up to 24 h, the antigen is shed more slowly. When lung-stage schistoso-mula were radiolabelled, however, the 38-kd antigen could not be detected although the 32-kd antigen was still available for radiolabelling. PAYARES et al. (1985) also found that the 20- and 15-kd antigens could also be detected on lung-stage schistoso-mula. The finding that these antigens can be labelled in lung-stage schistosomula seems to contradict the observed lack of immune antibody binding of this parasite form. It is possible, however, that although the antigens are present there is steric hin-drance that prevents antibody access. Alternatively, the amount of these antigens exposed on lung-stage schistosomula may be too small to be detectable by fluores-cent labelling techniques. Nevertheless, these results indicate that the expression of surface antigens may not be coordinated and that some antigens are lost from the developing schistosomula before others. This result was also reported by SIMPSON et al. (1984a). They demonstrated that when schistosomula were cultured for 48 h and then surface labelled, the 38- and 17-kd antigens could not be detected although the 32-, 20- and 15-kd antigens were all readily labelled. Thus there appear to be two groups of surface antigens, those that are retained for at least 2 days on the developing schistosomula and those that are rapidly lost. (Interestingly, those that are retained correspond to the surface antigens recognized by a single immunization with irradi-ated cercariae whereas the 17- and 38-kd antigens are only recognized following a second exposure to irradiated cercariae or chronic infection). These latter antigens were not exposed on 48-h schistosomula whether or not serum was included in the incubation media. The loss of these antigens, therefore, does not depend upon masking by host-derived molecules but is rather part of the parasite's developmental processes. The disappearance of the 38-and 17-kd antigens from the schistosomulum surface does not appear to be due to their being shed. When freshly transformed schi-stosomula were radiolabelled and then incubated in vitro for up to 48 h, a major pro-portion of all the antigens were found to be retained by the schistosomula. Thus the inability to detect the 38-kd and 17-kd antigens on older schistosomula appears to be due to their sequestration to a site within the tegument rather than antigen shedding or masking. The exact mechanism of this process and its significance to the pro-tection of the parasite remains to be established. The 180-kd antigen described by ARONSTEIN et al. (1983) was found to be present on lung-stage organisms but only available for antibody binding after extraction of lipids and glycolipids with ethanol (ARONSTEIN and STRAND 1983). This again suggests that the lack of antigen expres-sion on developing parasites is not due to shedding but to sequestration or masking.

In addition to the two groups of polypeptide surface antigens described above, those retained by the schistosomula and those rapidly lost, another set of antigens appear on the schistosomula surface during maturation. Prominent amongst these is a doublet of approximately 45 kd which was proposed by RUMJANEK et al. (1983) to be a receptor for host lipoproteins. Interestingly, these molecules were expressed only after the parasites had been exposed to serum whereas an 11-kd antigen that appeared during culture was expressed whether or not serum was present in the culture medium (SIMPSON et al. 1984a). Furthermore, when developing schistosomes recovered from mice were examined antigens of 95- and 25-kd were also detected on 5-day-old and 3-week-old worms of which only the latter was precipitated by serum from immune animals (PAYARES et al. 1985). The 95-kd antigen appears to be a major surface component of lung-stage parasites in particular and may account, at least in part, for the binding of the antisurface membrane antibodies to this developmental stage. Thus the modulation of surface antigens during the early stages of schistosomulum maturation involves not only the loss and retention but also the acquisition of polypeptides. Furthermore, some events are independent and some dependent of the presence of host serum. The understanding of the importance of the differently expressed groups of surface antigens will depend upon obtaining antibodies specific for each component. It should be noted, however, that monoclonal antibodies have been obtained against the 38-kd antigen which is rapidly lost and the 20-kd antigen which is retained on the schistosomulum surface. Both of these mediated in vitro killing of freshly transformed schistosomula.

Although polypeptide surface antigens are strongly expressed on schistosomula and apparently thereafter lost from the parasite surface it is not the schistosomula themselves that stimulate the antibody response against these surface antigens during a chronic schistosome infection in mice. Antibodies against schistosomula surface antigens are not detected in the serum of animals 6 weeks after infection but are present in high concentrations 12 weeks postinfection (DISSOUS and CAPRON 1981; SIMPSON et al. 1983a). These antibodies precipitate the 38-, 32-, 20- and 17-kd schistosomula surface antigens but not the 15-kd antigen. Thus it is the mature (and invulnerable) adult worms that stimulate the production of antibodies against antigens exposed on the vulnerable schistosomula. This observation is consistent with the concept of concomitant immunity originally proposed by SMITHERS and TERRY (1969) in which hosts develop immunity against freshly invading schistosomula but are unable to kill the adult worms which provoke the immunity. Antibodies provoked by an all-male infection or an all-female infection are identical to those produced by a chronic infection (SIMPSON et al. 1983a and unpublished observations). Thus both sexes have the potential to stimulate antibodies against schistosomulum surface antigens and the presence of eggs is not required. Indeed although injection of eggs into mice produces antibody that binds to the schistosomulum surface this does not immunoprecipitate any of the polypeptide surface antigens (BICKLE et al. 1983 and unpublished observations).

Immunization of rabbits with purified adult worm surface membranes was found to produce antibodies against the 38-, 32- and 20-kd schistosomula surface antigens (PAYARES et al. 1985). This result demonstrated that although not exposed on the adult worm surface, antigens identical with or cross-reacting with the schistosomulum surface antigens are associated with the adult surface membrane. As mentioned above, the adult surface membrane is very labile and is thought to turn over rapidly (WILSON and BARNES 1977). Thus it is possible that shed membrane fragments or vesicles are the major antigenic stimulus to antibodies against schistosomulum polypeptide surface antigens during chronic infection. However, no antibodies against the 17-kd antigen, strongly recognized during chronic infection, were detected in

antimembrane serum. Antibodies against the 17-kd antigen may therefore be stimulated by another adult worm structure.

To examine the antigens within the adult membrane that are immunogenic,
membrane components released by incubation in phosphate-buffered saline (PBS)
have been radiolabelled, solubilized and immunoprecipitated (DISSOUS and CAPRON
1983; PAYARES et al. 1985). The 32- and 20-kd antigens were detected as on the surface
of schistosomula and an additional antigen of 25-kd was also precipitated by chronically infected mouse sera (PAYARES et al. 1985). Utilizing a monoclonal antibody that
recognizes the 38-kd antigen on the schistosomula surface, DISSOUS and CAPRON
(1983) demonstrated that an antigen of approximately 115 kd was immunoprecipitated from the solubilized adult membrane, indicating that this cross-reacting
antigen may be responsible for the induction of anti-38-kd antibodies by purified
membranes and also perhaps during chronic infection.

Interestingly, it has recently been shown that immunization with a denuded adult
worm homogenate also induces antibodies against schistosomulum surface antigens,
suggesting that these antigens, or molecules that cross-react with them, are also
distributed within the parasite (F. HACKETT, unpublished observations). In this context it has been demonstrated that adult worm surface membranes and internal membranes appear to contain the same major polypeptide antigens, i.e. the 32- and 20-kd
polypepties (C. KELLY, unpublished observations).

Although adult worms provide the major antigenic stimulus during chronic infection of mice the time course of the antibody response shows that this is not the case
during S.mansoni infection of rats. In this host precipitating antibodies were detected
4 weeks after infection (DISSOUS et al. 1981), indicating that it is the schistosomula that
stimulate antibody. The reason for this difference may be the size of the infecting
dose. Rats are usually exposed to 500–1000 cercariae, since the worms are rejected
before they mature and lay eggs. In contrast, chronic infection of mice is achieved
with about 20–30 cercariae. Thus it may be that a small number of schistosomula
cannot stimulate detectable antibody production but the same organisms can when
they develop into the much larger adult stage of the parasite. Evidence to support this
view is provided by the immunization of mice with cercariae attenuated by exposure
to irradiation. Depending on the level of irradiation, these cercariae die in the skin,
lungs, or liver allowing the mice to be infected with several hundred larvae. In such an
infection antibody against schistosomula surface antigens was readily detected
within 4 weeks, indicating that this larger number of schistosomula are capable of
inducing a detectable antibody response in mice (SIMPSON et al. 1983a). Interestingly,
in those situations where the schistosomula stimulate an antibody response against
themselves (i.e. rat infection and immunization with highly irradiated cercariae) the
antibody response to surface antigens is restricted compared with the response
during a chronic infection. Rat infection only induces antibody against the 38- and 32-
kd antigens (DISSOUS et al. 1981; SIMPSON et al. 1985a) and the immunization of mice
with a single infection of highly irradiated cercariae induces antibody principally
against the 32-kd antigen but also detectably against the 20- and 15-kd antigens
(SIMPSON at al. 1983a, 1985a). The antibody response to a single immunization with
irradiated cercariae appears to reflect the differential expression of surface antigens
on schistosomula as described above. The 38- and 17-kd antigens that are rapidly lost
from the surface do not stimulate antibody whereas antibody is detectable against the

32-, 20- and 15-kd antigens that persist on the schistosomulum surface. The antibody response in rats infected with normal cercariae cannot be explained on this basis, however, although it may indicate that schistosomula develop differently in this host and modulate their surface antigens differently. Interestingly, however, rats immunized twice with highly irradiated cercariae show the same antibody response as mice immunized in the same way (SIMPSON et al. 1985a).

The work discussed above demonstrates that a group of schistosomulum polypeptide antigens are recognized by both rats and mice, the most commonly utilized experimental hosts of *S.mansoni* infection. The study of their expression and recognition has provided a detailed molecular insight into these host-parasite relationships, and monoclonal antibodies against two of the antigens have shown that they may play a crucial role in the mediating immune protection. Since schistosomiasis is a human disease, however, a further important stage of research is to establish to what extent the results obtained with rodent models correspond with human infection.

DISSOUS et al. (1981) were the first to show that both the 38- and 32-kd as well as some low molecular weight antigens were precipitated by the sera from two undefined schistosome patients and in recent work the human antibody response against these antigens has been investigated in more detail (DISSOUS et al. 1984). This work concentrated on the 38-kd antigen by making use of the monoclonal antibody previously raised against it (DISSOUS et al. 1982). It was demonstrated that this antigen was recognized by antibodies from 97% of schistosome patients from Brazil and that the amount of antibody present against this molecule varied with age and generally correlated with intensity of schistosome infection, the latter being maximal in the 2nd decade of life. This finding also appeared to extend to recognition of the entire 32- to 38-kd complex. This result is of great importance since it demonstrates that the 32- to 38-kd antigens are expressed by wild schistosome populations in Brazil and that they are recognized during human infection, indicating their possible involvement in protective immunity. A similar study was conducted by SIMPSON et al. (1985c) using serum samples collected from St. Lucia. It was shown again that although individuals showed a very heterogeneous antibody response as revealed by the immunoprecipitation of cell-free translation products all those tested precipitated the same ^{125}I-labelled schistosomulum surface antigens. In addition to 32- to 38-kd complex these patients also precipitated strongly the 20-kd antigens and less strongly the 17-kd antigen. The antibody response was thus very similar to that of chronically infected mice, indicating the relevance of this model and suggesting that, as in the rodent model, adult worms rather than schistosomula or cercariae provide the major immunogenic stimulus for antischistosomulum surface antigen antibodies during human infection. Patients from both South America and the Caribbean were thus found to recognize the same surface antigens as those expressed on laboratory-maintained parasite populations and indeed it was further shown that patients from serveral parts of Africa also have antibodies against the same surface antigens and that they do recognize the part of the molecule exposed on the schistosomulum surface (SIMPSON et al. 1985c). These data thus argue strongly against antigenic diversity within the species *S. mansoni* and suggest that the slow acquisition of human immunity against schistosomiasis is not due to the necessity of raising antibody against a variety of antigenic types. Furthermore, it suggests that if a vaccination using these antigens were possible it would be universally applicable.

Analysis of the human antibody response has also demonstrated that polypeptide surface antigens of *S.mansoni* that can be labelled with ^{125}I using Iodogen are species specific. Most *S.mansoni* antigens are precipitated by anti-*S.haematobium* antibodies and indeed antibodies from *S.haematobium* patients bind to *S.mansoni* schistosomula, indicating the overall antigenic similarity of these organisms (SIMPSON et al. 1985c). These antibodies do not, however, immunoprecipitate the polypeptide *S.mansoni* schistosomulum surface antigens. Thus these particular molecules are antigenically distinct within a background of antigenic similarity, perhaps indicating a particularly important immunological role for these molecules. These observations also clearly demonstrate that the polypeptides which can be labelled using Iodogen are not the only antigens of the schistosomulum surface and may indeed constitute minor components of the overall antigenicity of the schistosomulum surface. Similar results have also been found by DISSOUS et al. (1984), but curiously although most *S.mansoni* surface antigens were found to be species specific the 38-kd antigen was found to cross-react with *S.haematobium*. The reason for this discrepancy between the two studies has not yet been established.

Recently, polypeptide antigens on the surface of *S.haematobium* schistosomula have been described using indentical techniques to those used with *S.mansoni* (SIMPSON et al. 1985b). Interestingly, these antigens have similar molecular characteristics to those described on *S.mansoni*. The major antigenic feature is a broad smear of 24–35 kd together with another low molecular weight antigen of 17 kd. Again these antigens are species specific in terms of recognition by human sera while most antigens of a ^{125}I-labelled whole worm homogenate are cross-reactive. Antibodies from mice immunized with highly irradiated *S. mansoni* cercariae also failed to immunoprecipitate ^{125}I-labelled *S.haematobium* surface antigens but most interestingly antibodies from mice immunized by chronic *S.mansoni* infection strongly precipitated some of the dominant antigens with molecular weight 35–30 kd. Thus although the schistosomulum surface antigens themselves do not cross-react, antigens expressed either by adult worms or eggs of *S.mansoni* do cross-react with the *S.haematobium* surface antigens. This latter observation correlates with observed experimental immunity since mice immunized with highly irradiated cercariae exhibit species-specific immunity (BICKLE et al. 1979; CHEEVER et al. 1983; A. AGNEW, personal communication) whereas chronically infected rodents are risistant to heterologous challenge (SMITH et al. 1976). It has been suggested that the cross-specific immunity induced by chronic infection may be due to disease-induced pathology rather than direct immune effector mechanisms but the observations of cross-reaction described with surface antigens provide a possible molecular basis for a genuine cross-specific immune response.

2.5 Host-like Antigens on the Schistosome Surface

The schistosomulum surface expresses parasite antigens and presents a target of immune attack. As mentioned above, as the parasite matures parasite antigens become undetectable and instead a variety of host-like antigens are presented. It is not clear to what extent the acquisition of host antigens reflects an active and deliberate process by the parasite or an unavoidable consequence of intravascular existence. Furthermore, it is not certain that the expression of host antigens forms part of the parasite's defence against the immune system although the proposal that such molecules mask important antigens forms an appealing hypothesis (SMITHERS et al. 1969). Nevertheless the expression of host-like antigens is the dominant feature of the schistosome surface during all but the very earliest stages of its parasitic existence. Indeed there is evidence for host antigen acquisition by schistosomula even during skin penetration (SMITH and KUSEL 1979).

The existence of antigenic determinants shared by host and parasite was first described by DAMIAN (1964, 1967) and CAPRON et al. (1965). These workers held the view originally proposed by SPRENT (1959) that such antigens were synthesized by the parasite and mimicked host molecules, thus reducing the parasite's overall antigenicity. Indeed, DAMIAN et al. (1973) showed that an adult worm surface component synthesized by the parasite cross-reacts with mouse α^2-macroglobulin. In general, however, host antigens on the schistosome surface are thought to be acquired directly from the host and not synthesized by the parasite (SMITHERS et al. 1969). In the case of acquired major histocompatibility complex antigens this has now been demonstrated (SIMPSON et al. 1983c, see below).

The characterization of host antigens associated with the schistosome surface has indicated the presence of ABH and Lewis blood group substances as well as Forssman antigens (GOLDRING et al. 1976; DEAN and SELL 1972). All of these determinants can exist in the form of glycolipids and GOLDRING et al. (1977b) provided direct evidence that this is the form in which they are acquired by the parasite. Antigens that exist only in the form of glycoproteins such as MN, rhesus and Duffy antigens and Thy-1, Ly-1 and C3 are not acquired by the parasite (GOLDRING et al. 1976; SHER et al. (1978). Thus it is possible that host antigens are generally taken up by the intercalation of exogenous lipids within the surface membrane as originally proposed by CLEGG (1972). In this context it has recently been shown that there is extensive interchange of surface lipids between the schistosome and its environment (RUMJANEK and McLAREN 1981).

One interesting exception on the restriction of schistosome host antigens to glycolipids is the acquisition of major histocompatibility complex antigens which have been detected on the surface of schistosomula and adult worms recovered from mice (SHER et al. 1978; GITTER and DAMIAN 1982; GITTER et al. 1982). Taking the H-2K antigen as a model it was demonstrated that these antigens are acquired from the host in the form of intact glycoproteins by surface labelling of lung-stage schistosomula recovered from mice (SIMPSON et al. 1983c). Furthermore these authors demonstrated the absence of genes within the schistosome genome that could code for MHC antigens, thus ruling out the possibility of antigenic mimicry accounting for the presence of these molecules on the schistosome surface. It was the view of these authors that the MHC antigens were acquired by the schistosome associated with lipid vesicles since H-2 molecules are known to be shed from spleen cells in this form (EMERSON and CONE 1981). Interestingly, however, it has now been demonstrated using schistosomula recovered from the lung of bone marrow chimaeras that the class I and class II MHC antigens acquired by schistosomula originate from different tissues within the mouse (SHER et al. 1984). The class I antigens were derived from a non-haemapoietic tissue source and the class II antigens from bone marrow-derived donor cells. This result supports the hypothesis that the MHC product acquisition by schistosomes involves selective and specific interactions with host tissue but suggests that in the case of class I antigens the endothelium may be a major site of host molecular uptake for the parasite.

In addition immunoglobulin has also been found in association with schistosomes recovered from either mice or rats (SOGANDARES-BERNAL 1976; KEMP et al. 1976, 1977; YONG and DAS 1983). These antibodies were found to be heterospecific, i.e. not specific for the worm's tegument (KEMP et al. 1977). Furthermore, it was thought that the immunoglobulin was bound via their Fc portions to Fc receptors on the parasite surface (KEMP et al. 1980).

Although there is evidence that host glycolipids, alloantigens and immunoglobulins are acquired by schistosomes, there is not unequivocal proof that such host molecules act to protect schistosomes against the immune response. SMITHERS et al (1969) showed that transplantation of adult schistosomes from mice to monkeys immunized with mouse red blood cells resulted in the rapid killing of the parasites. This demonstrated the potential vulnerability of adult worms to immune attack and showed that under normal conditions adult parasites must present an antigenically inert surface to the host in order to survive. However, it does not demonstrate that host antigens are necessary to facilitate immune evasion. Certainly on younger, lung-stage schistosomula, host antigens are not the only mechanism of defence. When antibody binding to this form of the parasite is artificially induced immune killing is still avoided. Thus labelling the surface with trinitrophenyl (TNP) groups and exposure to anti-TNP antibody plus complement or eosinophils killed freshly transformed schistosomula but not lung-stage parasites (MOSER et al. 1980). Likewise, exposure to anti-host red blood cell antisera and eosinophils led to adherence and degranulation of the cells but did not lead to death of lung-stage schistosomula (MCLAREN and TERRY 1982); the same procedure did kill adult worms, however. Thus acquired host antigens are not essential for the survival of lung-stage schistosomula but they may help to protect older worms.

2.6 Conclusions and Perspectives

The surface of the schistosome is very complex and changes profoundly during maturation. Transiently exposed on the surface of young schistosomula are polypeptide antigens which have been shown to form at least a part of the immune target of the organism. The importance of the identification of these molecules is that it gives a focus to both analysis of the immune response to the parasite and the parasite's evasion of immunity. Furthermore it is now possible to investigate rationally the possibility of developing a schistosome vaccine using these antigens in either a purified or semipurified form. Since polypeptides have been identified on the schistosomulum surface which can act as the target of protective immunity, this creates the possibility of obtaining large amounts of these important antigens using the powerful techniques of modern biotechnology. Genes for protective polypeptide antigens have been cloned from many pathogenic organisms and have been utilized to direct the expression of the antigen in recombinant microorganisms. It is highly likely that this will soon be accomplished with the protective polypeptide antigens of *S.mansoni*. Already basic studies of the structure of the schistosome genome, its expression as well as the expression of schistosome antigens in vitro have been reported (TENNISWOOD and SIMPSON 1982; SIMPSON et al. 1982; GRAUSZ et al. 1983; BUGRA et al. 1983;

KNIGHT et al. 1984; TAYLOR et al. 1983). Furthermore genes that may be of use in schistosome identification and immunodiagnosis have already been cloned (CORDINGLY et al. 1984; SIMPSON et al. 1984c; McCUTCHAN et al. 1984). When genes from protective antigens are cloned and expressed, this will lead to a completely new phase in the study of the molecular structure of schistosome antigens and their role in immunity and hopefully a significant advance in our ability to diagnose and prevent schistosome infection.

3 Egg Antigens

During schistosome infection, many of the eggs laid by the female worms become trapped in the tissues; the liver is particularly affected in *S.mansoni* and *S.japonicum* infections, whilst the bladder and ureters are the main organs of egg deposition by *S.haematobium*. As the major factor in the pathogenesis of schistosomiasis is the host's granulomatous response to tissue trapped eggs (see SMITHERS and DOENHOFF 1982), it is not surprising that many attempts have been made to identify the egg antigens which are responsible for inducing this reaction. A further reason for studying egg antigens is that they may prove to be useful immunodiagnostic reagents. Unfractionated *S.mansoni* egg homogenate is claimed to be more statisfactory for the serodiagnosis of human infection than either cercarial or worm antigenic preparations (McLAREN et al. 1978).

  The granulomatous reaction induced by *S.mansoni* eggs has characteristics more typical of a cell-mediated immune response than that of an immediate (antibody-mediated) reaction (WARREN et al. 1967). The granulamatous-inducing antigens are secreted through ultramicroscopic pores in living eggs (RACE et al. 1969). These soluble egg antigens (SEAs) are conveniently prepared for study from homogenized eggs after discarding the 100 000-g pellet (BOROS and WARREN 1970). An intraperitoneal injection of SEA from *S.mansoni* eggs will sensitize mice to produce an accelerated and enhanced granulomatous response to an intravenous injection of *S.mansoni* eggs, which become trapped in the lungs (WARREN and DOMINGO 1970a). Sensitization is stage specific insofar as neither prepatent infections nor cercarial or worm constituents alone are capable of similar presensitization. Moreover, the antigens are relatively schistosome species specific, with neither *S.japonicum* nor *S.haematobium* eggs being capable of presensitizing mice to give enhanced *S.mansoni* granulomas (WARREN and DOMINGO 1970b).

  Between 6 and 9 µg SEA is sufficient to sensitize a mouse. The most interesting property of *S.mansoni* SEA is its ability to induce a delayed cellular response in the absence of humoral antibody. When mice were sensitized by an intraperitoneal injection and challenged homologously 1–2 weeks later by footpad injection of 10 µg SEA, the animals developed a delayed footpad swelling in the absence of detectable antibodies (BOROS and WARREN 1970).

3.1 Glycoprotein Antigens of *S.mansoni* Eggs

Soluble egg antigen is a very heterologous mixture and its antigenic components must be isolated from each other before investigation of the host's immunological response to specific egg antigens can be made. It has been known for some time that carbohydrate-containing moieties are prominent immunogens in SEA of *S.mansoni* (SMITHERS and WILLIAMSON 1961) and several investigators have therefore used lectin affinity chromatography as the main procedure of purification. PELLY et al. (1976) produced the first significant step in identifying the antigens responsible for the granulomatous reaction. Initially, serum from chronically infected mice (CIS) was used in simple immunodiffusion to identify three major serological antigens designated MSA_1, MSA_2 and MSA_3. Then, by a combination of concanavlin A (ConA) affinity chromatography and diethylaminoethanol (DEAE) ion exchange chromatography, MSA_1 and MSA_2 were isolated in purified form, whilst MSA_3, although purified, remained significantly contaminated with the other two antigens.

CARTER and COLLEY (1979), using a similar method, found that antigens recognized by antibodies in CIS were also present in the fraction which did not bind to ConA, but only the bound antigenic fraction was capable of eluting a T-cell response. The extraction of antigens from SEA by immunoaffinity chromatography using antibodies present in CIS gave similar results to antigen purification by ConA affinity chromatography (HARRISON et al. 1979).

Further studies on the MSA antigens have shown that MSA_1 is stable at 4 °C, whilst MSA_2 has a half-life of about 1 month when stored at this temperature (PELLEY et al. 1976). An antigen-competitive radioimmunoassay demonstrated that MSA_1 is completely stage specific; its activity fails to be inhibited by either cercariae or adult worm antigen. In contrast, MSA_2 and MSA_3 show a degree of cross-reactivity with cercarial antigen. Although all three MSA antigens are present in the fluid which escapes when the miracidia hatches, MSA_1 but not MSA_2 or MSA_3 is probably absent in immature eggs (HAMBURGER et al. 1976).

In further antigen inhibition studies using SEA from *S.japonicum* and *S.haematobium* eggs, HAMBURGER et al. (1976) showed that MSA_1 from *S.mansoni* had little or no cross-reactivity with the other species. In a different study, however, a monoclonal antibody to MSA_1, was shown to form a circumoval preciptin (COP) reaction with *S.mansoni* eggs and also with the eggs of *S.haematobium* and *S.japonicum* (HILLYER and PELLEY 1980).

More recent work has suggested that egg glycoprotein antigens may differ both quantitatively and qualitatively within a species. HAMBURGER et al. (1982) purified a glycoprotein from SEA of an Egyptian strain of *S.mansoni* by procedures similar to those used to purify MSA_1 from a Puerto Rican strain (PELLEY et al. 1976). This antigen reacted as a single precipitation line against CIS and ran as a single band in SDS-PAGE. Its serological activity was stable at 100 °C for 30 min and was resistant to HC1, NaOH and TCA treatment. Some loss of serological activity, however, followed treatment with pronase, and periodate oxidation resulted in a substantial loss of serological activity, suggesting the importance of both carbohydrate and peptide moieties for its serological activity.

Cross-reaction studies with this antigen, however, showed considerable differences from the properties of the MSA antigens previously isolated from a Puerto Rican strain of *S.mansoni* SEA (HAMBURGER et al. 1976), particularly in that a high degree of cross-reactivity occurred with SEA prepared from *S.haematobium* eggs. When the SEA of the Puerto Rican strain was compared with that of the Egyptian strain it was found that the Puerto Rican SEA was less cross-reactive with *S.haemato-*

bium SEA and that that the antigenic glycoprotein isolated in this study was four times more abundant in Puerto Rican SEA than in Egyptian SEA.

Some of the physicochemical characteristics of *S.mansoni* MSA antigens have been investigated by PELLEY and PELLEY (1976). MSA$_1$ is a glycoprotein which by gel filtration analysis has a mw of 138 000. Estimation of its mw by ultracentrifugation through sucrose gradients, however, gives a value of 90–100 kd, and when run on SDS gels in the reduced state it has a mw of 50 000. These last two estimates sugest that MSA$_1$ may be a dimer composed of two isoelectric polypeptide chains of 50 kd each.

MSA$_1$ binds tightly to ConA, eluting at 0.1M sugar concentration α-methyl-*D*-mannose gradients, suggesting that the carbohydrate portion of the molecule has a high density of terminal sugar residues with –1,4 glucopyranoside linkages. MSA$_1$ has an isoelectric point between 3.5 and 4.5 and (consistent with a negative charge) it binds easily to DEAE.

MSA$_3$ has a mw, estimated by gel filtration, of 80 000. By staining with PAS, it appears to contain little detectable carbohydrate, which is consistent with its being loosely bound to ConA Sepharose.

MSA$_2$ is a heterogeneous lipoglycoprotein and is possibly derived from the miracidial cell membranes. One predominant species, which has a mw of 450 000 by gel filtration, is almost invariably isolated from SEA. Two other species are occasionally observed but in very small concentrations (PELLEY and PELLEY 1976).

3.2 Glycoprotein Antigens of *S. japonicum* Eggs

Whilst *S. mansoni* and *S. japonicum* infections are similar in that adult worms of both species live within the portal venous system and both are associated with hepatosplenic disease, there are striking differences in the responses of the host to the eggs of both parasites and to their SEA preparations. For example, *S. mansoni* SEA induces and elicits a delayed hypersensitivity footpad reaction in *S. mansoni*-infected mice (BOROS and WARREN 1970). In contrast, injections of small amounts of an analogous *S. japonicum* SEA preparation into the footpads of *S. japonicum*-infected mice result in an immediate (15-min) inflammatory response (WARREN et al. 1975). Furthermore, injection of *S. japonicum* eggs intraperitoneally into mice will not induce sensitization to a secondary granulomatous response to *S. japonicum* eggs injected intravenously. Only when the parasite eggs or an equivalent amount of *S. japonicum* SEA are presented to the host by subcutaneous injection does a significant granulomatous response develop around eggs subsequently injected intravenously.

Studies with *S. mansoni* egg antigens have demonstrated that the glycoprotein antigen fraction which binds to ConA is primarily responsible for eliciting cell-mediated responses in mice (CARTER and COLLEY 1979). TRACY and MAHMOUD (1982) have shown that a similar ConA-binding glycoprotein fraction of *S. japonicum* SEA has an analogous biological activity. This factor when injected subcutaneously sensitizes mice to granuloma formation around *S. japonicum* eggs subsequently injected into the lungs and also elicits an immediate response when injected into the footpads of *S. japonicum*-infected mice.

Two other kinds of biological activity have been shown to be associated with the glycoproteins or proteoglycans of *S. japonicum* SEA. Allergenic components were isolated by OWHASHI and ISHII (1981) by gel filtration and ion-exchange chromatography. The allergens were absorbed by ConA and lentil lectin, and partially absorbed by wheat germ lectins. Allergenic activity was lost after treatment with $0.1M$ periodate oxidation, but was resistant to heating (100 °C for 60 min), urea and guanidine.

A glycoprotein antigen with eosinophilic chemotactic activity was isolated by the same authors (OWHASHI and ISHII 1982) who suggest it may be partially responsible for the presence of eosinophils within the granuloma. This 900 000 mw antigen was isolated by gel filtration and ion-exchange chromatography. Again, the biological activities, i.e. its antigenicity and chemotactic activity, were apparently dependent on the carbohydrate moiety of the molecule as both activities were destroyed by periodate oxidation.

Several workers have attempted to compare the antigens already characterized in SEA of *S. mansoni* with those present in *S. japonicum* SEA. A two-phase immunoabsorbent system was used by CARTER and COLLEY (1981); a specific anti-egg antibody was prepared from sera of mice infected with *S. japonicum* by affinity chromatography using a *S. japonicum* SEA CNBr Sepharose column. The specific immunoglobulin was then coupled to Sepharose and used to isolate the major serological antigens. Their findings suggest that acute infection serum (7-week) identifies one major glycoprotein antigen mw 130 000, whilst 12-week antibody isolates a mixture of that antigen plus a carbohydrate antigen of lower mw.

LONG et al. (1981a) demonstrated by ConA Sepharose affinity chromatography that the absorbent (glycoprotein) fraction of *S. japonicum* SEA produces two intense precipitation lines upon immunodiffusion analysis with human CIS. The protein fraction produces two faint precipitation lines which do not cross-react with the glycoprotein fraction. However, it was found impossible to analyse the glycoprotein fraction further by SDS-PAGE, because the fractions displayed such great heterogeneity in mw and the glycoproteins were more difficult to separate by ion-exchange chromatography than those of *S. mansoni*. Further purification was effected by isoelectric focussing, demonstrating that the glycoprotein fraction contained a complex mixture of at least eight different molecules. Thus the *S. japonicum* egg antigen glycoprotein system differs considerably from that of *S. mansoni*. *S. japonicum* SEA does not contain an antigen analogous to MSA_1, and no single antigen is serologically dominant to the degree that MSA_1 is in *S. mansoni* infection. *S. japonicum* SEA, however, does contain large amounts of a high-mw heterodisperse glycoprotein that is remiscent but not identical to MSA_2.

Hydrophobic interaction chromatography (phenyl Sepharose) was utilized to separate further the *S. japonicum* SEA glycoproteins into two groups designated JAG I and II complex and JAG III and IV complex (LONG et al. 1981b). The highly acidic hydrophilic JAG I and II contained more carbohydrate than protein; the less acidic more hydrophobic JAG III and IV were predominantly protein. About five times as much antibody in human CIS was directed towards JAG III and IV compared with JAG I and II and in fact it is these hydrophobic glycoproteins which are the major serological antigens of *S. japonicum* eggs. They confer to the COP test, its specificity and sensitivity.

JAG II in many ways resembles MSA_2 of *S. mansoni* eggs. It induces a faint diffuse precipitation line consistent with its heterodisperse nature and only low amounts of antibody are directed towards it in natural infections; it is completely cross-reactive with *S. mansoni*. MSA$_2$ is also heterodisperse on PAGE and is highly cross-reactive with *S.japonicum* eggs.

In a subsequent study in which these purified antigens of *S.japonicum* SEA were used in the ELISA (LONG et al. 1982) it was found, surprisingly, that the sensitivity of the ELISA employing crude *S.japonicum* SEA was so high and the specificity so good that the increased immunological sensitivity afforded by partially purified antigens made little improvement. This is not the case with *S. mansoni* ELISA, where crude antigens have decidedly inferior sensitivity and specificity (HILLYER and GOMEZ DE RIUS 1979; HILLYER et al. 1979).

Paradoxically, in ELISA there was far less *S. mansoni* cross-reactivity in the JAG II hydrophilic fraction than expected, and it was suspected that because of its hydrophilic nature JAG II was not absorbing to the solid phase and the ELISA reactivity of that fraction could be accounted for by contaminant hydrophobic glycoproteins which would preferentially bind to the solid phase. It is as well to remember that the plastic plate used in ELISA may act like an in situ hydrophobic interaction column, preferentially absorbing the minor amounts of hydrophobic glycoproteins.

3.3 Polysaccharide Egg Antigens

The presence of polysaccharide antigens in SEA has been demonstrated by BOCTER et al. (1982), who treated *S. mansoni* SEA with 44% aqueous phenol. The crude extract was subjected to ConA Sepharose 4B affinity chromatography to provide a bound and unbound fraction, which were both precipitated by CIS. The unbound fraction was subsequently shown to bind to wheat germ agglutinin. Neither fraction ran on SDS-PAGE or could be shown to contain protein material, thus differing from the MSA antigens. Both fractions contained a small percentage of amino acids and were shown to have different sugar compositions.

3.4 Non-soluble Egg Antigens

Soluble egg antigen is a water-soluble extract of schistosome egg amenable to fractionation by lectin affinity chromatography. It is unlikely, however, that the host's response to schistosome eggs is restricted to those antigens that are aqueous soluble, and it is possible that important membrane antigens, for example, which are insoluble under conditions used to prepare SEA, are overlooked when investigating SEA. In attempting to identify a highly potent antigen for use in ELISA, TSANG et al. (1981) have compared the antigenic activities and cross-reactivities in ELISA of *S. mansoni* SEA with the 8 *M* urea extract of the residual pellet left after SEA preparation. These authors found that three times as much antigenic activity was present in the urea extract compared with SEA and the cross-reactivity of the former against a *Trichinella spiralis* antiserum was considerably lower. SDS-PAGE showed that the fractions extracted in urea were distinct from those in SEA; there was not PAS +ve bands and the antigens contained mostly proteinaceous components of extremely large MWs (1.5×10^6). Essentially similar results were obtained with *S.japonicum* SEA (TSANG et al. 1982).

3.5 Egg Proteinase Antigens

Proteolytic activity, which may aid in the penetration of eggs through the host's intestinal tissue, has been reported in secretions of *S. mansoni* eggs by SMITH (1974). ASCH and DRESDEN (1979) have purified an acidic thiol-dependent proteolytic enzyme from *S. mansoni* egss using AcA54 or amino-phenyimercuricacetate-Sepharose and elution with acetate buffer. This enzyme appears to be antigenic during infection as a monoclonal antibody has been isolated from infected mice which inhibits its activity (DRESDEN et al. 1983). The egg proteinase does not bind to ConA Sepharose and has a relatively low mw of 25 000–27 000.

3.6 Hepatotoxic Egg Antigen

A potentially useful immunodiagnostic egg antigen has been isolated and characterized from a study initially designed to investigate granuloma-induced liver pathology. T-cell-deprived *S. mansoni*-infected mice fail to develop granulomatous reactions around tissue-deposited eggs, but nevertheless they suffer from an acute hepatotoxicity reaction 6–7 weeks after infection (BYRAM et al. 1979). Treatment of infected deprived mice with *S. mansoni* CIS, however, prevents the development of liver damage (DOENHOFF et al. 1979) and the protective effect of CIS is egg-stage-specific (DOENHOFF et al. 1981). In order to identify the putative liver-damaging *S. mansoni* egg antigens, DUNNE et al. (1981) have raised mouse antisera against various fractions of SEA and tested these antisera for their ameliorative effect in T-cell-deprived infected mice.

Twelve *S. mansoni* egg antigens were identified by immunoelectrophoresis in agar using mouse CIS. One of these antigens, designated ω_1, which fails to migrate from the site of application, is detected in all mice with patent chronic infections and induces the most concentrated of the precipitating anti-egg antibodies in pooled CIS. Serum from mice infected with SEA fractions containing ω_1 were fully protective against liver damage in infected T-cell-deprived mice; sera which did not contain anti-ω_1 were not hepatoprotective (DUNNE et al. 1981).

Initially ω_1 was difficult to separate from an anodally migrating antigen designated α_1. The fraction obtained after cation exchange on CM-Sephadex (CEF 6) which contained both antigens, however, gave improved sensitivitiy and specificity in ELISA compared with unfractionated SEA when tested against sera from infected patients (McLAREN et al. 1981).

Antigen ω_1 has now been purified by a combination of solid-phase immunoabsorbence and cation exchange chromatography. Rabbit antiserum specific only for α_1 has been raised by immunizing with one of the peaks obtained by cation exchange chromatography of SEA. This antiserum was then used to absorb out α_1 from crude SEA. The absorbed material was then passed through a column of CM-Sepharose and purified ω_1 was eluted with IM NaCl (DUNNE and DONEHOFF 1983).

ω_1 contains no detectable carbohydrate; it has a mw of between 22 000 and 26 000 and a pI greater than 9. Clearly therefore it is unrelated to the MSA antigens. It is not detectable in homogenates of *S. mansoni* cercariae or adult worms, nor is it present in *S. bovis*, *S. japonicum* or *S. haematobium* eggs or worm homogenate (DUNNE and

DONEHOFF 1983). The specificity is interesting in the light of recent findings that T-cell-deprived mice heavily infected with *S. bovis* do not suffer hepatocyte damage despite the presence of large numbers of eggs in the livers and the absence of granuloma (DOENHOFF et al. 1982).

3.7 Conclusions

Progress in the characterization and purification of egg antigens has arisen from a dichotomy of objectives – to understand the aetiology and pathogenesis of schistosomes and to improve serodiagnosis. There has been sufficient signs of progress in both areas to encourage further work. Egg antigens have a great potential in immunodiagnosis and it is clear that in *S. mansoni* infections at least the use of more purified antigens increases the sensitivity and specificity of the method. It remains to be seen whether individual antigens will provide more specific information about infection, pathology or cure. Furthermore, the identification of unique aetiological antigens in *S. mansoni* SEA (MSA_1 and ω_1), which have no apparent analogy in other species, is a crucial step towards understanding the disease process.

4 Circulating Schistosome Antigens

An interest in circulating schistosome antigens has developed from the need to improve immunodiagnosis. The detection of schistosome antigen in the blood or urine of an individual should signify the presence of the living parasite, in contrast to immunodiagnostic methods based on antibody detection, which may not discriminate between past experience of the disease and a current infection. The ability to demonstrate free-circulating antigen would be useful therefore in evaluating the efficiency of treatment and the quantification of antigen may provide an estimation of infection intensity.

Although antigens in the urine of *S. japonicum* patients were detected by OKABE and TANAKA in 1958, more detailed studies began only after the finding by BERGGREN and WELLER (1967) of a circulating antigen in the serum and urine of mice and hamsters heavily infected with *S. mansoni*. On subsequent characterization, this antigen was shown to be a negatively charged polysaccharide which gave a characteristic anodic precipitation in immunoelectrophoresis (DEELDER et al. 1976) and it was therefore designated circulating anodic antigen or CAA. An antigen fraction called "GASP", purified by NASH et al. (1981) from a crude TCA-soluble polysaccharide fraction of adult *S. mansoni*, is now reported to be identical to CAA (COLLEY 1983).

A second circulating antigen was originally described from the urine and serum of *S. mansoni* patients and from the milk of infected mothers (CARLIER et al. 1975, 1978; SANTORO et al. 1977). It was originally called antigen M, but because this antigen has the characteristics of a polysaccharide with a cathodic migration at pH 8.2, it has been designated circulating cathodic antigen or CCA (DEELDER et al. 1976). CCA is not the same as antigen 4, which has also been described from the milk of infected mothers (SANTORO et al. 1977).

Recent studies have shown that both CAA and CCA are also found in adult worms of *S. japonicum* and that they circulate in *S. japonicum*-infected hosts (QIAN and DEELDER 1983). At least five other circulating polysaccharide antigens have been described (QIAN and DEELDER 1983) but, apart from PSAP, a glycoprotein of molecular weight ranging from 90 000 to 13 000 (NASH et al. 1981, 1983), only CAA and CCA have so far been characterized in any detail.

Both CAA and CCA are extracted from adult worms in the TCA-soluble fraction (NASH 1974; CARLIER et al. 1978; DEELDER et al. 1980). They are heat stable and resistant to proteolytic enzymes, DNase and RNase, but are destroyed by periodate treatment (GOLD et al. 1969; NASH 1974; CARLIER et al. 1978). Purification of CAA on DEAE chromatography has revealed that the antigen contains large amounts of carbohydrate, primarily *N*-acetylgalactosamine and *D*-glucoronic acid, but also galactose, glucose, *N*-acetylglucosamine and trace amounts of other sugars; amino acids account for about 11% of its weight. The antigen therefore, appears to have the characteristics of a proteoglycan (NASH et al. 1977).

CAA can be separated from CCA in TCA-soluble extract by ion-exchange chromatography using DE 52 (DEELDER et al. 1980). Although antibodies to both antigens are produced in natural infections (CARLIER et al. 1975; NASH 1978; DEELDER et al. 1980) isolated CAA is only weakly antigenic and must be complexed with a carrier such as methylated bovine serum albumin (BSA) to induce high titres of antibodies in rabbits and sheep. It has not been possible to raise antibodies to CCA in rabbits, even when the antigen is coupled to methylated BSA. So far, antibodies to CCA have been obtained only from the serum of infected hamsters or man.

Indirect fluorescent antibody studies have shown that CAA and CCA are derived from the cells lining the schistosome gut (NASH 1974; CARLIER et al. 1980b; DEELDER et al. 1980) and almost certainly enter the blood stream in the worm vomitus. Free CAA can be detected in the serum of heavily infected (1500 cercariae) hamsters from 28 days postinfection and CCA in the serum and urine by day 45. It has been calculated that in heavily infected hamsters a steady-state production of 50 ng CCA/worm is reached (DEELDER et al. 1978).

As both antigens induce an antibody response during infections of animals and man, it is not surprising that they are found in circulating immune complexes (DEELDER et al. 1980) and that their deposition in the kidneys of mice, hamsters (DEELDER et al. 1980; VAN MARCK et al. 1981; CARLIER et al. 1980a) and man (MORIEARTY and BRITO 1977) has been demonstrated. The retention of circulating antigens in the renal glomeruli of patients with *S. mansoni* is a possible factor in the initiation of kidney injury in human schistosomiasis (BRITO et al. 1979).

If circulating antigens of schistosomes are to be of value in immunodiagnosis, they must be detectable at very low levels. A number of sensitive techniques have been developed for measuring CAA, including the complement fixation reaction (BAWDEN and WELLER 1974), indirect haemagglutination (IHA) (DEELDER and EVELEIGH 1978) and various immunosassays using fluorescent conjugates (DEELDER et al. 1978) or peroxidase conjugates (DEELDER et al. 1980). The ELISA, IHA and DASS techniques will all detect CAA levels as low as 10–100 ng CAA/ml serum (DEELDER et al. 1980). Generally no correlation between the circulating CAA level and the worm burden has been found, although in rabbits lightly infected with *S. japonicum* a linear relationship between CAA level and worm burden was seen (QIAN and DEELDER 1983). In these animals, levels of immune-complexed CAA were very low and did not interfere with accurate determinations of free CAA levels as appears to occur in the sera of heavily infected hamsters. HIRATA et al. (1977), however, studying *S. japonicum*-infected rabbits claim to demonstrate a correlation between CAA levels and worm burden up to 41 weeks postinfection. ABDEL-HAFEZ et al. (1983) used an inhibition ELISA to assay the presence of crude schistosome antigens added to PBS or normal human serum. Using an appropriate titre of infected mouse serum, less than 10 ng/ml antigen could be detected. The technique also showed a higher level of circulating soluble egg antigen (SEA) in mice exposed to 120 cercariae than in those exposed to 60 or 30 cercariae and a rapid elimination of circulating SEA after treatment.

A radioimmunoprecipitation assay combined with PEG precipitation (RIPEGA) has been developed for detecting circulating antigens in the serum of human patients (SANTORO et al. 1978). Using RIPEGA, SANTORA et al. (1980) showed a direct correlation between the level of circulating antigens and the number of eggs per gram (egp.) of faeces. Circulating immune complexed and C3d serum levels were similarly correlated. In addition antigen "4" was detected in the serum only from patients with more than 500 epg faeces, suggesting a correlation between this circulating antigen and the intensity of infection. CARLIER et al. (1975) also noted a correlation between M antigen (= CCA) in the urine of patients and the number of *S. mansoni* eggs in the faeces.

Thus, there may be grounds for believing that circulating antigen measurements could be used to evaluate the level of worm burden. These latter studies in patients however showed a wide variation in the relationship between circulating antigen and epg faeces, and conclusions about the relationship of these parameters were made only on group studies. At the present time, the measurement of circulating antigen in an individual cannot be relied on as an assessment of worm burden.

DEELDER et al. (1980) have suggested that CCA appears to be the most suitable antigen for detection as it is readily found in the urine of infected animals and man by relatively insensitive immunoprecipitation techniques. The development of a qualitative assay such as ELISA for this antigen however is hampered by the difficulty of producing a strong antiserum.

Nash and his co-workers have had more success in differentiating the clinical status of patients by measuring antibody levels to highly purified circulating glycoprotein antigens. They have demonstrated major differences in antibody responses to "PSAP" and to "GASP" (= CAA). IgM and IgG responses to GASP are elevated in acutely infected patients. In contrast, PSAP initially evokes a predominantly IgM response, but in heavily infected chronically exposed patients there is an elevated response to PSAP which is mainly IgG; in the same chronically infected patients IgM and IgG responses to GASP are depressed (NASH et al. 1983). Antibody levels to a crude TCA-soluble extract of adult worms fail to differentiate between early infected and chronically infected patients and Nash and his colleagues suggest that only measurements of antibodies recognizing purified schistosome antigens are likely to unmask clinically relevant correlations.

4.1 Conclusions

The future improvement and refinement of immunodiagnosis may depend more on the measurement of antibody to certain defined antigens rather than the detection of free circulating antigens which will inevitably become complexed with antibody and metabolized at rates which differ in individuals. However, the recent discovery that the schistosome produces ecdysteroids (TORPIER et al. 1982; NIRDE et al. 1983) could prove to be a major advance in the development of serodiagnosis. Ecdysteroids of *S. mansoni* are released into the sera and urine of infected hosts, where they can be quantified by radioimmunoassay from the 6th day of infection. The ecdysteroid concentration in the urine of *S. haematobium* patients decreases markedly after chemotherapy (NIRDE et al. 1984). Thus, this moulting hormone, which does not become complexed to antibody, could prove to be a more valuable reagent in the diagnosis of active infections and in monitoring the efficiency of chemotherapy than circulating schistosome antigens.

References

Abdel-Hafez SK, Phillips SM, Zodda DM (1983) *Schistosoma mansoni:* detection and characterization of antigens and antigenemia by inhibition enzyme-linked immunosorbent assay (1 ELISA). Exp Parasitol 55:219–232

Aronstein WS, Strand M (1983) Lung stage expression of a major schistosome surface antigen. J Parasitol 69:1027–1032

Aronstein WS, Norden AP, Strand M (1983) Tegumental expression in larval and adult stages of a major schistosome structural glycoprotein. Am J Trop Med Hyg 32:334–342

Asch HL, Dresden MH (1979) Acidic thiol proteinase activity of *Schistosoma mansoni* egg extracts. J Parasitol 65:543–549

Asch HL, Read CP (1975) Membrane transport in *Schistosoma mansoni:* transport of amino acids by adult males. Exp Parasitol 38:123–135

Bawden MP, Weller TH (1974) *Schistosoma mansoni* circulating antigen: detection by complement fixation in sera from infected hamsters and mice. Am J Trop Med Hyg 23:1077–1084

Berggren WL, Weller TH (1967) Immunoelectrophoretic demonstration of specific circulating antigen in animals infected with *Schistosoma mansoni*. Am J Trop Med Hyg 16:606–612

Bickle QD, Ford MJ (1982) Studies on the surface antigenicity and susceptibility to antibody-dependent killing of developing schistosomula using sera from chronically infected mice and mice vaccinated with irradiated cercariae. J Immunol 128:2101–2106

Bickle QD, Taylor MG, James ER, Nelson GS, Hussein MF, Andrews BJ, Dobinson AR, de C-Marshall TF (1979) Further observations on immunization of sheep against *Schistosoma mansoni* and *S. bovis* using irradiation-attenuated schistosomula of homologous and heterologous species. Parasitology 78:185–193

Bickle QD, Ford MJ, Andrews BT (1983) Studies on the development of antischistosomula surface antibodies by mice exposed to irradiated cercariae, adults and/or eggs of *S. mansoni*. Parasit Immunol 5:499–511

Boctor FN, Cheever AW, Higashi GI (1982) *Schistosoma mansoni:* fractionation of polysaccharide egg antigens by lectin affinity chromatography. Immunology 46:237–245

Boros DL, Warren KS (1970) Delayed hypersensitivity type granuloma formation and dermal reaction induced and elicited by a soluble factor isolated from *Schistosoma mansoni* eggs. J Exp Med 132:488–507

Brink LH, Krueger K, Harris C (1980) Stage specific antigens of *Schistosoma mansoni*. In: Van den Bossche H (ed) Procedings of the 3rd international symposium on biochemistry of parasites and host parasite relationships. The Host Invader Interplay. Elsevier/North-Holland, Amsterdam, pp 393–404

Brito E, Santoro F, Rocha H, Dutra M, Capron A (1979) Immune complexes in schistosomiasis. VI. Circulating IC levels in patients with and without nephropathy. Rev Inst Med Sao Paulo 21:119–125

Bugra K, Tanaka RD, Boyle WJ, MacInnis AJ (1983) Isolation of polyA(T)RNA from *Schistosoma mansoni* and immunoprecipitation of its in vitro translation products. J Parasitol 69:486–490

Butterworth AE, Taylor DW, Weith MC, Vadas MA, Dessain A, Sturrock RF, Wells E (1982) Studies on the mechanisms of immunity in human schistosomiasis. Immunol Rev 61:5–39

Byram JE, Doenhoff MJ, Musallam R, Brink LH, von Lichtenberg F (1979) *Schistosoma mansoni* infections in T cell-deprived mice, and the ameliorating effect of administering homologous chronic infection serum. II Pathology. Am J Trop Med Hyg 28:274–285

Capron A, Biguet J, Rose F, Vernes A (1965) Les antigenes de *Schistosoma mansoni*. II Etude immuno-electrophoretique comparee de divers stade labaires et des adultes de deux sexes. Aspects immuno-logiques des relations hote-parasite de la cercaire et de l'adulte de *S. mansoni*, Annls Inst Pasteur (Paris) 109:798–810

Carlier Y, Bout D, Bina JC, Camus D, Figueiredo JFM, Capron A (1975) Immunological studies in human schistosomiasis. I Parasitic antigen in urine. Am J Trop Med Hyg 24:949–954

Carlier Y, Bout D, Capron A (1978) Further studies on the circulating M antigen in human and experi-mental *Schistosoma mansoni* infections. Ann Immunol (Paris) 129C:811–818

Carlier Y, Bout D, Capron A (1980a) Detection of *Schistosoma mansoni* M antigen in circulating immune-complexes and in kidneys of infected hamsters. Trans R Soc Trop Med Hyg 74:534–538

Carlier Y, Bout D, Strecker G, Bebray H, Capron A (1980b) Purification, immunochemical and biological characterization of the schistosoma circulating M antigen. J Immunol 124:2442–2450

Carter CE, Colley DG (1979) Partial purification and characterization of *Schistosoma mansoni* soluble egg antigen with Con-A Sepharose chromatography. J Immunol 122:2204–2209

Carter CE, Colley DG (1981) *Schistosoma japonicum* soluble egg antigens: separation by Con-A chromato-graphy and immunoaffinity purification. Mol Immunol 18:219–225

Cesari IM, Simpson AJG, Evans WH (1981) Properties of a series of tegumental membrane-bound phospho-hydrolase activities of *Schistosoma mansoni*. Biochem J 198:467–473

Cheever AW, Heiny S, Duvall RH, Sher A (1983) Lack of resistance to *Schistosoma japonicum* in mice immunized with irradiated *S. mansoni* cercariae. Trans R Soc Trop Med Hyg 77:812–814

Clegg JA (1972) The schistosome surface in relation to parasitism. Functional aspects of parasite surfaces. In: Taylor AER, Muller R (eds) Symposia of the British Society for Parasitology, vol 10 Blackwell Scientific, Oxford, pp 23–40

Colley DG (1983) Schistosome antigen laboratory workshop: summary of a meeting sponsored by the Edna McConnell Clark Foundation. J Parasitol 69:45–48

Cordingly JB, Taylor DW, Dunn DW, Butterworth AE (1984) Clone banks of cDNA from the parasite, *Schistosoma mansoni*: isolation of clones containing a potentially immunodiagnostic antigen gene. Gene 26:25–39

Cruise KM, Mitchell GF, Garcia EG, Tiu WU, Hocking RE, Anders RF (1983) Sj23, the target antigen in *Schistosoma japonicum* adult worms of an immunodiagnostic hydridoma antibody. Parasite Immunol 5:37–46

Cutter BD, Damian RT (1982) Murine alloantigen acquisition by schistosomula of *Schistosoma mansoni*: further evidence for the presence of K, D and I region products on the tegumental surface. Parasite Immunol 4:363–367

Cutter BD, McCormick SL, Damian RT (1982) Murine alloantigen acquisition by *Schistosoma mansoni*: presence of H-2k determinants on adult worms and failure of allogeneic lymphocytes to recognize acquired MHC gene products on schistosomula. J Parasitol 68:513–518

Damian RT (1964) Molecular mimicry: antigen sharing by parasite and host and its consequences. Am Naturalist 98:129–149

Damian RT (1967) Common antigens between *Schistosoma mansoni* and the laboratory mouse. J Parasitol 53:60–64

Damian RT, Greene ND, Hubbard WJ (1973) Occurrence of mouse α2-macroglobulin antigenic deter-minants on *Schistosoma mansoni* adults, with evidence on their nature. J Parasitol 59:64–73

Dean DA, Sell KW (1972) Surface antigens on *Schistosoma mansoni*. II Adsorption of a Forssman-like host antigen by schistosomula. Clin Exp Immunol 12:525–540

Dean DA, Bukowski MA, Cheever AW (1981) Relationship between acquired resistance, portal hyper-tension and lung granulomas in ten strains of mice infected with *Schistosoma mansoni*. Am J Trop Med Hyg 30:806–814

Deelder AM, Eveleigh PC (1978) An indirect haemagglutination reaction for the demonstration of *Schisto-soma mansoni* circulating anodic antigen. Trans R Soc Trop Med Hyg 72:178–180

Deelder AM, Klappe HTM, Van den Aardweg GJMJ, Van Meerbeke EHEM (1976) *Schistosoma mansoni:* demonstration of two circulating antigens in infected hamsters. Exp Parasitol 40:189-197

Deelder AM, Van Dalen DP, Van Egmond JG (1978) *Schistosoma mansoni:* microfluorometric determination of circulating anodic antigen and antigen-antibody complexes in infected hamster serum. Exp Parasitol 44:216-224

Deelder AM, Kornelis D, Van Marck EAE, Eveleigh PC, Van Egmond JG (1980) *Schistosoma mansoni:* characterization of two circulating polysaccharide antigens and the immunoglobulin response to these antigens in mouse, hamster and human infections. Exp Parasitol 50:16-32

Dessein AJ, Samuelson C, Butterworth AE, Hogan M, Sherry BA, Vadas MA, David JR (1981) Immune evasion by *Schistosoma mansoni:* loss of susceptibility to antibody or complement-dependent eosinophil attack by schistosomula cultured in medium free of macromolecules. Parasitology 82:357-374

Dissous C, Capron A (1981) Isolation of surface antigens from *Schistosoma mansoni* schistosomula. In: Peeters H (ed) Protids of the Biological Fluias (Proceedings of the 29th Colloquium). Pergamon pp 179-182

Dissous C, Capron A (1983) *Schistosoma mansoni:* antigenic community between schistosomula surface and adult worm incubation productions as a support for concomitant immunity. FEBS Lett 162:355-359

Dissous C, Dissous C, Capron A (1981) Isolation and characterization of surface antigens from *Schistosoma mansoni* schistosomula. Mol Biochem Parasitol 3:215-225

Dissous C, Gryzch J-M, Capron A (1982) *Schistosoma mansoni* surface antigen defined by a rat monoclonal IgG2a. J Immunol 129:2232-2234

Dissous C, Prata A, Capron A (1984) Human antibody response to *Schistosoma mansoni* surface antigens defined by protective monoclonal antibodies. J Infect Dis 149:227-233

Doenhoff M, Musallam R, Bain J, McGregor A (1979) *Schistosoma mansoni* infections in T-cell deprived mice, and the ameliorating effect of adminstering homologous chronic infection serum. I. Pathogenesis. Am J Trop Med Hyg 28:260-273

Doenhoff MJ, Pearson S, Dunne DW, Bickle Q, Lucas S, Bain J, Musallam R, Hassounah O (1981) Immunological control of hepatoxicity and parasite egg excretion in *Schistosoma mansoni* infections: stage specificity of the reactivity of immune serum in T-cell deprived mice. Trans R Soc Trop Med Hyg 75:41-53

Doenhoff M, Harrison R, Sabah A, Morare H, Dunne D, Hassounah O (1982) Schistosomiasis in the immunosuppressed host: studies on the host-parasite relationship of *Schistosoma mansoni* and *S. bovis* in T-cell deprived and hydrocortisone-treated mice. In: Owen (ed) Animals models in parasitology. Macmillan, London, pp 155-169

Dresden MH, Sung CK, Deelder AM (1983) A monoclonal antibody from infected mice to a *Schistosoma mansoni* egg proteinase. J Immunol 130:1-3

Dunne DW, Doenhoff MJ (1983) *Schistosoma mansoni* egg antigens and hepatocyte damage in infected T-cell deprived mice. Contrib Microbiol Immunol 7:22-29

Dunne DW, Lucas S, Bickle Q, Pearson S, Madgwick L, Bain J, Doenhoff MJ (1981) Identification and partial purification of antigen (ω_1) from *Schistosoma mansoni* eggs which is putatively hepatotoxic in T-cell deprived mice. Trans R Soc Trop Med Hyg 75:54-71

Emerson SG, Cone RE (1981) 1-K^k and H-2^k antigens are shed as supramolecular particles in association with membrane lipids. J Immunol 127:482-485

Gitter BD, Damian RT (1982) Murine alloantigen acquisition by schistosomula of *Schistosoma mansoni:* further evidence for the presence of K, D and I region gene products on the tegumental surface. Parasite Immunol 4:383-393

Gitter BD, Damian RT (1982) Murine alloantigen acquisition by schistosomula by *Schistosoma mansoni:* presence of H-2K determinants on adult worms and failure of allogenic lymphocytes to recognise acquired MHC gene products on schistosomula. J Parasitol 68:513-518

Gold R, Rosen FS, Weller TH (1969) A specific circulating antigen in hamsters infected with *Schistosoma mansoni.* Detection of antigen in serum and urine, and correlation between antigenic concentration and worm burden. Am J Trop Med Hyg 18:545-551

Goldring OL, Clegg JA, Smithers SR, Terry RJ (1976) Acquisition of human blood group antigens by *Schistosoma mansoni.* Clin Exp Immunol 26:181-187

Goldring OL, Kusel JR, Smithers SR (1977a) *Schistosoma mansoni:* origin of in vitro host-like surface antigens. Exp Parasitol 43:82-93

Goldring OL, Sher A, Smithers SR, McLaren DJ (1977b) Host antigens and parasite antigens of murine *Schistosoma mansoni.* Trans R Soc Trop Med Hyg 71:144-148

Grausz D, Dissous C, Capron A, Roskam W (1983) Messenger RNA extracted from *Schistosoma mansoni* larval forms codes for parasite antigens when translated in vitro. Mol Biochem Parasit 7:293–301

Gryzch J-M, Capron M, Bazin H, Capron A (1982) In vitro and in vivo effector function of rat IgG$_{2a}$ monoclonal anti-*S. mansoni* antibodies. J Immunol 129:2739–2743

Hamburger J, Pelley RP, Warren KS (1976) *Schistosoma mansoni* soluble egg antigens: determination of the stage and species specificity of their serologic reactivity by radioimmunoassay. J Immunol 117:1561–1566

Hamburger J, Lustigman S, Siongok TKA, Ouma JH, Mahmoud AAF (1982) Characterization of a purified glycoprotein from *Schistosoma mansoni* eggs: specificity, stability, and the involvement of carbohydrate and peptide moieties in its serologic activity. J Immunol 128:1864–1869

Harn DA, Mitsuyana M, David JR (1984) *Schistosoma mansoni:* anti-egg monoclonal antibodies protect against cercarial challenge in vivo. J Exp Med 159:1371–1387

Harrison DJ, Carter CE, Colley DG (1979) Immunoaffinity purification of *Schistosoma mansoni* soluble egg antigens. J Immunol 122:2210–2217

Harrison DJ, Carter CE, Colley DG (1979) Immunoaffinity purification of *Schistosoma mansoni* soluble egg antigens. J Immunology 122:2210–2217

Hayunga EG, Murrell KD, Taylor DW, Vannier WE (1979a) Isolation and characterisation of surface antigens from *Schistosoma mansoni*. 1. Evaluation of techniques for radioisotope labelling of surface proteins from adult worms. J Parasitol 65:488–496

Hayunga EG, Murrell KD, Taylor DW, Vannier WE (1979b) Isolation and characterisation of surface antigens from *Schistosoma mansoni*. II. Antigenicity of radiolabelled proteins from adult worms. J Parasitol 65:497–506

Hayunga EG, Vannier WE, Chesnut RY (1981) Partial characterization of radiolabelled antigens from adult *Schistosoma haematobium*. J Parasitol 67:589–591

Hayunga EG, Sumner MP, Stek M (1982) Isolation of a concanavalin A-binding glycoprotein from adult *Schistosoma japonicum*. J Parasitol 68:960–961

Higgins HG, Harrington KJ (1959) Reaction of amino acids and proteins with Diazonium compounds. II. Spectra of protein derivatives. Arch Biochem Biophys 85:409–425

Hillyer GV, Gomez de Rois I (1979) The enzyme linked immunosorbent assay (ELISA) for the immuno-diagnosis of schistosomiasis. Am J Trop Med Hyg 28:237–241

Hillyer GV, Pelley RP (1980) The major serological antigen (MSA$_1$) from *Schistosoma mansoni* eggs in a "circumoval" precipitinogen. Am J Trop Med Hyg 29:582–585

Hillyer GV, Ruiz-Tiben E, Knight WB, Gomez de Rios I, Pelley RP (1979) Immunodiagnosis of infection with *Schistosoma mansoni:* comparison of ELISA, radioimmunoassay, and precipitation tests per-formed with antigen from eggs. Am J Trop Med Hyg 28:661–669

Hirata M, Takamori K, Tsutsumi M (1977) Circulating antigen in animals infected with *Schistosoma japonicum*. III. Detection of circulating antigen by counter immunoelectrophoresis. Kurume Med J 24:139–146

Hockley DJ (1973) Ultrastructure of the tegument of *Schistosoma*. Adv Parasitol 11:233

Hockley DJ, McLaren DJ (1973) *Schistosoma mansoni:* changes in the outer membrane of the tegument during development from cercariae to adult worms. Int J Parasitol 3:13–25

Horowitz S, Tarrab-Hazdai R, Eshhar Z, Arnon R (1983) Anti-schistosome monoclonal antibodies of different isotypes – correlation with cytotoxicity. EMBO J 2:193–198

Kelly C, Payares G, Simpson AJG, Smithers SR (1985) Analysis of antigenic and nonantigenic surface polypeptides of *Schistosoma mansoni* by two dimensional electrophoresis. Exp Parasitol (in press)

Kemp WM, Damian RT, Green ND (1976) Immunocytochemical localisation of IgG on adult *Schistosoma mansoni* tegumental surfaces. J Parasitol 62:830–832

Kemp WM, Merritt SC, Bogucki MS, Roster JG, Seed JR (1977) Evidence for adsorption of heterospecific host immunoglobulin on the tegument of *Schistosoma mansoni*. J Immunol 119:1849–1854

Kemp WM, Brown PR, Merritt SC, Miller RE (1980) Tegument-associated antigen modulation by adult male *Schistosoma mansoni*. J Immunol 124:806–811

Knight M, Simpson AJG, Payares G, Chaudhri M, Smithers SR (1984) Cell-free synthesis of *Schistosoma mansoni* surface antigens: stage specificity of their expression. EMBO J 3:213–219

Kusel JR (1972) Protein composition and protein synthesis in the surface membranes of *Schistosoma mansoni*. Parasitology 65:55–59

Long W, Lewert RM, Pelley RP (1981a) Characterization of the glycoprotein antigens which mediate the *Schistosoma japonicum* circumoval precipitin reaction. Infect Immun 34:389–396

Long W, Lewert RM, Pelley RP (1981b) Fractionation of *Schistosoma japonicum* soluble egg antigen glycoproteins by hydrophobic interaction chromatography. Infect Immun 34:397–406

Long GW, Yogore MG, Lewert RM, Blas BL, Pelley RP (1982) Efficacy of purified *Schistosoma japonicum* egg antigens for ELISA serodiagnosis of human schistosomiasis japonica: specificity and sensitivity. Am J Trop Med Hyg 31:1006–1014

McCutchan TF, Simpson AJG, Mullins A, Sher A, Nash TE, Lewis F, Richards C (1984) Differentiation of schistosomes by species, strain and sex using cloned DNA markers. Proc Natl Acad Sci USA 81:889–893

McLaren DJ (1980) *Schistosoma mansoni:* the parasite surface in relation to host immunity. In Trop Med Res Stud ed K.N. Brown, vol 1./John Wiley, Chichester

McLaren DJ, Hockley DJ (1976) Blood flukes have a double outer membrane. Nature 269:147–149

McLaren DJ, Terry RJ (1982) The protective role of acquired host antigens during schistosome maturation. Parasite Immunol 4:129–148

McLaren M, Draper CC, Roberts JM, Minter-Goedbloed E, Lighthart GS, Teesdale CH, Amin MA, Omer AHS, Bartlett A, Voller A (1978) Studies on the enzyme-linked immunosorbent assay (ELISA) test for *Schistosoma mansoni* infections. Ann Trop Med Parasitol 72:243–253

McLaren ML, Lilleywhite JE, Dunne DW, Doenhoff M (1981) Serodiagnosis of human *Schistosoma mansoni* infections: enhanced sensitivity and specificity in ELISA using a fraction containing S. *mansoni* egg antigens ω1 and α1. Trans R Soc Trop Med Hyg 75:72–79

Mitchell GF, Cruise KM, Garcia EG, Anders RF (1981) Hybridoma-derived antibody with immunodiagnostic potential for schistosomiasis japonicum. Proc Natl Acad Sci USA 78:3165–3169

Moriearty PL, Brito E (1977) Elution of renal antischistosome antibodies in human schistosomiasis mansoni. Am J Trop Med Hyg 26:717–722

Moser G, Wassom D, Sher A (1980) Studies of the antibody-dependent killing of schistosomula of *Schistosoma mansoni* employing haptenic target antigens. I. Evidence that the loss in susceptibility to immune damage undergone by developing schistosomula involves a change unrelated to the masking of parasite antigens by host molecules. J Exp Med 152:41–53

Murrell KD, Taylor DW, Vannier WE, Dean DE (1978) *Schistosoma mansoni:* analysis of surface membrane carbohydrates using lectins. Exp Parasitol 46:247–255

Nash TE (1974) Localization of the circulating antigen within the gut of *Schistosoma mansoni.* Am J Trop Med Hyg 23:1085–1087

Nash TE (1978) Antibody response to a polysaccharide antigen present in the schistosome gut I. Sensitivity and specificity. Am J Trop Med Hyg 27:938–943

Nash TE, Nasir-Ud-Din, Jeanloz RW (1977) Further purification and characterization of a circulating antigen in schistosomiasis. J Immunol 119:1627–1633

Nash TE, Lunde MN, Cheever AW (1981) Analysis and antigenic activity of a carbohydrate fraction derived from adult *Schistosoma mansoni.* J Immunol 126:805–810

Nash TE, Garcia-Coyco C, Ruiz-Tiben E, Nazario-Lopez HA, Vazquez G, Torres-Borges A (1983) Differentiation of acute and chronic schistosomiasis by antibody responses to specific schistosome antigens. Am J Trop Med Hyg 32:776–784

Nirde P, De Reggi ML, Torpier G, Capron A (1983) Ecdysone and 20 hydroxyecdysone: new hormones for the human parasite *Schistosoma mansoni.* FEBS Lett 151:223–227

Nirde P, De Reggi ML, Tsoupras G, Torpier G, Fressancourt P, Capron A (1984) Excretion of ecdysteroids by schistosomes as a marker of parasite infection. FEBS Lett 168:235–240

Nordon AP, Aronstein WS, Strand M (1982) *Schistosoma mansoni:* identification, characterization and purification of spine glycoprotein by monoclonal antibody. Exp Parasitol 54:432–442

Okabe K, Tanaka T (1958) A new urine precipitation reaction for schistosomiasis japonica, a preliminary report. Kurume Med J 5:45–52

Owhashi M, Ishii A (1981) Fractionation and characterization of allergens extracted from eggs of *Schistosoma japonicum.* Int Arch Allergy Appl Immunol 64:146–156

Owhashi M, Ishii A (1982) Purification and characterization of a high molecular weight eosinophil chemotactic factor from *Schistosoma japonicum* eggs. J Immunol 129:2226–2231

Payares G, McLaren DJ, Evans WH, Smithers SR (1985) Antigenicity and immunogenicity of the tegumental outer membrane of adult *Schistosoma mansoni.* Parasite Immunol 7:45–61

Pelley RR, Pelley RJ (1976) S. *mansoni* soluble egg antigens. IV. Biochemistry and immunochemistery of major serological antigens with particular emphasis on MSA1. In: Van den Bossche H (ed) Proceedings of 2nd International Symposium on the biochemistry of parasites and host-parasite relationships, Beerse Belgium. North Holland, Amsterdam, pp 283-290

Pelley RR, Pelley RJ, Hamburger J, Peters PA, Warren KS (1976) *Schistosoma mansoni* soluble egg antigens. I. Identification and purification of three major antigens and the employment of radioimmunoassay for their further characterization. J Immunol 117:1553-1566

Philipp M, Rumjanek FD (1984) Antigenic and dynamic properties of helminth structures. Mol Biochem Parasitol 10:245-268

Podesta RB, McDiarmid SS (1982) Enrichment and partial enzyme characterization of ATPase activity associated with the outward-facing membrane complex and inward-facing membrane of the surface epithelial syncytium of *Schistosoma mansoni*. Mol Biochem Parasitol 4:225-235

Qian ZL, Deelder AM (1983) *Schistosoma japonicum:* immunological characterization and detection of circulating polysaccharide antigens from adult worms. Exp Parasitol 55:168-178

Race GJ, Michaels RM, Martin JH, Larsh JE Jr, Matthews JL (1969) *Schistosoma mansoni* eggs: an electron microscope study of shell pores and microbarbs. Proc Soc Exp Biol Med 130:990-992

Ramasamy R (1979) Surface proteins on schistosomula and cercariae of *Schistosoma mansoni*. Int J Parasitol 9:491-493

Roberts SM, Aitken R, Vojvodic M, Wells E, Wilson RA (1983) Identification of exposed components on the surface of adult *Schistosoma mansoni* by lactoperoxidase-catalysed iodination. Mol Biochem Parasitol 9:129-143

Rogers SH, Bueding E (19757 Anatomical localization of glucose uptake by *Schistosoma mansoni*. Int J Parasitol 5:369-371

Rotmans JP, Mooji GW (1982) KCL-extractable surface antigens of *Schistosoma mansoni:* immunological characterization and applicability in immunodiagnosis. Z Parasitenkd 68:211-226

Rumjankek FD, Curiel M (1983) The occurrence of proteolipid antigens in *Schistosoma mansoni*. Mol Biochem Parasitol 7:183-195

Rumjanek FD, McLaren DJ (1981) *Schistosoma mansoni:* modulation of schistosomular lipid composition by serum. Mol Biochem Parasitol 3:239-252

Rumjanek FD, McLaren DJ, Smithers SR (1983) Serum induced expression of a surface protein in schistosomula of *Schistosoma mansoni:* a possible receptor for lipid uptake. Mol Biochem Parasitol 9:337-350

Ruppel A (1978) A study of surface proteins in the parasite *Schistosoma mansoni*. PhD thesis, University of Freiburg, GFR.

Samuelson JC, Caulfield JP (1982) Loss of covalently labelled glycoproteins and glycolipids from the surface of newly transformed schistosomula of *Schistosoma mansoni*. J Cell Biol 94:363-369

Samuelson JC, Sher A, Caulfield JP (1980) Newly transformed schistosomula spontaneously lose surface antigens and C3 acceptor sites during culture. J Immunol 124:2055-2057

Samuelson JC, Caulfield JP, David JR (1982) Schistosomula of *Schistosoma mansoni* clear concanavalin A from their surface by sloughing. J Cell Biol 94:355-362

Santoro F, Borojevic R, Bout D, Tachon P, Bina JC, Capron A (1977) Mother-child relationship in human schistosomiasis mansoni. I. Parasitic antigens and antibodies in milk. Am J Trop Med Hyg 26:1164-1168

Santoro F, Vandemeulebrouke B, Capron A (1978) The use of the radioimmunoprecipitation-PEG assay (RIPEGA) to quantify circulating antigens in human and experimental schistosomiasis. J Immunol Methods 24:229-237

Santoro F, Prata A, Castro CN, Capron A (1980) Circulating antigens, immune complexes and C3d levels in human schistosomiasis: relationship with *Schistosoma mansoni* egg output. Clin Exp Immunol 42:219-225

Shah J, Ramasamy R (1982) Surface antigens on cercariae, schistosomula and adult worms of *Schistosoma mansoni*. Int J Parasitol 12:451-461

Sher A, Hall BF, Vadas MA (1978) Acquisition of murine major histocompatibility complex gene products by schistosomula of *Schistosoma mansoni*. J Exp Med 148:46-57

Sher A, Sacks D, Simpson AJG, Singer A (1984) Dichotomy in the tissue origins of schistome acquired Class I and Class II histocompatability complex antigens. J Exp Med 159:952-957

Simpson AJG, McLaren DJ (1982) *Schistosoma mansoni:* tegumental damage as a consequence of lectin binding. Exp Parasitol 53:105-116

Simpson AJG, Smithers SR (1980) Characterization of the exposed carbohydrates on the surface membrane of adult *Schistosome mansoni* by analysis of lectin binding. Parasitology 81:1-15

Simpson AJG, Schryer MD, Cesari IM, Evans WH, Smithers SR (1981a) Isolation and partial characterization of the tegumental outer membrane of adult *Schistosoma mansoni*. Parasitology 83:163–177

Simpson AJG, Rumjanek FD, Payares G, Evans WH (1981b) Glycosyl transferase activities are associated with the surface membrane in adult *Schistosoma mansoni*. Parasitology 83:163–177

Simpson AJG, McCutchan F, Sher A (1982) The genome of *Schistosoma mansoni:* isolation of DNA, its size, bases and repetitive sequences. Mol Biochem Parasitol 6:125–137

Simpson AJG, James SL, Sher A (1983a) Identification of surface antigens of schistosomula of *Schistosoma mansoni* recognized by antibodies from mice immunized by chronic infection and by exposure to highly irradiated cercariae. Infect Immun 41:591–597

Simpson AJG, Correa-Oliveira R, Smithers SR, Sher A (1983b) The exposed carbohydrates of schistosomula of *Schistosoma mansoni* and their modification during maturation in vivo. Mol Biochem Parasitol 8:191–205

Simpson AJG, Singer D, McCutchan TF, Sacks DL, Sher A (1983c) Evidence that schistosome MHC antigens are not synthesized by the parasite but are acquired from the host as intact glycoproteins. J Immunol 131:962–963

Simpson AJG, Payares G, Walker T, Knight M, Smithers SR (1984a) The modulation of expression of polypeptide surface antigens on developing schistosomula of *Schistosoma mansoni*. J Immunol 133:2725–2730

Simpson AJG, Knight M, Payares G, Smithers SR (1984b) Identification and biosynthesis of schistosome surface antigens. In: August T (ed) Proceedings of the Third John Jacob Abel Symposium on drug development: molecular parasitology Academic, London, pp 177–184

Simpson AJG, Dame JB, Lewis FA, McCutchan TF (1984c) The arrangement of ribosomal RNA genes in *Schistosoma mansoni:* identification of polymorphic structural variants. Eur J Biochem 139:41–45

Simpson AJG, Hackett F, Walker T, Payares G, de Rossi R, Smithers SR (1985a) The antibody response against schistosomula surface antigens following immunization with highly irradiated cercariae of *Schistosoma mansoni* Parasite Immunol 7:133–152

Simpson AJG, Knight M, Hagan P, Hodgson J, Wilkins HA, Smithers SR (1985b) The schistosomulum surface antigens of *Schistosoma haematobium*. Parasitology 90:499–508

Simpson AJG, Hackett F, Kelly C, Knight M, Payares G, Ali P, Lillywhite J, Fleck SL, Smithers SR (1985c) The recognition of *Schistosoma mansoni* surface antigens by antibodies from patients infected with *S. mansoni* and *S. haematobium*. Trans Roy Soc Hyg Trop Med. In press

Smith HV, Kusel JR (1979) The acquisition of antigens in the intercellular substance of mouse skin by schistosomula of *Schistosoma mansoni*. Clin Exp Immunol 36:430–435

Smith M (1974) Radioassays for the proteolytic enzymes secreted by living eggs of *Schistosoma mansoni*. Int J Parasitol 4:681–683

Smith MA, Clegg JA (1979) Different levels of immunity to *Schistosoma mansoni* in the mouse: the role of variant cercariae. Parasitology 78:311–321

Smith MA, Clegg JA, Webbe G (1976) Cross immunity to *Schistosoma mansoni* and *S. haematobium* in the hamster. Parasitology 73:53–64

Smith MA, Clegg JA, Snary D, Trejdosiewicz AJ (1982) Passive immunization of mice against *Schistosoma mansoni* with an IgM monoclonal antibody. Parasitology 84:83–91

Smith MA, Clegg JA, Snary D (1984) Monoclonal antibodies demonstrate similarity of surface antigens on different clones of *Schistosoma mansoni* schistosomula. Trans Soc Trop Med Hyg 78:187–189

Smithers SR, Doenhoff MJ (1982) Schistosomiasis. In: Cohen S, Warren KS (eds) Immunology of parasitic infections. Blackwell, London, pp 527–607

Smithers SR, Terry RJ (1969) Immunity in schistosomiasis. Ann NY Acad Sci 160:826–840

Smithers SR, Williamson J (1961) Antigenic polysaccharide material in cercariae and eggs of *Schistosoma mansoni*. Trans R Soc Trop Med Hyg 55:308–309

Smithers SR, Terry RJ, Hockley DJ (1969) Host antigens in schistosomiasis. Proc R Soc Lond (Biol) 171:483–494

Snary D, Smith MA, Clegg JA (1980) Surface proteins of *Schistosoma mansoni* and their expression during morphogenesis. Eur J Immunol 10:573–575

Sogandares-Bernal F (1976) Immunoglobulins attached to and in the integument of adult *Schistosoma mansoni* Sambon 1907, from first infection of CF_1 mice. J Parasitol 62:222–226

Sprent JFA (1959) Parasitism, immunity and evolution. In: Leeper GW (ed) The evolution of living organisms. Melbourne University, Melbourne pp 149–165

Tavares CAP, de Rossi R, Payares G, Simpson AJG, McLaren DJ, Smithers SR (1984) A monoclonal antibody raised against adult *Schistosoma mansoni* which recognizes a surface antigen on schisto-somula. Z Parasitenkd 70:189–197

Taylor DW, Butterworth AE (1982) Monoclonal antibodies against surface antigens of schistosomula of *Schistosoma mansoni*. Parasitology 84:65–82

Taylor DW, Hayunga EG, Vannier WE (1981) Surface antigens of *Schistosoma mansoni*. Mol Biochem Parasitol 3:157–168

Taylor DW, Cordingly JS, Butterworth AE (1984) Immunoprecipitation of surface antigen precursors from *Schistosoma mansoni* messenger RNA in vitro translation products. Mol Biochem Parasitol 10:305–318

Tenniswood MPR, Simpson AJG (1982) The extraction, characterization and in vitro translation of RNA from adult *Schistosoma mansoni*. Parasitology 84:253–261

Torpier G, Hirn M, Nirde P, de Reggi M, Capron A (1982) Detection of ecdysteroids in the human trematode *Schistosoma mansoni*. Parasitology 84:123–130

Tracey JW, Mahmoud AAF (1982) Isolation of *Schistosoma japonicum* egg glycoprotein antigens which sensitize mice to lung granuloma formation and elicit an immediate hypersensitivity response. Am J Trop Med Hyg 31:1201–1212

Tsang VCW, Tao Y, Maddison SE (1981) Systematic fractionation of *Schistosoma mansoni* urea-soluble egg antigens and their evaluation by the single-tube kinetic-dependent, enzyme-linked immunosorbent assay (k-ELISA). J Parasitol 67:340–350

Tsang VCW, Tao Y, Qui L, Xue H (1982) Fractionation and quantitation of egg antigens from *Schistosoma japonicum* by the single-tube kinetic dependent enzyme-linked immunosorbent assay (k-ELISA); higher antigenic activity in urea soluble than in aqueous soluble fractions. J Parasitol 68:1034–1043

Warren KS, Domingo EO (1970a) *Schistosoma mansoni:* stage specificity of granuloma formation around eggs after exposure to irradiated cercariae, unisexual infections or dead worms. Exp Parasitol 27:60–66

Warren KS, Domingo EO (1970b) Granuloma formation around *Schistosoma mansoni*, *S. haematobium* and *S. japonicum* eggs. Size and rate of development, cellular composition, cross sensitivity and rate of egg destruction. Am J Trop Med Hyg 19:292–304

Warren KS, Domingo EO, Cowan RBT (1967) Granuloma formation around schistosome eggs as a manifestation of delayed hypersensitivity. Am J Pathol 51:735–756

Warren KS, Boros DL, Hang LM, Mahmoud AAF (1975) The *Schistosoma japonicum* egg granuloma. Am J Pathol 80:279–294

Wilson RA, Barnes PE (1977) The formation and turnover of the membranocalyx of the tegument of *Schistosoma mansoni*. Parasitology 74:61–71

Wilson RA, Coulson PS, McHugh SM (1983) A significant part of the 'concomitant immunity' of mice to *Schistosoma mansoni* is a consequence of a leaky hepatic portal system, not immune killing. Parasitol Immunol 5:595–601

Van Marck EAE, Deelder AM, Gigase PLJ (1981) *Schistosoma mansoni:* anodic polysaccharide antigen in glomerular immune deposits of mice with unisexual infections. Exp Parasitol 52:62–68

Yong WK, Das PK (1983) Acquisition of host proteins by the tegument of *Schistosoma mansoni* recovered from rats. Z Parasitenkd 69:53–60

Zodda DM, Phillips SM (1982) Monoclonal antibody-mediated protection against *Schistosoma mansoni* infection in mice. J Immunol 129:2326–2328

Trematodes, Excluding Schistosomes with Special Emphasis on *Fasciola*

D.L. HUGHES

1 Introduction

The digenetic trematodes with which we are concerned have an indirect life cycle with at least one intermediate host involved before the final mammalian or bird host becomes infected.

Excluding the blood flukes of man and his domestic animals (*Schistosoma* spp.), which have separate sexes, we are left with a handful of hermaphrodite flukes of economic importance which can and do cause serious disease. Genera of most importance are *Paramphistomum, Clonorchis, Opisthorchis, Dicrocoelium, Metagonimus, Paragonimus, Fasciola, Fasciolopsis* and *Fascioloides*. These digenetic trematodes are fairly specific in their choice of intermediate host or hosts, one of which is a mollusc, but generally infect a wide variety of mammalian species. We are not concerned with those members which infect birds.

The liver fluke *Fasciola hepatica,* for example, will infect only a few species of molluscs, whereas most mammals including man can be infected although some species are much more resistant than others. *Clonorchis sinensis,* the human liver fluke of China and the Orient (Japan, Korea, Vietnam and India), will also infect other mammals, such as rabbits, rats and mice. Its first intermediate host is a snail, (*Parafossarulus, Bulimus* or *Alocinma* spp.) and the second a fish most commonly belonging to the Cyprinidae family. Man becomes infected by eating these infected raw fish.

AFRC Institute for Research on Animal Diseases, Compton, Newbury, Berkshire RG16 0NN, United Kingdom

Our concern with parasitic antigens in protection, diagnosis and escape is confined to host/parasite relationships in mammals. The fact that many of the trematodes with which we are concerned are able to infect a wide variety of definitive hosts appears at first sight a distinct advantage from the point of view of the experimental immunoparasitologist until the difficulties of interpretation and dangers of extrapolation from one species to another become apparent. The difficulties of verifying the applicability of experimental results obtained from animals to man are obvious. Somewhat less obvious are the problems of work in species and strains of inbred laboratory animals when attempting to relate results to the situation in larger domestic animals. We have no inbred strains of sheep or cattle. In this situation any attempts at cell transfer, for example, can only be effectively carried out in monozygotic twins, animals made tolerant experimentally, or chimaeric twins which will have shared the same placenta.

It is one of the stated aims of the volume to examine antigens as defined biochemical entities. This is obviously not possible with the trematodes with which we are concerned here. Although we will concentrate on *Fasciola* this should not be taken as any indication that much is known about defined fluke antigens, but rather as an indication that we know a little more about the parasite, its biology, structure and the antigenicity of some of the stages because of the vast amount of literature that has accrued. This arises from its importance as a parasite of domestic animals and probably because of the ease of obtaining some of its stages, especially the large adult, for experimental work and analysis.

We know very little about the parasite in terms of a defined antigen for use as an immunising or diagnostic agent or in terms of parasite protection. An approach to parasite antigen analysis and the concept of host-parasite interactions with the various "host effector" and "parasite avoidance" mechanisms that are brought into play, or theoretically could be brought into play, have been reviewed in depth by a number of authors, notably ANDERS et al. (1982) and MITCHELL (1979, 1982).

Fasciola hepatica, the liver fluke, is a parasite of veterinary importance mainly in ruminants, with a marked effect on production and the ability of the animal to realise its full potential. It can and frequently does kill the animal although this is more often associated with the infection in sheep and goats rather than in cattle. Fluke is cosmopolitan in its distribution and the grazing animal is liable to become heavily infected and develop disease unless frequent anthelmintic dosing is carried out in endemic areas. There are as yet no immunological means of preventing this disease in ruminants. This is perhaps not surprising when we look at the situation generally. To quote MITCHELL (1982) "there is currently no example of a molecular (i.e. antigenically defined) vaccine which is capable of inducing protective immunity in natural hosts against a metazoon or protozoan parasite of medical or economic importance".

Before looking at protection, diagnosis and escape and what we know about the antigens involved, it will help if we examine briefly the biology and life cycle of this parasite.

1.1 Life Cycle of *Fasciola hepatica*

The infective stages or metacercariae are normally found as small white cysts on the grass or herbage on low-lying damp ground, which is the usual habitat of the intermediate snail host. Infected snails shed a free-swimming cercarial stage which attaches to and then encysts on herbage with the loss of its tail. Once ingested by a mammal the cyst hatches in the intestine and the emerging young fluke burrows across the wall to gain entry into the abdominal cavity. It browses on intestinal serosa and adjacent tissues and eventually reaches and penetrates the liver capsule. After a period of migration in the liver parenchyma the young fluke enters the bile ducts, becomes sexually mature and commences to lay eggs. Individual flukes which are hermaphrodite can lay fertile eggs without cross-fertilisation taking place. The eggs pass out with the faeces and after embryonation the free-swimming miracidium infects a snail and the 5- to 6-month cycle begins again.

1.2 Experimental Infections and the Production of Parasite Material

The most reliable source of infective metacercariae is from snails infected experimentally and obtained from a well-maintained laboratory colony of snails. Small laboratory mammals or ruminants can be infected by a number of different routes and with the various stages as Table 1 shows.

Infected animals can be used as a source of parasitic material by recovering flukes from the body cavity, liver or bile ducts. The large size of the adult (approximately 3 cm long) enables many grams of material to be collected from say a single sheep (approximately 15 g wet weight for 100 flukes) whereas it is tedious to obtain large quantities of liver stages quickly and free of host tissues. Newly excysted flukes obtained from cysts using the method of SEWELL and PURVIS (1969) are very reliable for producing parenteral infections and for use in in vitro culture, etc. It is useful to remind ourselves that, although the parasite easily infects most mammalian hosts, the course of the disease and time scale of the different stages in the body cavity, liver, ect. varies in the different hosts. In the mouse, one or two flukes can and frequently do kill this host in about 3 weeks and only in a small number of mice does the parasite

Table 1. Routes of infection

Parasite stage	Route[a]
Metacercariae	Oral
	Intraperitoneal
	Subcutaneous
	Intrarumenal[a]
Newly excysted juveniles (NEJ)	Intraperitoneal
	Subcutaneous
	Into gut loops
Liver stages	Intraperitoneal
	Intrabiliary[a]
Adults	Intraperitoneal
	Subcutaneous
	Intrabiliary

[a] Intrabiliary and intrarumenal routes have only been used in ruminants

reach sexual maturity in the bile duct. Rats and rabbits are better hosts, with only a very small percentage dying after oral infection (eg. approximately 10% in rats given 20–30 infective cysts). Although sheep and cattle also die with heavy fluke infections, experimental infections of 500 infective cysts are seldom fatal. Susceptibility and the response of the liver to infection vary in the domestic animals. In the series sheep-cattle-pig, sheep are the most susceptible whereas pigs are the most resistant and show marked fibrotic changes in the liver. Any work on resistance must bear in mind the possible role of the fibrotic and damaged liver acting as a physical barrier to reinfection.

2 Antigens in Protection, Evasion and Diagnosis

It is probably true to say of parasitic helminths that nearly all the antigenic material used has been rather crude, ill-defined material often consisting of whole homogenised parasites, soluble extracts or different fractions isolated in various ways. These antigens can certainly not be regarded as defined biochemical entities. Trematodes are no exception, in that the isolation of defined antigens with known functions still presents a great problem and as yet this goal has not been realised.

2.1 Protection and Its Induction

Resistance to reinfection in *F. hepatica* has been demonstrated in rats (HAYES et al. 1972) and in cattle (DOYLE 1971) whereas it has not been clearly demonstrated in mice, rabbits and sheep. The rat then shows resistance to reinfection after a primary infection, and this ability to kill an incoming challenge, either of adults introduced via the intraperitoneal challenge route or orally dosed metacercariae, is lost as the infection ages (HUGHES et al. 1977; DOY 1980). In this situation rats with a long-standing primary infection in the bile ducts are susceptible to reinfection. The early invasive forms of the challenge infection which are normally the targets for this concomitant immunity are therefore no longer affected. In this situation it would appear to be a loss of host effector factors rather than evasive tactics on the part of the parasite, because reinfection of the rats which have lost their ability to kill a challenge have this facility quickly restored.

When rats are sensitised initially with orally administered metacercariae, it is found that a subsequent challenge, with either metacercariae, newly in vitro excysted juveniles (NEJ), juvenile stages from the liver or adult flukes from donor animals is effectively dealt with. The fact that challenge flukes of different stages are killed indicates that antigen/s common to all stages are functional in immunity. This has been further shown by immunising rats with materials derived from adults (OLDHAM and HUGHES 1982) or Ag/Ab complexes from newly excysted flukes maintained in vitro (HOWELL and SANDEMAN 1979). The sloughing off of these complexes from newly excysted juveniles maintained in vitro indicates that the surface is probably an important source of relevant antigens.

The immunological nature of such resistance has been demonstrated in rats by passive serum transfer and also by cell transfer (ARMOUR and DARGIE 1974; CORBA et al. 1971) but it is worth noting that cell donors infected for only 4 weeks were not capable of protecting recipients whereas 8-week-infected donors were. This is somewhat surprising because resistance to reinfection develops very quickly after an initial infection. It could indicate that not sufficient antibody-producing cells are present at 4 weeks postinfection.

In cattle although there is good evidence for resistance to reinfection (DOYLE 1971, 1973) and also some evidence obtained from a few animals indicating that cells will transfer resistance in monozygotic twin calves (CORBA et al. 1971), there is little unequivocal information to show that resistance to reinfection is immunologically mediated rather than due to damaged or fibrotic livers. We also know from work using intrabiliary challenge that there is a barrier at the bile duct wall in cattle previously infected and having fibrotic and partly calcified bile duct walls (HUGHES and HARNESS 1975). Cattle and rats unlike any other hosts, with the possible exception of man und some deer, eliminate a primary fluke infection with time. It is unlikely that the mechanisms are the same in both rats and cattle and although the rat has been used extensively to extend our knowledge of host/parasite relationships it may not be a good model for the cow. Preliminary results in cattle using body cavity recovery techniques to enumerate challenge flukes (DOY and HUGHES, 1984a, 1984b) and migration of newly excysted flukes placed in gut loop preparations in previously infected calves (BURDEN and BLAND 1983) would also indicate that it is not a good model. Mechanisms of resistance have been examined using athymic nude (*rnu/rnu*) rats and their hairy heterozygous litter mates (*rnu/+*) (DOY and HUGHES 1982a). This has shown that there appears to be two distinct mechanisms of resistance to reinfection with *F. hepatica,* the first being a T-independent system effective at the gut wall and the second effective once the migrating parasite has penetrated the gut wall and which is T dependent and seen to act in the body cavity.

The apparently T-independent mechanism operating at the level of gut lumen/gut wall is possibly non-specific as work with *Nippostrongylus brasiliensis* gives similar results to that in nude rats. In this case rats previously infected with *N. brasiliensis* are resistant when challenged orally, but not if challenged with NEJs directly into the body cavity. Further evidence for a possible non-specific effector mechanism is found if adult flukes are placed into the body cavity of previously uninfected nude (*rnu/rnu*), hairy (*rnu/+*) and PVG rats. In both the nude and its hairy litter mate > 60% of the transferred flukes are killed within a few hours with marked cellular involvement (HUGHES and DOY 1983). This is in marked contrast to the situation in our inbred PVG rats in which both adults and immature flukes transferred by the intraperitoneal route remain unaffected and live for many months or years, unless of course the rats have experienced a previous infection. Obviously the question of rat strain differences on fluke resistance needs further study. The way in which flukes are killed in an immune host and the possible cells involved have been examined by a number of workers (BENNETT et al. 1980; DAVIES and GOOSE 1981; BURDEN et al. 1983a) and although the eosinophil appears to be one of the important cells in this relationship there is only circumstantial evidence that it is involved in killing the fluke. Attempts to reconstitute the necessary factors involved in killing in an in vitro situation have not been successful (DOY and HUGHES 1982b; DUFFUS and FRANKS 1980) although as might be expected eosinophil basic protein kills young flukes in vitro (DUFFUS et al. 1980). Similar results were found by CHAPMAN and MITCHELL (1982a), who were unable to produce damage to NEJs in vitro with serum and cells. Although conventional rats infected with *F. hepatica* contain in their sera a factor which readily promotes adherence of peritoneal cells which are mainly eosinophils (> 80%), serum from infected nude rats does not promote adherence; yet these animals are resistant to challenge. What do we know about homologous and heterologous antigens used in protection experiments? There is little information in biochemical terms of homologous antigens used and generally they have given only partial protection when compared with the high levels of resistance produced after infection with living parasites. The antigen used by HOWELL and SANDEMAN (1979), for example, was a precipitate formed when newly excysted juvenile flukes were incubated in rat serum. The precipitate was thought to be a complex of metabolic antigen and rat immunoglobulin, possibly IgG, and when it was used in conjunction with Freund's complete adjuvant gave a 50% reduction in worm burden. Other workers notably LANG and HALL (1977) have claimed to produce resistance in mice using excretory/secretory antigens from immature worms maintained in vitro. It should be noted that the demonstration of resistance in mice has not been demonstrated in other laboratories (CHAPMAN and MITCHELL 1982a) including our own.

Resistance has been produced when mature flukes contained in diffusion chambers have been implanted in rats (HAROUN et al. 1980) and also by subcutaneous transplantation of living flukes (ERIKSEN and FLAGSTAD 1974; ANDERSON et al. 1975). HAROUN et al. (1982) also showed in further work that the serum transfer of resistance in the rat using cattle serum could be ablated if the serum was first adsorbed with adult fluke culture fluid. The use of S/E products from immature flukes produced no significant resistance in experiments reported by BURDEN and HAMMET (1980) in contrast to the findings of RAJASEKARIAH et al. (1979). DAVIES et al. (1979) were unable to detect any resistance in rats given E/S products obtained from in vitro maintained newly excysted flukes. These somewhat conflicting results no doubt demonstrate different techniques and antigen processing, but the rat strain may be very important. The balance in favour of the host may be easily tipped with the minimum of antigenic stimulus in some situations and with some strains of rat.

References has been made earlier to the importance of the fluke surface and the host cells which are found on this surface during the death of flukes in previously infected rats. It will be useful if we briefly summarise what is known about the nature and altered antigenicity of the fluke surface as the fluke develops. During the course of development of *Fasciola hepatica* after entry into the final host there is seen in its tegument a sequential appearance of morphologically different granules (BENNETT and THREADGOLD 1973, 1975). These granules have been designated T0, T1 und T2. T0-type granules are associated with early stages – metacercariae and NEJ, the T1 type mostly with liver migratory stages and T2 granules with pre-bile duct and bile duct stages. The bile duct dwelling adult parasites having mainly T2 with T1 granules present to a lesser extent.

Tegumental antigens appear to be expressed at different stages of development and their expression appears to be associated with the presence in the tegument of the different granules (BENNETT 1978; HANNA 1980b). The profiles of various specific antibodies which appear in response to sequentially expressed antigens during the development of *F. hepatica* in sheep and cattle have been described by HANNA (1980b) and HUGHES et al. (1981). Indirect fluorescent antibody labelling was used for this work and the test antigens were sections of different aged flukes embedded in JB4 plastic embedding resin. A section of adult fluke would show little fluorescence against sera obtained a few weeks after infection because mainly T2 granules are present in the adult. In contrast, sera obtained about 15 weeks after·infection when all granules had been expressed showed marked fluorescence to T2 antigens in the adult. The reverse situation is seen with sections of young flukes containing T1 granules which show bright fluorescence with early sera, but not with sera obtained much later in infection. That the antibody response is associated with granule expression has been further shown by strategic anthelmintic dosing. A response to the later T2 granules can be prevented by killing 3-week-old flukes in experimentally infected sheep at a time before T2 granules are expressed (HANNA et al. 1982).

What is the relevance of these granule antigens in immunity? During the course of a normal infection all granule types are expressed and these can be monitored by measuring the appropriate antibody response. Rats given a normal infection are resistant to challenge 2–3 weeks later (HUGHES et al. 1977) whereas expression of T2 granules in significant numbers does not occur until just before invasion of the bile ducts (4–5 weeks postinfection in the rat) (HANNA 1980b). The protection demonstrated using sloughed glycocalyx of juvenile flukes as antigen (HOWELL and SANDEMAN 1979) would therefore also implicate T0/T1 as potential candidate antigens. The development of tegumental granules of normal and γ-irradiated flukes during infection of rats and mice has been described in detail by BURDEN et al. (1983b). These authors showed that an irradiation dose of 4 Kr prevented the development of T1 and T2 granules in rat flukes whereas 3 Kr had a variable effect, with some flukes developing normally.

When groups of rats were infected with metacercariae irradiated at either 4 or 3 Kr and then challenged, only rats given the 3-Kr dose were protected (HUGHES et al. 1982a). It would appear that the sensitising fluke infection must reach a stage where T1 and/or T2 granules are produced before immunity develops. A challenge infection with normal metacercariae is killed either in the gut or body cavity before the appearance of T2 granules and this would therefore leave T1 as a potential candidate granule. Rather surprisingly there is an antibody response to T1 granules seen in rats given 4-Kr attenuated flukes which are known not to have such granules present. It had been suggested that T0 and T1 may contain similar antigenic determinants by HANNA (1980b), and HUGHES et al. (1982b) verified this. Monoclonal antibodies against T0 and T1 granules were found to be cross-reactive, confirming the sharing of a common functional antigenic determinant (HANNA and TRUDGETT 1983). The T1 granules may therefore act as a marker for protection antigens produced at the same time and prevented from functioning by γ-irradiation, or γ-irradiation may have altered the functional antigens in the developing fluke (BURDEN et al. 1983b). If this is so the very early irradiated stages of the fluke are still recognised antigenically because rats given an initial infection of normal metacercariae are able to kill a challenge with metacercariae fully attenuated by γ-irradiation. The role of granules is therefore not clear. Further work on granule antigens will obviously provide valuable information and recent work by HANNA and TRUDGETT (1983) shows with the aid of monoclonal antibodies that the T1 antigen occurs in metacercarial tegument, gut cells and excretory ducts of juvenile and adult flukes. These same authors also found that when isolated by immunoadsorption and separated electrophoretically under reducing conditions T1-type antigens consisted of a polypeptide (MW 50 000) possibly linked to smaller entities.

A number of workers have tried to isolate protective antigens from fluke but generally these have been unsuccessful in producing immunity. LEHNER and SEWELL (1979, 1980) isolated antigens as S/E products from both juvenile and adult flukes, but were unable to protect mice, rats and rabbits against a challenge when these products were used. These antigens were found to have molecular weights within two ranges: 10 000–25 000 and 100 000–500 000. KORACH and BENEX (1966a) isolated a lipoprotein from adult *F. hepatica* which failed to produce protection (KORACH and BENEX 1966b). HILLYER (1980) has isolated tegumental antigens from *F. hepatica* which were genus specific and also protected mice from challenge with *S. mansoni*. Cross-reacting antigens between different genera of helminths have been described by CAPRON et al. (1968) and more recently Hillyer and colleagues have used purified *F. hepatica* antigen to produce resistance to *S. mansoni*. The techniques used by Hillyer and collagues have taken the isolation and purification of *F. hepatica* antigens further than most workers, but the antigen has been mainly examined for its ability to protect against *S. mansoni* in mice (HILLYER and SAGRAMOSO DE ATECA 1979; DEMAREE and HILLYER 1982; HILLYER and SERRANO 1982; HILLYER 1979). The mouse is not a good host for demonstrating immunity to *F. hepatica* and its use for examining the efficacy of antigens from fluke for use against *S. mansoni* in man and domestic animals remains to be proved as does the nature of resistance engendered by these fluke antigens to *S. mansoni*.

Cross-reacting antigens with *Fasciola* are also found in *Paragonimus, Clonorchis, Paramphistomum* as well as several species of Schistosomes. The possibility that cross-reacting antigens could be used to protect man and his domestic animals from various parasitic diseases has been a hope for many years. Liver fluke is no exception and many organisms have been tried to prevent this disease. *Nippostrongylus brasiliensis* as mentioned earlier, when given to rats, protected them from oral challenge with metacercariae but not against an intraperitoneal challenge with NEJs and although it was suggested by DOY et al. (1981) that induced intestinal eosinophilia may have been the reason for this resistance, sera from *N. brasiliensis* rats did not induce cell adherence to NEJs. The involvement of the eosinophil in effector killing mechanisms of *Schistosoma* is well documented and the fact that *F. hepatica* is such a potent inducer of eosinophilia even in athymic nude rats (DOY and HUGHES 1982c) is worthy of further study in connection with cross-immunity to *Schistosoma*. The effect of fluke infections and fluke antigens on the circulatory system of the liver in experimental animals and cattle is also relevant with respect to cross-immunity work.

Schistosoma bovis has been used to infect both sheep and cattle prior to challenge with *F. hepatica* and MONRAD et al. (1981) and SIRAG et al. (1981) claim that resistance is developed in both these hosts respectively but some of the results are open to criticism. There was a particularly poor "take" in the controls in the sheep work. A mean of only 16.4 flukes was found in the controls after a dose of 400 metacercariae. In the calves, previous infections with *S. bovis* 10 weeks prior to challenge with 900 metacercariae of *F. hepatica* resulted in a mean of 154 ± 37 flukes compared with 217 ± 33 in the controls ($P < 0.025$).

Reciprocal cross-protection has also been demonstrated in mice using infection of *F. hepatica* and *S. mansoni* (CHRISTENSEN et al. 1980; HILLYER 1981). One of the more exciting cross-protection results was that claimed by the Australian workers CAMPBELL et al. (1977). These authors showed that when sheep were infected with eggs of the dog tapeworm *Taenia hydatigena,* the resulting infection with the liver-migrating metacestodes (*Cysticercus tenucollis*) induced resistance to *F. hepatica.* This protection was said to last for at least 9 months after the initial tapeworm infection (*Dineen* et al. 1978) and to be associated with the presence of long-surviving tapeworm cysts. This work was not confirmed by HUGHES et al. (1978) in goats, sheep or cattle and these authors suggested that the anthelmintic levamisole which had been used to rid Campbell et al.'s experimental sheep of nematodes may have been responsible due to its possible immunomodulatory effects. However, further work with levamisole (HUGHES et al. unpublished) in grazing sheep failed to induce resistance. In contrast MITCHELL and ARMOUR (1981) claimed to have produced some resistance using this compound in sheep which also had nematodes as well as experimental tapeworm infections before challenge with *F. hepatica.* Finally, the Australian workers have not been able to confirm their earlier results and it would appear that *T. hydatigena* is not capable of preventing fluke infections.

Monoclonal antibodies to fluke will obviously be much used in attempts to isolate specific antigens for use as immunising agents. A major problem is that to assay any of these antigens for protection small animal models will be used and unfortunately we do not know at present if these small animals give any indication of "protection-inducing antigens" functional in ruminants.

Another approach to the production of *F. hepatica* antigens has been that of HOWELL (1981), who has reported the formation of hybrid cells between *F. hepatica* and a rat fibroblast cell line. Although fluke-derived hypoxanthine-guanine phosphoribosyl transferase was found in the hybrid cells it was not possible to prove that fusion had taken place because no obvious difference was found in chromosome studies. The only difference seen was an extra chromosome possibly derived from *F. hepatica* found in the hybrid cell. No *Fasciola* antigens were produced, but it is a most encouraging start in this new area.

2.2 Evasion Mechanisms

Host antigen acquisition by the liver fluke does not appear to be a survival mechanism in this parasite. Attempts to detect an effect on young (18-day-old) or adult flukes transferred into hosts immunised against the donors' blood tissues were unsuccessful (HUGHES and HARNESS 1973a, b), an observation since confirmed in other laboratories. However, BEN-ISMAIL et al. (1982) reported that cattle flukes were able to synthesise A,H and Lewis blood group antigens and suggest that as these surface markers are clearly defined in immunological and biochemical terms their study could shed further light on tegument development and its alteration during migration in different host regions.

CHAPMAN et al. (1981) have examined the effect of different clones of *F. hepatica* in the immune situation. They found that there was no difference if homologous or heterologous clones were used to challenge previously infected animals. A clone is defined as those metacercariae obtained from a snail infected with a single miracidium.

The possibility that flukes produce substances that prevent immunocompetent effector cells functioning or specific antibodies from attaching has been reported. GOOSE (1978) showed that E/S products of flukes were capable of killing lymphocytes in vitro and suggested flukes could circumvent the host defences in regions such as the liver and bile ducts where a high concentration of these products would be in close proximity to the flukes. This would also apply to the situation when flukes become surrounded by a fibrous cyst when placed in the body cavity of rats. The nature of such substances has however not been defined and as these E/S products were obtained by allowing living flukes to expel caecal contents into dry tubes, more refined techniques should prove useful in relation to the in vivo situation. Proteolytic cleavage of immunoglobulins, by papain or cathepsin-B-like proteolytic enzymes released from *F. hepatica*, have been described by CHAPMAN and MITCHELL (1982b). The Ig cleavage is said to be partial in contrast to that found with *Schistosoma mansoni* protease (AURIOULT et al. 1980), but nevertheless could be effective against host antibody to the parasite. CHAPMAN and MITCHELL (1982b) also suggested that passive transfer could possibly be explained by the presence of non-specific protease inhibitors (maybe antienzyme antibodies) rather than antiparasitic antibodies in the serum. This is worthy of further examination.

Glycocalyx turnover or replacement as a possible mechanism for protection against host immunity has been suggested by HANNA (1980a) and the replacement glycocalyx is found to be antigenically similar to its predecessor. Further work by DUFFUS and FRANKS (1980, 1981) also supports the view that glycocalyx replacement is a protective device and the reason why specific antibody and antibody-mediated eosinophil adherence fails to damage NEJs in vitro.

The antigenic nature of the different granules (T0, T1, T2) and the antibody response they engender has been described earlier although their function in the context of immunity is not clear. The contribution of the granules to the formation and replacement of glycocalyx is somewhat clearer especially in terms of the "sloughing off" of outer surface to which either antibody or cells, or both, may be attached. The large stores of T0 granules found in metacercarial tegumental cells (BENNETT and THREADGOLD 1975) have also been suggested as a preadaptation which might help the emerging and vulnerable young fluke to survive attack in previously infected hosts (HANNA 1980b). Once the T0 granules are replaced by T1 granules the latter are present in great numbers thoughout liver migration and are still found, although to a much lesser extent, in the bile duct stage. The bile duct stage has T2 granules predominantly and Hanna suggests that the appearance of these granules at about the time the parasite enters the bile duct is a preadaptation to the altered mode of life necessary for living in bile.

HUGHES et al. (1981) compared the antibody profiles to the various granules, in both sera and bile using bile duct cannulated cows. There was a much lower concentration of immunoglobulins in bile than serum although IgA appeared to be preferentially concentrated in the bile. No antibodies were detected in bile to T2 granules and these authors suggested that this might provide further protection since the dominant granule in bile-dwelling flukes is T2.

Recent work by HANNA and TRUDGETT (1983) with monoclonal antibodies prepared against antigen present in T1 granules and fluke glycocalyx showed that these antigens are indeed powerful in that all the monoclonals appeared to bind to the same epitope. The shared antigenicity of T0, T1 and gut was also confirmed. If the response in fluke infection is mainly, if not exclusively, to a single antigen early in infection, and if this antigen is not relevant in immunity, this could be yet another mechanism used by fluke whereby many cells are switched on to produce irrelevant antibody early in infection. It was suggested by HANNA and TRUDGETT (1983) that the functional epitope was probably a polypeptide.

The liver has been suggested as a sequestered site for parasites as have the bile ducts. There is little evidence for flukes being removed from either the liver or bile ducts by immunological mechanisms. The loss of flukes in long-standing infections from the bile ducts of both rats and cattle however does occur but this could be also explained by nutritional problems encountered by the parasite. It could be due to fluke size in the bile ducts and local damage produced in rats, and in the cattle, extra fibrosis and calcification could prevent its normal feeding activities. The possibility that the thin bile duct walls found in long-standing infections in rats may allow constituents of effector mechanisms to gain entry cannot however be ruled out (HUGHES et al. 1976).

2.3 Immunodiagnosis

The need for and usefulness of immunodiagnostic tests for trematodes is probably much greater for man than his domestic animals. Their potential uses in individual patients either in endemic areas or after visiting such areas, before specific therapy or surgery can be prescribed, are obvious, as is the potential of such tests in epidemiological surveys. In contrast animals generally are dealt with on a herd or flock basis, and diagnosis for adult parasites is quickly carried out using conventional egg flotation techniques. Diagnosis of immature infection can be confirmed by slaughter and postmortem examination of one or two animals if considered necessary, although treatment with the appropriate anthelmintic of the whole group would normally be carried out.

Immunodiagnostic tests for animal parasites are likely to be of most use in the experimental situation, to assess the presence of living parasites, or to follow the course of primary and challenge infections and monitor anthelmintic activity.

Recent reviews of immunodiagnosis and recent advances together with relevant techniques and reagent availability have been published (VOLLER and BIDWELL 1982; KAGAN 1980; KAGAN and HILLYER 1981).

Specificity is a problem in anthelmintic diagnostic reagents and many authors have attempted to purify such reagents to eliminate cross-reactivity. In some situations the detection of circulating antigen will be of greater value than the detection of antibodies.

Cross-reactions of *F. hepatica* antigens and other helminths such as *Schistosoma, Clonorchis, Paragonimus* and *Paramphistomum* occur (CAPRON et al. 1968; CHOI and LEE 1979).

HILLYER and CAPRON (1976) found that a crude extract of *F. hepatica* antigen cross-reacted with sera from humans with various parasitic infections. Partial purification of this *F. hepatica* antigen using Sephadex G-200 eliminated most of this cross-reactivity although it was not clear if cross-reactivity with amoebiasis and cysticercosis sera was eliminated because of possible concurrent infection with *F. hepatica* in these patients.

The literature on immunodiagnosis of both natural and experimental infections in cattle, sheep and some small laboratory animals has been extensively reviewed by VAN TIGGELE (1978). In natural infections it was found that skin tests were generally judged useful for diagnosis in practice, but were not specific enough. This criticism was also brought against the complement fixation test and double immunodiffusion although immunoelectrophoresis was found to be more sensitive. HILLYER and DE WEIL (1977, 1979, 1981) partially purified *F. hepatica* antigen for use in immuno-diagnosis of experimental infections mainly in rats and also used it in the enzyme-linked immunosorbent assay (ELISA) test. Antibody titres dropped dramatically after treatment, indicating chemotherapeutic success. The indirect fluorescent anti-body test used by HANNA et al. (1982) does not appear to be suitable for serodiagnosis of fluke in the field because of the rise and fall of antibody titre associated with the sequential granule expression in ruminants. BURDEN and HAMMET (1978) used the supernatant from lyophilised adult flukes in phosphate buffer saline after elution overnight at 4 °C as their antigen in the ELISA test. In experimentally infected calves increased values were noted 5 weeks after infection. Although sera from *Ostertagia*-infected calves did not cross-react with the antigen, later work (Burden, personal communication) showed that in older cattle there was a high background in non-fluke-infected animals which invalidated the test for use in the field. OLDHAM (1983) developed a standardised assay which was able to detect antibodies 3 weeks after infection in experimentally infected calves. He used partially purified *F. hepatica* antigens obtained by chromatography of a soluble phosphate-buffered saline (PBS) extract on Sephadex G-200. The antibody response to the seven fractions he obtained showed a different time course and he suggested that some of these antigens may re-late to maturation and others were common to all stages of fluke. The relevance of this assay to the field situation has not been investigated.

3 Other Trematodes

The different migration patterns and the predilection sites of the various adult flukes of different species should provide a fruitful source for comparative work in the fields of immunology and serology. *Clonorchis, Opisthorchis* and *Dicrocoelium* all hatch from metacercariae in the intestine and ascend the bile ducts from the duodenum. They all appear to be able to live for many years in this environment (*Clonorchis* up to 15 years, for example).

Intestinal flukes such as *Fasciolopsis, Metagonimus* and *Echinostoma* inhabit the small bowel whereas *Gastrodiscoides* lives in the caecum and colon. *Fasciolopsis buski* is said to live for only 6 months in the gut.

Extensive migration via the gut, peritoneum and diaphragm occur before *Para-gonimus* reaches the lung and *Fasciola* also migrates extensively in gut, body cavity and liver parenchyma before attaining sexual maturity in the bile ducts.

Experimental hosts for flukes can and often do give different results or appear to have different mechanisms operating compared with natural hosts. A similar situation is seen in the field when the large liver fluke *Fascioloides magna* infects different species of ruminant. In the white-tailed deer there is little liver damage and the sexually mature adults lay their eggs within small liver cysts which connect up to the bile ducts via channels and so allow the eggs to attain the outside world. In sheep infected with this parasite there is massive liver damage and often deaths, whereas in cattle there is complete encapsulation within the liver, no eggs are produced and the life cycle is not completed (FOREYT and TODD 1976). A few of the more important of these trematodes will now be briefly discussed under separate headings.

3.1 *Paragonimus*

There appears to be little published on immunity to this parasite. Antigenic analysis of adult worm extract of *Paragonimus westermani* and the antibody response to these fractionated antigens has been carried out by IMAI (1979). The antibody response in rats sensitised with adult worm antigen has also been examined by HAMAJIMA et al. (1977) and the effect of prednisolone on antibody response in rats for another species *P. iloktsuenensis* was reported by LEE et al. (1976).

The IgE response to *Paragonimus ohirai* has been followed in some detail by IKEDA and FUJITA (1980), who compared the responses in orally infected rats with those given immature flukes by surgical transplantation into the peritoneum and other sites. They found that the indirect haemagglutination antibody response was not influenced by infection route whereas the IgG response was affected by both the site of metacercarial inoculation and also the age of the inoculated worms. The migration of the parasite in the peritoneal cavity and liver was found to be important in the production of specific IgE. When worms were transplanted into the peritoneum the IgE response was significantly lower and correlated with the age of these worms. The authors point out that as reported by OGILVIE (1964) for *N. brasiliensis*, only living parasites were found to induce this IgE response and that continuous stimulation as well as the correct site was essential for its production. It was also thought that 4-week-old worms placed in the body cavity induced a suppression of IgA when placed in the body cavity with metacercariae. An extension to other trematodes using this approach could prove profitable. The extraction of antigens from *Paragonimus* and their use in diagnosis and also their cross-reactivity has been dealt with in detail by YOKOGAWA (1964), who reviews *Paragonimus* and paragonimiasis in great depth with over 600 references. Further information is given by MIYAZOKI (1964) in the same volume and this author quotes MANNOJI (1952) as finding rats resistant to reinfection with *P. ohirai*. While it appears that some acquired immunity develops in animals infected with lung flukes further experiments are needed. The few earlier studies that have been made are reviewed by YOKOGAWA et al. (1960). More recent reviews by YOKOGAWA (1965, 1969) cover seroimmunological diagnosis in great detail with a comparison of various antigenic preparations for use in the intradermal, precipitin, flocculation and complement fixation test and other tests; involving the use of extracts of faeces or urine from infected patients which were found to induce precipitation with rabbit anti-*P. westermani* immune serum. Cross-reactivity was found with both *Clonorchis*- and *Schistosoma*-infected patients in some of these tests. Yokogawa does not discuss immunity to *Paragonimus* here, but considers that in rats the duration of infection with *P. okirai,* for example, is influenced by the worm burden as is the ability of the parasite to produce lung cysts. No lung cysts were found in infected rats given only one metacercaria.

3.2 *Clonorchis*

As is the case with *Paragonimus* there is little information on immunity to *Clonorchis sinensis,* the human liver fluke, although it is capable of infecting a wide range of experimental laboratory animals. Work has, however, been carried out on the effect of various antisera on the parasite in vitro and the use of whole fluke extracts and metabolic products as antigens. SUN (1969) points out that about one-quarter of the population of Hong Kong are infected, with reinfection a common feature, indicating little or no immunity. This author found precipitate around adult flukes when placed in antisera and these parasites did not survive as long as those placed in normal control sera. SUN and GIBSON (1969a, b) also examined antigens of *C. sinensis* in experimental and human infections and looked at the site of antigen formation in the different developmental stages of this parasite. They found that metabolic products appeared to be the main source of antigens during infection. Attempts to protect rabbits and guinea pigs immunised with whole fluke extract were unsuccessful. They point out that in man a probable infective dose eaten at one meal would be in the region of 26 metacercariae per infected fish portion. Because many thousands of worms are found in infected individuals, reinfection must take place many times over and this gives no indication of a naturally acquired immunity. Concerning the site of antigen formation, specific fluorescence was seen in the cuticle and around the suckers as well as in the caeca, with only weak fluorescence seen on the testes and ovary. The eggs gave no reaction. This parasite spends most of its life inside thickened bile ducts of the liver and unless antibodies found in the sera are found at this site the relevance of in vitro tests is questionable. Some work has been carried out examining the immune response in rats (LEE and SONG 1977) to infection and also the fate of degranulated mast cells responding to the homologous antigen (AHN and SOH 1977). The possible role of the bile duct as a sequestered site is obviously relevant to this parasite but there appears to be little further information on parasite evasion mechanisms. Experiments using a closely related parasite *Opisthorchis viverrini* were reported by FLAVELL et al. (1980) and these authors claimed partial success, as measured by reduced egg counts, in adoptively transferring immunity in the hamster with spleen cells and serum. Attempts at immunodiagnosis using various antigenic preparations have been reported and these have been reviewed by KOMIYA (1966). More recently AHN et al. (1975) have examined different antigens for their specificity, but much of this work is reported in Korean.

3.3 *Paramphistomum*

Paramphistomiasis of domestic ruminants and the various genera and species involved are reviewed by HORAK (1971), who considers that immunity develops in the field. Multiple experimental infections with *Paramphistomum microbothrium* result in partial immunity to reinfection in sheep, but the challenge worms are able to excyst and attach to the small intestine. The situation in cattle appears different in that the immunity is virtually complete after multiple infections and worms are eliminated from subsequent infections. In immunisation experiments using live metacercariae (approximately 40 000/animal), Horak was able to protect sheep, goats and cattle. This immunity manifested itself in reduced worm burdens and the ability of the immunised animals to withstand a lethal challenge. Although no attempts were made to isolate functional antigens, it was found that immunity in sheep seemed to depend on the continued presence of worms. If these were removed the animals became susceptible and, furthermore, adult worms given orally and which produced an infection did not give rise to immunity. Horak considered that the normal life cycle in the final hot must be completed for immunity to develop. X-irradiated metacercariae were also able to produce immunity and these cysts hatched normally and the worms attached to the small intestine. Large numbers were lost however on their migration to or arrival at the rumen site. The growth rate of paramphistomes in immune animals is retarded. An interesting observation was that in cattle a high proportion of a residual challenge infection in the immune animal was attached behind the first 3 m of the small intestine. This is in marked contrast to the situation in normal animals, where the first 3 m would be heavily infected. This may indicate different effects of the gut regions after initial infection or that attachment in the lower section may render the paramphistome incapable of growth and/or migration. The use of purified antigen introduced into the various gut regions and an examination of its effect on subsequent challenge could prove rewarding. Attempts to diagnose the infection serologically have been mainly concerned with intradermal tests, complement fixation and the ability of infected serum to produce precipitates around living parasites. Antigenic extracts prepared by HORAK (1967) were not specific in that cross-reactions occurred with nematodes, *Fasciola* and *Schistosoma*.

3.4 *Dicrocoelium*

Dicrocoelium dendriticum, the lancet fluke of ruminants, is similar to *Clonorchis* in that it spends the vast majority of its life in the mammalian host's bile ducts. There is little evidence of any acquired resistance. The serological response to this parasite has been examined by BODE and GEYER (1981) in both golden hamsters and rabbits. An attempt was made to correlate worm burdens in hamsters with the humoral immune response as measured by CFT, ELISA, double diffusion and immuno- and counter-immunoelectrophoresis. CFT and ELISA were found to be the most sensitive for early detection of response. In rabbits exposed to 500, 1000 and 3000 metacercariae per animal a persisting antibody response (using precipitation tests) was only demonstrated after the highest infective dose of 3000 cysts and the initial response was first observed on day 63. This initial response is much later than that found with *F. hepatica* in rabbits in which double diffusion in gel detects a response 2 or 3 weeks after infection, when the parasites are migrating in the liver. *Dicrocoelium* does not have this tissue-migrating stage in its mammalian cycle.

In summary it can be stated that some progress have been made in obtaining information about those antigens involved in protection, evasion or diagnosis of the trematode infections discussed here. There is, however, little solid evidence available to define and characterise such antigens in biochemical terms. Hopefully the new techniques and the newer technologies will help unravel the many unsolved problems associated with these complex and well-adapted metazoan parasites.

References

Ahn YK, Soh CT (1977) Fate of degranulated mast cells and extruded granules in response to homologous antigen in rats infected with *Clonorchis sinensis*. Korean J Parasitol 15:93–99 (in Korean, summary in English) Helm Abs OHO48-00264

Ahn YK, Soh CT, Han JK (1975) Comparison of dermal reaction with VBS and KST antigens of *Clonorchis sinensis* in reference to sensitivity and specificity. New Med J 18:1195–1207

Anders RF, Howard RJ, Mitchell GF (1982) Parasitic antigens and methods of analysis. In: Cohen S, Warren KS (eds) Immunology of parasitic infections, 2nd edn. Blackwell Scientific, Oxford pp 28–73

Anderson JC, Hughes DL, Harness E (1975) The immune response of rats to subcutaneous implantation with *Fasciola hepatica*. Br Vet J 131:509–517

Armour J, Dargie JD (1974) Immunity to *Fasciola hepatica* in the rat: successful transfer of immunity by lymphoid cells and serum. Exp Parasitol 35:318

Auriault C, Ouiassi MA, Torpier G, Eisen H, Capron A (1980) Proteolytic cleavage of IgG bound to the receptor of *Schistosoma mansoni* schistosomula. Parasite Immunol 3:33–44

Ben-Ismail R, Mulet-Clamagirand C, Carme B, Gentilini M (1982) Biosynthesis of A, H and Lewis blood group determinants in *Fasciola hepatica*. J Parasitol 68:402–407

Bennett CE (1978) The identification of soluble adult antigen on the tegumental surface of juvenile *Fasciola hepatica*. Parasitology 77:325–332

Bennett CE, Threadgold LT (1973) Electronmicroscopy studies of *Fasciola hepatica*. XIII. Fine structure of newly excysted juveniles. Exp Parasitol 34:85–99

Bennett CE, Threadgold LT (1975) *Fasciola hepatica:* development of tegument during migration in mouse. Exp Parasitol 38:38–55

Bennett CE, Hughes DL, Harness E (1980) *Fasciola hepatica:* changes in tegument during killing of adult flukes surgically transferred to sensitised rats. Parasite Immunol 2:39–55

Bode L, Geyer E (1981) Experimental dicrocoeliasis: the humoral immune response to golden hamsters and rabbits to primary infection with *Dicrocoelium dendriticum*. Z Parasitenkd 66:167–178

Burden DJ, Bland AP (1983) *Fasciola hepatica:* the fate of challenge flukes in naive and previously infected rats and cattle. Parasitology 87(2):lxx–lxxi

Burden DJ, Hammet NC (1978) Microplate enzyme-linked immunosorbent assay for antibody to *Fasciola hepatica* in cattle. Vet Rec 103:158

Burden DJ, Hammet NC (1980) *Fasciola hepatica:* attempts to immunise rats using fluke eggs and in vitro culture products. Vet Parasitol 7:51–57

Burden DJ, Bland AP, Hammet NC, Hughes DL (1983a) *Fasciola hepatica:* migration of newly excysted juveniles in resistant rats. Exp Parasitol 56:277–288

Burden DJ, Bland AP, Hughes DL, Hammet NC (1983b) *Fasciola hepatica:* development of the tegument of normal and γ-irradiated flukes during infection in rats and mice. Parasitology 86:137–145

Campbell NJ, Kelly JD, Townsend RB, Dineen JK (1977) The stimulation of resistance in sheep to *Fasciola hepatica* by infection with *Cysticercus tenuicollis*. Int J Parasitol 7:347–357

Capron A, Biguet J, Vernes A, Afchain D (1968) Structure antigenique des helminthes. Aspects immunologiques des relations hote-parasite. Pathol Biol 16:121–138

Chapman CB, Mitchell GF (1982a) *Fasciola hepatica:* comparative studies on fascioliasis in rats and mice. Int J Parasitol 12:81–91

Chapman GB, Mitchell GF (1982b) Proteolytic cleavage of immunoglobulin by enzymes released by *Fasciola hepatica*. Vet Parasitol 11:165–178

Chapman CB, Rajasekariah GR, Mitchell GF (1981) Clonal parasites in the analysis of resistance to reinfection with *Fasciola hepatica*. Am J Trop Med Hyg 30:1039–1042

Choi WY, Lee OR (1979) Immunoelectrophoresis for *Fasciola hepatica*. Korean J Parasitol 17:73–80

Christensen NO, Monrad J, Nansen P, Frandsen F (1980) *Schistosoma mansoni* and *Fasciola hepatica:* cross-resistance in mice. Exp Parasitol 46:113–120

Corba J, Armour J, Roberts RJ, Urquhart GM (1971) Transfer of immunity to *Fasciola hepatica* infection by lymphoid cells. Res Vet Sci 12:292–295

Davies C, Goose J (1981) Killing of newly excysted juveniles of *Fasciola hepatica* in sensitised rats. Parasite Immunol 3:81–86

Davies C, Rickard MD, Smyth JD, Hughes DL (1979) Attempts to immunise rats against infection with *Fasciola hepatica* using in vitro culture antigens from newly excysted metacercariae. Res Vet Sci 26:259–260

Demaree RS, Hillyer GV (1982) Immunoperoxidase localisation of *Fasciola hepatica* worm tegument antigen by electron microscopy. Int J Parasitol 12:179–183

Dineen JK, Kelly JD, Champbell NJ (1978) Further observations on the nature and characteristics of cross protection against *Fasciola hepatica* produced in sheep by infection with *Cysticercus tenuicollis*. Int J Parasitol 8:172–176

Doy TG (1980) Examination of the immune response of rats to infection with *Fasciola hepatica*. PhD thesis Brunel University

Doy TG, Hughes DL (1982a) Evidence for two distinct mechanisms of resistance in the rat to reinfection with *Fasciola hepatica*. Int J Parasitol 12:357–361

Doy TG, Hughes DL (1982b) In vitro cell adherence to newly excysted *Fasciola hepatica*: failure to affect their subsequent development in rats. Res Vet Sci 32:118–120

Doy TG, Hughes DL (1982c) The role of the thymus in the eosinophil response of rats infected with *Fasciola hepatica*. Clin Exp Immunol 47:74–76

Doy TG, Hughes DL (1984a) Early nugiration of immature *Fasciola hepatica* and associated liver pathology in cattle. Res Net Sci 37:219–222

Doy TG, Hughes DL (1984b) *Fasciola hepatica:* site of resistance to reinfection in cattle. Exp Parasitol 57:274–278

Doy TG, Hughes DL, Harness E (1981) The heterologous protection of rats against a challenge with *Fasciola hepatica* by prior infection with the nematode *Nippostrongylus brasiliensis*. Parasite Immunol 3:171–180

Doyle JJ (1971) Acquired immunity to experimental infection with *Fasciola hepatica* in cattle. Res Vet Sci 12:527–534

Doyle JJ (1973) The relationship between the duration of a primary infection and the subsequent development of an acquired resistance to experimental infections with *Fasciola hepatica*. Res Vet Sci 14:97–103

Duffus WPH, Franks D (1980) In vitro effect of immune serum and bovine granulocytes on juvenile *Fasciola hepatica*. Clin Exp Immunol 41:430–440

Duffus WPH, Franks D (1981) The interaction in vitro between bovine immunoglobulins and juvenile *Fasciola hepatica*. Parasitology 82:1–10

Duffus WPH, Thorne L, Oliver R (1980) Killing of juvenile *Fasciola hepatica* by purified bovine eosinophil proteins. Clin Exp Immunol 40:336–344

Ericksen L, Flagstad T (1974) *Fasciola hepatica:* influence of extrahepatic adult flukes on infections and immunity in rats. Exp Parasitol 35:411–417

Flavell DJ, Pattanapanyasat K, Flavell SU (1980) *Opisthorchis viverrini:* partial success in adoptively transferring immunity with spleen cells and serum in the hamster. J Helminthol 54:191–197

Foreyt WJ, Tood AC (1976) Development of the large American liver fluke, *Fascioloides magna*, in white-tailed deer, cattle and sheep. J Parasitol 62:26–32

Goose J (1978) Possible role of excretory/secretory products in evasion of host defences by *Fasciola hepatica*. Nature 275:216–217

Hamajima E, Fujimo T, Koga M (1977) Time course of production of antibodies in rats sensitised with adult worm antigen of *Paragonimus westermani*. Jpn J Parasitol 26:32 (in Japanese) Helm Abs OHO47-00555 Title only).

Hanna REB (1980a) *Fasciola hepatica:* glycocalyx replacement in the juvenile as a possible mechanism for protection against post immunity. Exp Parasitol 50:103–114

Hanna REB (1980b) *Fasciola hepatica:* an immunofluorescent study of antigenic changes in the tegument during development in the rat and sheep. Exp Parasitol 50:155–170

Hanna REB, Trudgett AG (1983) *Fasciola hepatica:* the development of monoclonal antibodies and their use to characterise a glycocalyx antigen in migrating flukes. Parasite Immunol 5 (4):409–425

Hanna REB, Hughes DL, Taylor SM (1982) *Fasciola hepatica:* antibody levels in sheep serum before and after treatment with anthelmintic. Res Vet Sci 33:328–332

Haroun EM, Hammond JA, Sewell MMH (1980) Resistance to *Fasciola hepatica* in rats and rabbits following implantation of adult flukes contained in diffusion chambers. Res Vet Sci 29:310–314

Haroun EM, Hammond JA, Sewell MMH (1982) Absorption of protective components from serum of cattle infection with *Fasciola hepatica.* Res Vet Sci 33:263–264

Hayes TJ, Bailer J, Mitrovic M (1972) Immunity in rats to superinfection with *Fasciola hepatica.* J Parasitol 58:1103–1105

Hillyer GV (1979) *Schistosoma mansoni:* reduced worm burdens in mice immunised with isolated *Fasciola hepatica* antigens. Exp Parasitol 48:287–295

Hillyer GV (1980) Isolation of *Fasciola hepatica* tegument antigens. J Clin Microbiol 12:695–699

Hillyer GV (1981) Effect of *Schistosoma mansoni* infections on challenge infections with *Fasciola hepatica* in mice. J Parasitol 67:731–733

Hillyer GV, Capron A (1976) Immunodiagnosis of human fascioliasis by counter-electrophoresis. J Parasitol 62:1011–1013

Hillyer GV, de Weil NS (1977) Partial purification of *Fasciola hepatica* antigen for the immunodiagnosis of fascioliasis in rats. J Parasitol 63:430–433

Hillyer GV, de Weil NS (1979) Use of immunologic techniques to detect chemotherapeutic success in infections with *Fasciola hepatica.* II. The enzyme linked immunosorbent assay in infected rats and rabbits. J Parasitol 65:680–684

Hillyer GV, de Weil NS (1981) Serodiagnosis of experimental fascioliasis by immuno-precipitation tests. Int J Parasitol 11:71–78

Hillyer GV, Sagramoso de Ateco (1979) Immunity in *Schistosoma mansoni* using antigens of *Fasciola hepatica* isolated by concanavalin. An affinity chromatography. Infect Immun 26:802–807

Hillyer GV, Serrano AE (1982) Cross protection in infections due to *Schistosoma mansoni* using tegument antigens of *Fasciola hepatica.* J Infect Dis 145:728–732

Horak IG (1967) Host parasite relationships of *Paramphistomum microbothrium* Fischoeder 1901 in experimentally infected ruminants with particular reference to sheep. Onderstepoort J Vet Res 34:457–540 (quoted in Horak 1971)

Horak IG (1971) Paramphistomiasis of domestic ruminants. Adv Parasitol 9:33–72

Howell MJ (1981) An approach to the production of helminth antigens in vitro: the formation of hybrid cells between *Fasciola hepatica* and a rat fibroblast cell line. In J Parasitol 11:235–242

Howell MJ, Sandeman RM (1979) *Fasciola hepatica:* some properties of a precipitate which forms when metacercariae are cultured in immune rat serum. Int J Parasitol 9:41–45

Hughes DL, Doy TG (1983) Resistance to the liver fluke *Fasciola hepatica.* In: Proc 11th Scand Symp Parasitol Just Parasitol Åbo Academic Finland, Information 17:76

Hughes DL, Harness E (1973a) Attempts to demonstrate a 'host antigen' effect by the experimental transfer of adult *Fasciola hepatica* into recipient animals immunised against the donor. Res Vet Sci 14:151–154

Hughes DL, Harness E (1973b) The experimental transfer of immature *Fasciola hepatica* from donor mice and hamsters to rats immunised against the donors. Res Vet Sci 14:220–222

Hughes DL, Harness E (1975) Some immunological aspects of *Fasciola hepatica* infections in animals. 2nd European Multicolloquy of Parasitology, Trogir (Jugoslavia) Drusvo Parazitologa Jugoslavije Beograd, Bulerarja 18:60–61

Hughes DL, Harness E, Doy TG (1976) The extablishment and duration of *Fasciola hepatica* infections in two strains of rat and the development of acquired resistance. Res Vet Sci 20:207–211

Hughes DL, Harness E, Doy TG (1977) Loss of ability to kill *Fasciola hepatica* in sensitised rats. Nature 267:(5611), 577–578

Hughes DL, Harness E, Doy TG (1978) Failure to demonstrate resistance in goats, sheep and cattle to *Fasciola hepatica* after infection with *Cysticercus tenuicollis.* Res Vet Sci 25:356–359

Hughes DL, Hanna REB, Symonds HW (1981) *Fasciola hepatica:* IgG and IgA levels in the serum and bile of infected cattle. Exp Parasitol 52:271–279

Hughes DL, Doy TG, Burden DJ, Oldham G (1982a) Stage specific immunity in the rat as demonstrated by γ-irradiated *F. hepatica.* Parasitology 84:XIII

Hughes DL, Hanna REB, Doy TG (1982b) Antibody response in cattle, sheep and rats to infection with γ-irradiated metacercariae of *Fasciola hepatica.* Res Vet Sci 32:354–358

Ikeda T, Fujila K (1980) IgE in *Paragonimus ohirai*-infected rats: relationship between titre, migration route and parasite age. J Parasitol 66:197–204

Imai J (1979) Studies on the antigenic analysis in *Paragonimus westermani*. 2. Observations on antibody response against fractionated antigens from adult worm extract. Trop Med 21:2 45–55 (in Japanese, English summary)

Kagan IG (1980) Serodiagnosis of parasitic diseases. In: Rose NR, Friedman H (eds) Manual of clinical immunology. American Society for Microbiology, Washington DC, pp 573–604

Kagan IG, Hillyer GV (1981) Recent advances in serodiagnosis of parasitic diseases. In: Slursarski W (ed) Proceedings of fourth International Congress of Parasitology, 1978, Warsaw Poland. PWN-Polish Scientific Publishers, Warsaw

Komiya Y (1966) *Clonorchis* and clonorchiasis. Adv Parasitol 4:53–101

Korach S, Benex J (1966a) A lipoprotein antigen in *Fasciola hepatica*. I. Isolation, physical and chemical data. Exp Parasitol 19:193–198

Korach S, Benex J (1966b) A lipoprotein antigen in *Fasciola hepatica*. II. Immunological and immuno-chemical properties. Exp Parasitol 19:199–205

Lang BZ, Hall RF (1977) Host parasite relationships of *Fasciola hepatica* in the white mouse. VII. Successful vaccination with culture incubate antigens and antigens from sonic disruption of immature worms. J Parasitol 63:1046–1049

Lee SH, Song CY (1977) Studies on the immune response of albino rats infected with *Clonorchis sinensis*. Korean J Parasitol 15:55–56 (in Korean) (Helm. Abs. OHO47-04660 title only)

Lee SH, Song CY, Seo BS, Bae JH (1976) Studies on the lung fluke *Paragonimus iloktsuenensis*. VI. Effect of prednisolone injection on the immune responses of albino rats. Korean J Parasitol 14:172 (in Korean) (Helm Abs OHO46-05445 title only)

Lehner RP, Sewell MMH (1979) Attempted immunisation of laboratory animals with metabolic antigens of *Fasciola hepatica*. Vet Sci Com 2:337–340

Lehner RP, Sewell MMH (1980) A study of the antigens produced by adult *Fasciola hepatica* maintained in vitro. Parasite Immunol 2:99–109

Mitchell GF (1979) Response to infection with metozoan and protozoan parasites in mice. Adv Immunol 28:451–511

Mitchell GF (1982) Host-protective immune responses to parasites and evasion by parasites: generalisation and approaches to analysis. In: Symonds LEA, Donald AD, Dineen JK (eds) Biology and control of endoparasites. Academic, pp 331–341

Mitchell GBB, Armour J (1981) Stimulation of resistance to *Fasciola hepatica* infection in sheep by a regime involving the use of the immunomodulatory compound L tetramisole (levamisole). Res Vet Sci 30:343–348

Miyazaki I (1964) *Paragonimus* in Japan with special reference to *P. ohirai*, Miyazaki 1939, *P. iloktsuenensis* Chen 1940 and *P. kellicotti* Word 1908. In: Morishita K, Komiya Y, Matsubayashi H (eds) Progress of medical parasitology in Japan, vol 1 Meguro Parasitological Museum, Tokyo, pp 157–182

Monrad J, Christensen NO, Nansen P, Frandsen F (1981) Resistance to *Fasciola hepatica* in sheep harbouring primary *Schistosoma bovis* infections. J Helminthol 55:261–271

Ogilvie BM (1964) Reagin-like antibodies in animals immune to helminth parasites. Nature 204:91–92

Oldham G (1983) Antibodies to *Fasciola hepatica* antigens during experimental infections in cattle measured by ELISA. Vet Parasitol 13:151–158

Oldham G, Hughes DL (1982) *Fasciola hepatica:* immunisation of rats by intraperitoneal injection of adult fluke antigen in Freunds' adjuvant. Exp Parasitol 54:7–11

Rajasekariah GR, Mitchell GF, Chapman CB, Montague PE (1979) *Fasciola hepatica:* attempts to induce protection against infection in rats and mice by injection of excretory/secretory products of immature worms. Parasitology 79:397–400

Sewell MMH, Purvis GM (1969) *Fasciola hepatica:* the stimulation of excystation. Parasitology 59:4P

Sirag SB, Christensen NO, Nansen P, Monrad J, Frandsen F (1981) Resistance to *Fasciola* in calves harbouring primary patent *Schistosoma bovis* infections. J Helminthol 55:63–70

Sun T (1969) The in vitro action of antisera on the adults of *Clonorchis sinensis*. Trans R Soc Trop Med Hyg 63:582–590

Sun T, Gibson JB (1969a) Antigens of *Clonorchis sinensis* in experimental and human infections: on analysis by gel diffusion techniques. Am J Trop Med Hyg 18:241–252

Sun T, Gibson JB (1969b) The sites of antigen formation in different developmental stages of *Clonorchis sinensis* Jpn J Med Sci Biol 22:263–271

Van Tiggele LJ (1978) Host parasite relations in *Fasciola hepatica* infections. Immunopathology and diagnosis of liver fluke disease in ruminants. Thesis, Rijksuniversiteit te Leiden, The Netherlands.
Voller A, Bidwell D (1982) Immunodiagnosis of parasitic diseases. In: Mettrick DF, Desser SS (eds) Parasites – their world and ours. Elsevier Biomedical, Amsterdam, pp 381–386
Yokogawa M (1964) *Paragonimus* and paragonimiasis. In: Morishita K, Komiya Y, Matsubayashi H (eds) Progress of medical parasitology in Japan, vol 1. Meguro Parasitological Museum, Tokyo, pp 61–156
Yokogawa M (1965) *Paragonimus* and paragonimiasis. Adv Parasitol 3:99–153
Yokogawa M (1969) *Paragonimus* and paragonimiasis. Adv Parasitol 3:375–387
Yokogawa S, Cort WW, Yokogawa M (1960) *Paragonimus* and paragonimiasis. Exp Parasitol 10:139–205